A

Abdomen 69, 72
Abdominal flexor
  muscles 70
Acetabulum 80, 92
Acromium process 95
Actinophrys sol 12
Adductor longus 79
Adductor magnus 79
Adhesive gland 58
Adipose tissue 88
Aeciospores 19
African sleeping
  sickness 13
Agar .5
Air bladder 24
Air space 40
Algae
  blue green 6
  brown 24
  green 21-23
  in lichens 20
  yellow green 8
Allium
  mitosis 4
  root 38
Alveolar sacs, ducts 103
Alveoli 103
Ambulacral groove . . . 74-75
Ambulacral ridge 73-75
Amoeba 12
Amphibia 77-80
Amphioxus 76
Ampullae
  shark 98
  starfish 73-75
Anabena 6
Anaphase 3-4
Anemones (sea) 56
Angiospermae 38-49
Animalia 51-110
Annelida 69
Annulus
  fern 32
  mushroom 18
Antenna
  copepod 71
  crayfish 69
  grasshopper 72
Antennules 69
Anterior adductor
  muscle 64
Anterior chamber . . . 99-100
Anterior gray horn 96
Anterior interventricular
  groove 102
Anterior median
  fissure 96
Anterior mesenteric
  artery
  frog 78
  pig 84
Anterior retractor
  muscle 64
Anterior superior spine
  of ilium 95
Anterior vena cava
  pig 83, 85
  sheep 102
Anterior white
  columns 96
Anther 45-46, 48-49
Antheridia
  fern 33-34
  Fucus 24
  liverwort 25
  Oedogonium 23
  Vaucheria 8
Antheridial head 25
Antipodal cells 47
Anus
  Amphioxus 76
  clam 64
  earthworm 66
  sandworm 65
  starfish larva 110
Aorta 81, 83-84, 102
Aortic arches 67
Apical meristem
  seed 48
  stem 41
Apothecium
  lichen 20
  Peziza 17
Apple 49
Aqueduct of Silvius 97
Aqueus humor 100
Arcella 12
Archegonia
  fern 33-34
  liverwort 26
  moss 29
  pine 37
Archegonial head 25-26
Archenteron 110
Areolar connective
  tissue 88
Arrector pili 106
Arteries
  frog 78
  pig 84
  pulmonary 102
  x.s. 103
Ascaris 61-62
Aschelminthes 61-63
Ascomycetes 17
Ascospores 17
Ascus 17
Asteroidea 73-75
Astragulus 94
Atlas 92-94
Atriopore 76
Atrium 76, 81-83, 102
Aurelia 55
Auricle
  clam 64
  planaria 57
  sheep 102
Axillary artery 84
Axillary bud 41
Axillary vein 85
Axis 92-94
Axons 91, 96-97
Axopodia 12
Azygous vein 83

B

Bacillariophyceae 7
Bacilli 5
Bacteria 5
Barberry leaf 19
Bark 42
Basal disc 53
Basal layer 106
Basement membrane . . .108
Basidiomycetes 18-19
Basidiospores 18
Basidium 18
Basilar membrane 98
Basophil 101
Bean 48
Bipinnaria larva 110
Bipolar neurons 100
Bladderworms 60
Blastocoel 110
Blastopore 110
Blastostyle 55
Blastula
  hydra 54
  sponge 52
  starfish 110
Blepharoplast 13
Blind spot 99-100
Blood typing 101
Blood 101
Blue-green algae 6
Body of vertebra 94
Bones
  compact 95
  frog 80
  human 92-95
  spongy 95
  x.s. 89
Bowman's capsule 107
Brachial vein 85
Brachiocephalic
  artery 83-85
Bracket fungi 18
Bracts 46
Brain
  crayfish 69
  earthworm 67
  frog 80
  human 96
  sheep 97
Branchial heart 65
Bronchiole 103
Brown algae 24
Brownian movement 23
Bryophyta 25-29
Buccal cirri 76
Buccal nerve 68
Bud
  hydra 53
  Obelia 55
  yeast 17
Budding (yeast) 17
Bulb (hair) 106
Bulbourethral gland 86
Bulliform cells 43
Bundle cap 41
Bursa 63
Byssus 65

C

Caecum 59
Calcaneous 94
Calcium sulfate 23
Calyptera 27
Cambarus 69-71
Cambium 41-42
Canaliculi 89
Cap
  moss 27
  mushroom 18
Capitate 94
Capsella seed 48
Capsule 27-28
Carapace 69
Cardiac muscle 91
Cardiac stomach 71
Carotid arch 78
Carotid gland 78
Carpals 92, 94
Carpellate cones 35
Carpus 94
Cartilage (hyaline) 89
Caudal mesenteric
  artery 84
Caudal vena cava . . . 82, 85
Cell division 3, 4
Cell membrane 1, 3, 12
Cell plate 4
Cell wall . . . 1, 2, 4, 7, 21-22
Cells (cheek) 1
Central canal 96
Central vacuole 21
Centriole 3
Cephalic vein 85
Cephalochordata 76
Cephalopoda 65
Cephalothorax 69
Ceratium 7
Cercaria 59
Cerebellum
  frog 80
  human 96
  sheep 97
Cerebral ganglia 68
Cerebral hemisphere 80
Cerebrum
  human 96
  sheep 97
Cervical vertebrae 92-94
Cestoda 60
Cheek epithelium 1
Cheliped 69
Chinese liver fluke 59
Chlorophyta 21-23
Chloroplasts
  diatom 8
  Elodea 1
  Euglena 14
  Spirogy
  guard ce
Choanocyt
Chondrocyt
Chordata . . .
Choroid 99-100
Chromosomes 3, 4, 47
Chrysophyta 7
Cilia
  epithelium 87, 103
  Paramecium 9-11
  planaria 57-8
  rotifers 63
Ciliary muscle 99-100
Ciliary process 100
Ciliated pseudostratified
  columnar
  epithelium 104
Ciliophora 9-11
Circular muscles
  (earthworm) 68
Circulatory system
  (pig) 83-85
Circumesophageal
  connective 69
Circumpharyngeal
  connective 68
Cirri 66
Cladophora 23
Clam worm 66
Clam 64-65
Clavicle 80, 92, 95
Clitellum 66
Clonorchis 59
Closterium 23
Clown fish 56
Club fungi 18
Club moss 30
Clypeus 72
Cnidaria 53-56
Cnidoblasts 53-54
Cocci 5
Coccyx 92, 95
Cochlea 98
Cochlear duct 98
Cocoons (earthworm) 68
Coelenterata 53-56
Coeliac artery
  frog 78
  pig 82-84
Coeliaco-mesenteric
  artery 78
Coelom
  earthworm 68
  starfish 75
Coelomic sac 110
Coenocytic filament 8
Coenosarc 55
Collar cells 51-52
Collecting tubules . . 88, 107
Collenchyma 41-42
Colon
  frog 77-78
  pig 81, 83, 86
Columella
  fungi 15
  moss 29
Common carotid
  artery 83, 85
Common iliac vein 85
Compact bone 94
  x.s. 89
Companion cell 40
Composite flowers 46
Compound eye
  crayfish 70
  grasshopper 72
Conceptacles 24
Cones 100
Conidia 17
Conifers 35-37
Conjugation
  Paramecium 9
  Rhizopus 16
  Spirogyra 22
Conjugation fungi 15
Conjugation tube 22
Connective tissue
 12
  Paramecium 9-10
Conus arteriosus 78
Convoluted tubules 107
Copepods 71
Coprinus 18
Copulation
  (earthworm) 66
Coracoid process 95
Coracoid 80
Coral 56
Cork cells 1
Cork 42
Corn grain 48
Corn leaf 43
Cornea 99-100
Corneal epithelium . . 99-100
Corolla 46
Corona radiata 109
Corona 63
Coronal suture 93
Coronary artery & vein . . .84
Coronoid process 93
Corpora quadrigemina . . .97
Corpus albicans 109
Corpus callosum 97
Corpus luteum 109
Cortex
  fern 32
  kidney 107
  root 38-39
  stem 41-42
Corti, organ of 98
Coscinodiscus 8
Costal cartilage 92
Cotyledons 48-49
Coxa 72
Cranial mesenteric
  artery 83-84
Crayfish 69-71
Crop
  earthworm 66
  grasshopper 72
Crustacea 69-71
Crustose lichens 20
Cuboid 94
Cuneiforms 94
Cutaneous abdominis 79
Cutaneous artery 78
Cuticle
  Ascaris 62
  leaf 43-44
Cyanophyta 6
Cyclops 71
Cyst (trichina) 63
Cytoplasm 1, 2, 7,
  8, 21, 91
Cytoplasmic strands 21

D

Dactylozooids 56
Daughter colony 14
Deep circumflex artery . . .84
Deep femoral artery 84
Deltoid 79
Dendrites 91, 96
Dens 94
Dermal papulae 75
Dermal pores 52
Dermis 106
Desmids 23
Diaphragm 82
Diatoms 7, 8, 71
Dicot
  root 39
  stem (herbaceous) 41
  stem (woody) 33
Didinium 11
Digestive gland
  clam 64
  crayfish 69-71
  starfish 73
Dinoflagellates 7
Dinophysis 7
Disc flowers 46

- Distal convoluted tubules 107
- Dorsal aorta 78
- Dorsal blood vessel 67
- Dorsal fin 76
- Dorsal nerve cord 76
- Dorsoventral muscles 57-58
- Ductus arteriosus 83
- Ductus deferens 86

E

- Ear 98
- Echinodermata 73-75
- Ectoderm
  - hydra 53-54
  - starfish 110
- Egg
  - *Ascaris* 62
  - cat 109
  - crayfish 71
  - fern 33-34
  - flower 47
  - *Fucus* 24
  - grasshopper 72
  - *Marchantia* 26
  - moss 29
  - *Oedogonium* 23
  - pine 37
- Egg sac 71
- Ejaculatory duct 61-62
- Elaters 26
- *Elodea* 1
- Elongation region 38
- Embryo
  - seed 48-49
  - starfish 110
- Embryo sac 45, 47
- Embryonic region 38
- Endoderm
  - hydra 53-54
  - planaria 57-58
  - starfish 110
- Endodermis
  - pine needle 43
  - root 39
  - stem 42
- Endosperm 48
- Endothelium 99
- Eosinophil 101
- Epibranchial groove 76
- Epidermis
  - *Amphioxus* 76
  - earthworm 68
  - fern 32
  - hydra 53-54
  - leaf 19, 43-44
  - planaria 57-58
  - root 38-39
  - skin 106
  - stem 40-41
- Epididymis 86
- Epithalamus 97
- Epithelium
  - cheek 1
  - ciliated 87, 104
  - cuboidal 88
  - germinal 109
  - simple
    - columnar 87, 105
  - stratified squamous 87
- Erythrocytes 13, 101
- Esophagus
  - *Amphioxus* 76
  - clam 69
  - crayfish 69
  - fluke 59
  - frog 80
  - grasshopper 72
  - pig 83
  - starfish 110
- Ethmoid 93
- *Euglena* 14
- Euglenophyta 14
- Eustachian tube 80
- Excretory canals 61-62
- Excretory duct 60
- Excretory pore 59
- Excurrent siphon 64
- External acoustic meatus 93
- External carotid artery 78
- External iliac artery 82, 84
- External iliac vein 85
- External jugular vein 83, 85
- External malleolus 94
- External oblique 79
- Extrinsic muscle 99
- Eye
  - crayfish 69-70
  - frog 80
  - monkey 99-100
  - sheep 99
- Eye spot
  - *Euglena* 14
  - planaria 57

F

- Facet 94
- False ribs 92
- Fat bodies 78
- Fat droplet 88
- Femur
  - frog 80
  - grasshopper 72
  - human 92, 95
  - split 94
- Ferns 30-34
- Fetal pig 81-86
- Fibula 94
- Fiddlehead 30
- Filament 45-46, 49
- Fin rays 76
- Fission 9
- Flagella
  - bacteria 5
  - protozoa 13-14
  - sperm 108-109
- Flatworms 57-60
- Floating ribs 92
- Flowers 45-48
- Flukes 59
- Foliose lichens 20
- Follicle
  - graafian 109
  - hair 106
  - primoridal 109
  - secondary 109
- Food vacuole
  - *Amoeba* 12
  - *Paramecium* 9, 10, 14
- Foot
  - clam 64
  - human 94
  - *Marchantia* 26
  - rotifer 63
- Foraminifera 12
- Fornix 97
- Fourth ventricle 97
- Fovea 100
- *Fragilaria* 7
- Frog 77-80
- Frond 30
- Frons 72
- Frontal 93
- Fruiticose lichens 20
- *Frustulia* 7
- *Fucus* 24
- Fungi 15-20
- Funiculus 47, 49

G

- Gall bladder
  - frog 77
  - pig 81
- Gametangia 16
- Gametes
  - *Rhizopus* 16
  - *Spirogyra* 22
- Gametophyte
  - fern 30-34
  - flower 45, 47-48
  - liverwort 25-26
  - moss 27-29
  - pine 36-37
- Ganglion cells 100
- Gastric caeca 72
- Gastric mill 71
- Gastric muscle 70
- Gastrocnemius 79
- Gastrodermis 53-54
- Gastrovascular cavity
  - hydra 53-54
  - planaria 57-58
- Gastrozooids 56
- Gastrula 110
- Gemmae 25
- Gena 72
- Generative cell
  - flower 48
  - pine 36
- Generative nucleus 48
- Genital pore
  - *Ascaris* 61
  - crayfish 70
  - tapeworm 60
- Germinal epithelium 109
- Gill
  - clam 64-65
  - crayfish 70
  - mushroom 18
  - squid 65
- Gill bars 76
- Gill slits 76
- Girdle 7
- Gizzard 67
- Gladiolus 92
- Glenoid cavity 95
- Glenoid fossa 80
- *Gleocapsa* 6
- *Globigerina* 12
- Glochidia 65
- Glomerulus 107
- Glottis 80
- Glumes 46
- Gluteus 79
- Goblet cell 87, 104, 105
- Gonads
  - *Obelia* 55
  - nematode 63
  - starfish 73-75
- Gonangium 55
- Gonopore 55
- Gonotheca 55
- Gonozooids 56
- Gorgonian coral 56
- Graafian follicle 109
- Gracilis major 79
- Gracilis minor 79
- *Grantia* 50-52
- Grass flower 46
- Grasshopper 72
- Gravid proglottids 60
- Gray matter 96
- Greater trochanter 95
- Greater tuberosity 95
- Green gland duct 69
- Green gland 68-69
- Guard cells 44
- Gubernaculum 86
- Gullet 9
- Gymnospermae 35-37

H

- Hair cells 98
- Hair 106
- Hamate 94
- Hand 94
- Haversian system, canals 89
- Heart
  - clam 64
  - crayfish 70
  - earthworm 67
  - frog 77-80
  - human 102
  - squid 65
- Hepatic caecum 76
- Hepaticae 25
- Herbaceous dicot stem 41
- Heterocyst 6
- Hilum 48
- Hip 95
- Histology 87-91
- Holdfast 24
- Hookworm 63
- Horn of uterus 81
- Humerus
  - frog 80
  - human 92, 95
- Hydra 53-54
- Hydranth 55
- *Hydrodictyon* 23
- Hymenium 17
- Hyphae 15-18, 20
- Hypobranchial groove 76
- Hypocotyl 48
- Hypodermis 43
- Hypophysis 97
- Hypostome
  - hyrda 53
  - *Obelia* 55

I

- Iliac crest 95
- Ilium 80, 92, 95
- Incurrent canal 50-51
- Incurrent siphon 64
- Indusium 31-32
- Inferior articular facet 94
- Inferior nasal concha 93
- Inferior orbital fissure 93
- Infraorbital foramen 93
- Infundibulum 97
- Ink sac 65
- Innominate artery 102
- Insecta 72
- Integument
  - inner & outer 47
  - pine 37
  - seed 48-49
- Intercalated disc 91
- Intermediate mass 97
- Internal carotid artery 78
- Internal iliac vein 85
- Internal jugular vein 83, 85
- Internal malleolus 94
- Internal nares 80
- Internal thoracic artery 84
- Internal thoracic vein 85
- Interphase 3-4
- Interstitial cells 108
- Interthalamic adhesion 97
- Intertrochanteric line 95
- Interventricular septum 102
- Intervertebral discs 92
- Intestinal glands 105
- Intestine
  - *Amphioxus* 76
  - *Ascaris* 61-62
  - clam 64
  - crayfish 70
  - earthworm 67-68
  - flukes 59
  - grasshopper 72
  - pig 81
  - planaria 58
  - starfish larva 110
- Involuntary muscle 91
- Iris 99-100
- Ischium 80, 92, 95

J

- Jaws 66
- Jejunum 105

K

- Kidney
  - clam 64
  - frog 77-78
  - lamb 107
  - pig 81, 83, 86
  - rat 107
  - squid 65

L

- Labial palps
  - clam 64
  - grasshopper 72
- Lacrimal 93
- Lacteal 105
- Lacuna 87
- Lamina 94
- Large intestine 72
- Larva
  - Bipinnaria 110
  - Copepod 71
  - Trichina 63
  - Starfish 110
- Larynx 81-83
- Lateral lines 61-62
- Lateral malleolus 94
- Latissimus dorsi 79
- Leaf bud (primordium) 41
- Leaves 43-44
- Left atrium 102
- Left azygous vein 83
- Left brachiocephalic vein 85
- Left subclavian artery 83
- Left ventricle 102
- Lens 99-100
- Lesser trochanter 95
- Lesser tuberosity 95
- Leucocytes 101
- Leucoplasts 2, 21
- *Leucosolenia* 50
- Leukemia 101
- Lichens 20
- Ligamentum arteriosum 102
- Limestone 12
- Lima bean 48
- Lip cell 32
- Liver
  - *Amphioxus* 76
  - clam 64
  - frog 77-78
  - pig 81-83
- Liverworts 25
- Locule 47
- Longitudinal muscle
  - *Ascaris* 62
  - earthworm 67-68
- Loop of Henle 107
- Lumbar vertebrae 92, 94-95
- *Lumbricus* 66-68
- Lumen
  - *Ascaris* 62
  - mammal 105
- Lunate 94
- Lung
  - frog 77-78
  - pig 81-83
  - sheep 103
- Lymph vessel 103
- Lymphocytes 101

M

- *Macrocystis* 24
- Macronucleus
  - *Paramecium* 9
  - *Stentor* 11
- Madreporite 73-75
- Malar 93
- Malaria 13
- Malpighian layer 106
- Malpighian tubules 72
- Mamillary body 97
- Mamillary process 94
- Mandible
  - crayfish 69
  - grasshopper 72
  - human 93
- Mandibular (glenoid) fossa 93
- Mandibular condyle 93
- Mandibular muscle 69-70
- Mantle 64-65
- Manubrium
  - human 92
  - *Obelia* 55

A GUIDE TO BIOLOGY LAB

3rd Edition

THOMAS G. RUST, M.Ed., M.A.

TABLE OF CONTENTS

Cytology 1-2
Mitosis 3-4

KINGDOM MONERA

Schizophyta (Bacteria) 5
Cyanophyta (Blue-green algae) 6

KINGDOM PROTISTA

Pyrrophyta (Dinoflagellates) 7
Chrysophyta (Diatoms) 7-8
Protozoa 9-14
Euglenophyta 14

KINGDOM FUNGI

Zygomycetes (Conjugation fungi) 15-16
Ascomycetes (Sac fungi) 17
Basidiomycetes (Club fungi) 18-19
Lichens 20

KINGDOM PLANTAE

Chlorophyta (Green algae) 21-23
Phaeophyta (Brown algae) 24
Bryophyta (Liverworts, mosses) 25-29
Tracheophyta (Vascular plants)
Filicinae (Ferns) 30-34
Gymnosperms (Conifers) 35-37
Angiosperms (Flowering plants)
Roots 38-39
Stems 40-42
Leaves 43-44
Flowers, seeds, fruits 45-49

KINGDOM ANIMALIA

Porifera (Sponges) 50-52
Coelenterata (Cnidaria) (Hydra, jellyfish, sea anemones) ... 53-56
Platyhelminthes (Flatworms) 57-60
Aschelminthes (Nemathelminthes) (Roundworms, nematodes) ... 61-63
Mollusca (Clams, snails, squid) 64-65
Annelida (Segmented worms) 66-68
Arthropoda (Crustaceans, insects) 69-72
Echinodermata (Starfish) 73-75
Chordata
Amphioxus 76
Frog dissection 77-80
Fetal pig dissection 81-86
Histology (tissues) 87-91
Skeletal system 92-95
Nervous system 96-97
Ear 98
Eye (dissection) 99-100
Blood 101
Heart (dissection) 102
Circulatory system 103
Respiratory system 103-104
Digestive system 104-105
Skin 106
Excretory system 107
Reproductory system 108-109
Embryology 110

ISBN: 0-937029-01-7

Additional copies may be ordered from: Southwest Educational Enterprises
31400 IH-10 West
Boerne, Texas 78006
(210) 342-2297

A PHYLOGENETIC TREE

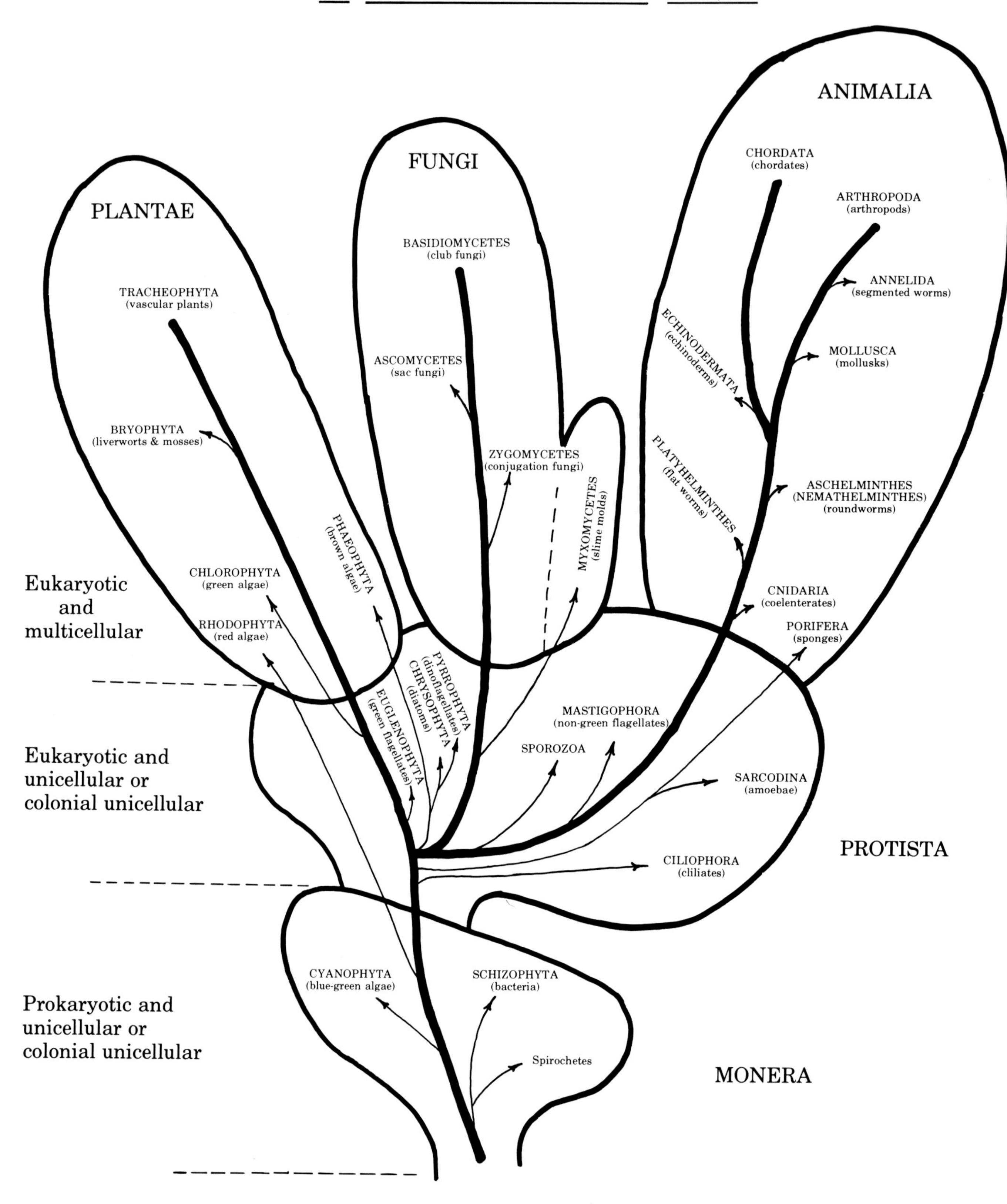

ORIGIN OF LIFE

The student should realize that there is no universally agreed upon classification system. The taxonomic system followed in this book will be largely compatible with the majority of current biology texts.

Due to new information from research in fields such as electron microscopy and biochemical analysis, changes and corrections to the system will inevitably result as our understanding of taxonomic and evolutionary relationships continues to improve.

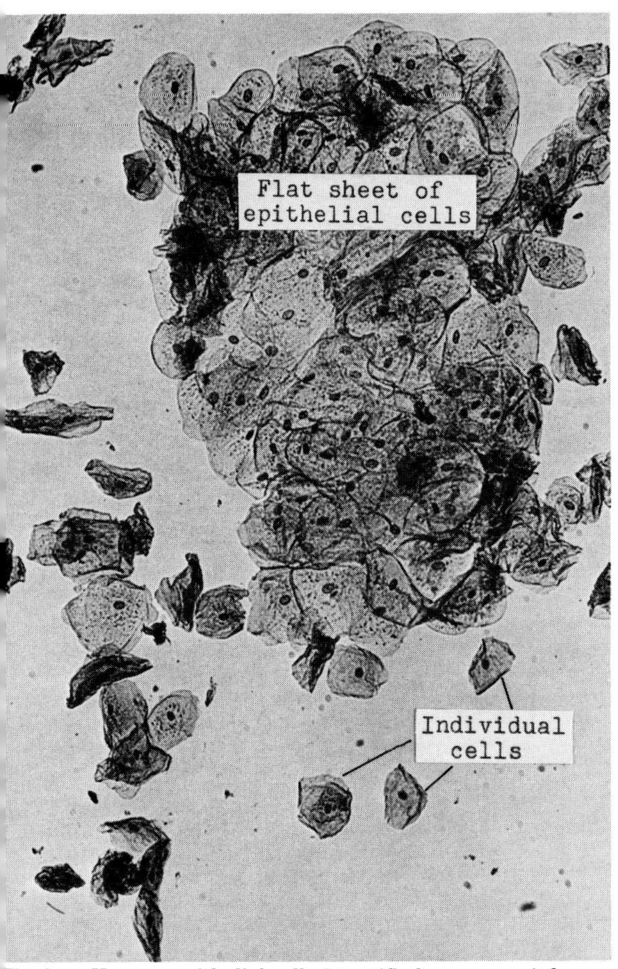

Fig. 1a Human epithelial cells (stratified squamous) from the lining of the oral cavity w.m. x100.

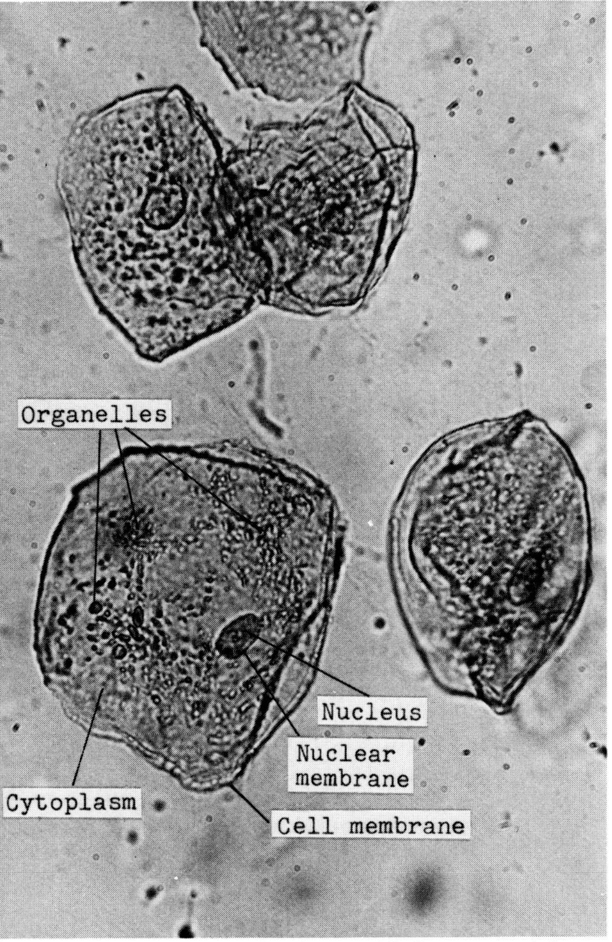

Fig. 1b Human epithelial cells (stratified squamous) from the lining of the oral cavity w.m. x430.

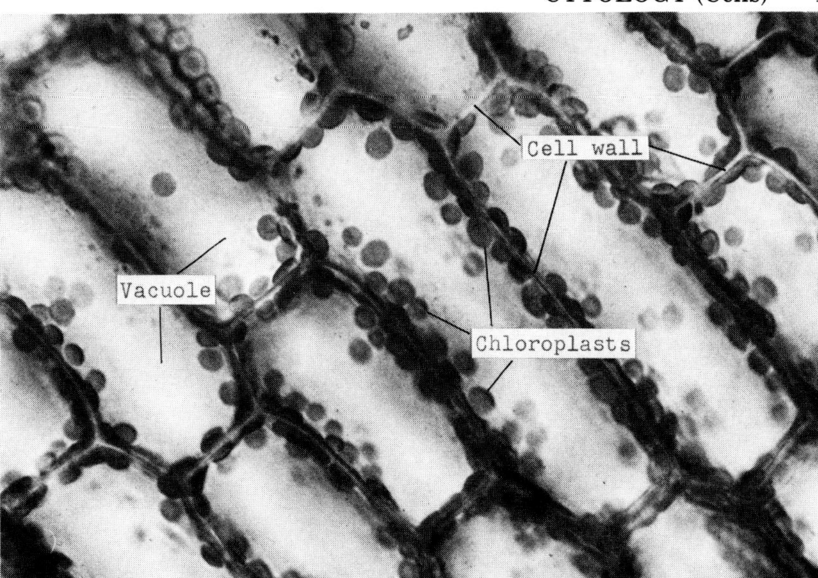

Fig. 1c Live Elodea leaf cells w.m. x430. Both the cell membrane and the vacuolar membrane are too thin to be seen.

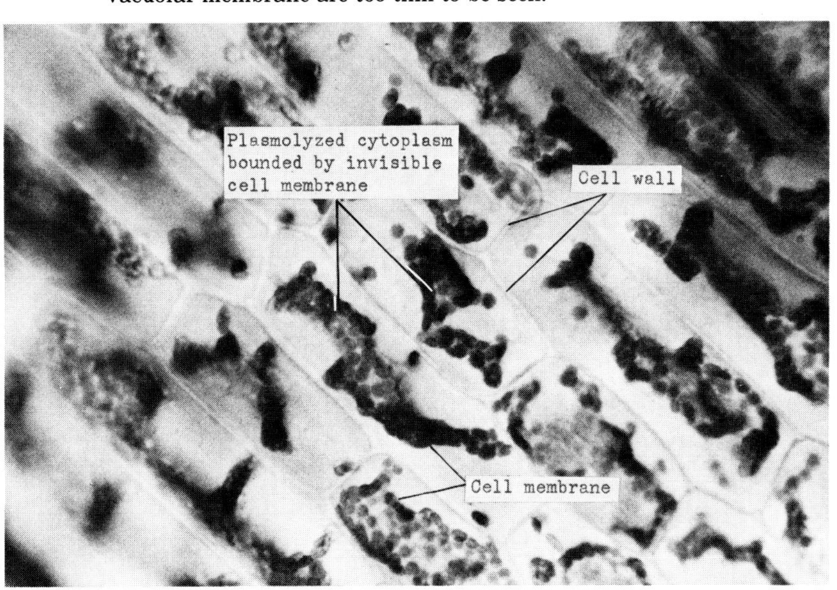

Fig. 1d Plasmolyzed Elodea leaf cells w.m. x430. Placed in a hypertonic 10% salt solution, the water in the cells diffused out of the cytoplasm (shrinking them) as it moved into the region of lower concentration of water outside the cell.

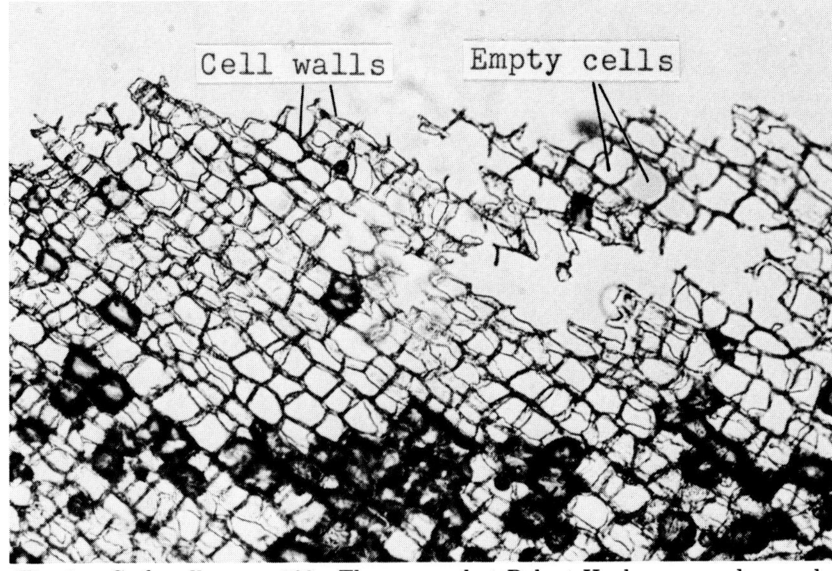

Fig. 1e Cork cells x.s. x100. These are what Robert Hooke saw and named "cells" in the mid 1600's. Ironically, these are only the empty boxes formed by the cell walls; the cells having long since died.

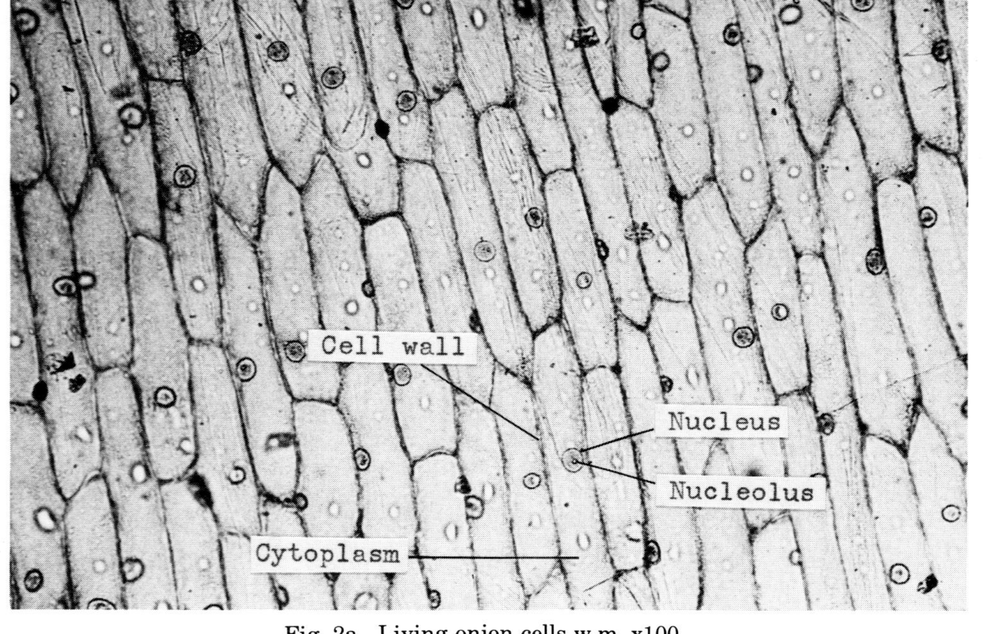

Fig. 2a Living onion cells w.m. x100.

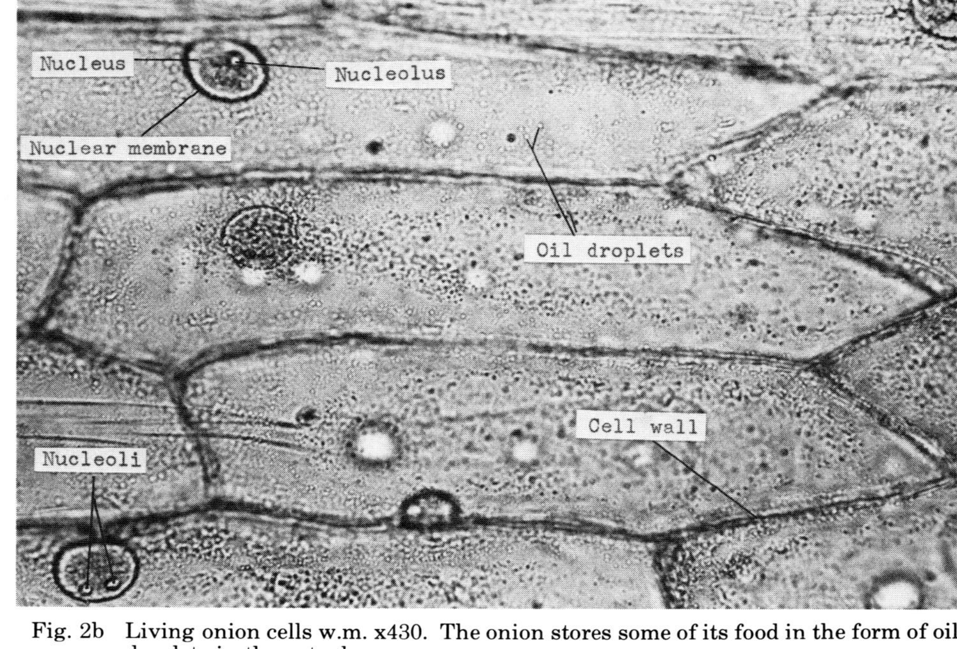

Fig. 2b Living onion cells w.m. x430. The onion stores some of its food in the form of oil droplets in the cytoplasm.

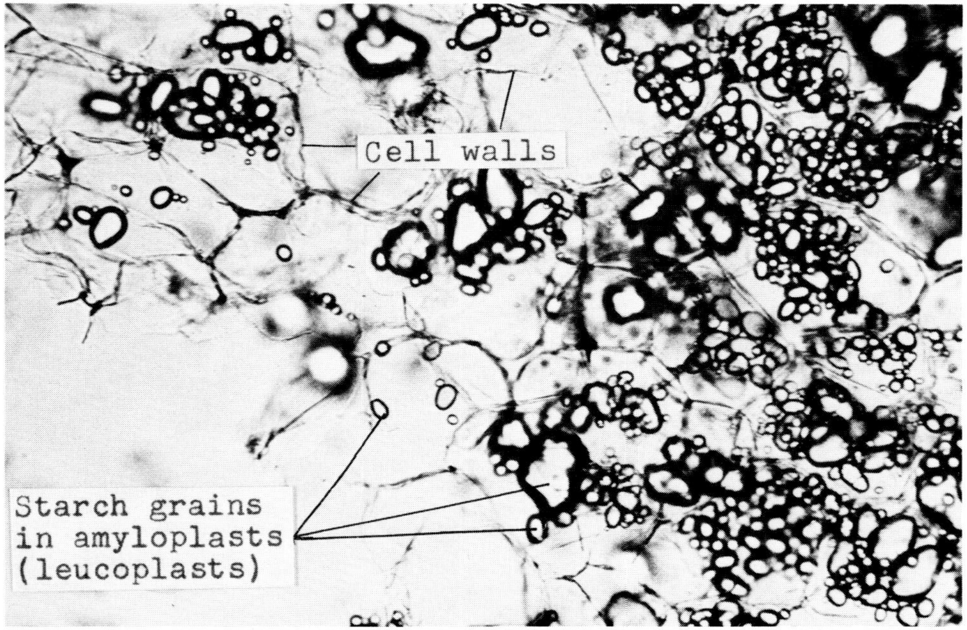

Fig. 2c Living potato cells w.m. x100. The potato stores food as starch grains in organelles called pyrenoids. These pyrenoids (leucoplasts) as well as other cellular organelles are present but not visible unless specially stained.

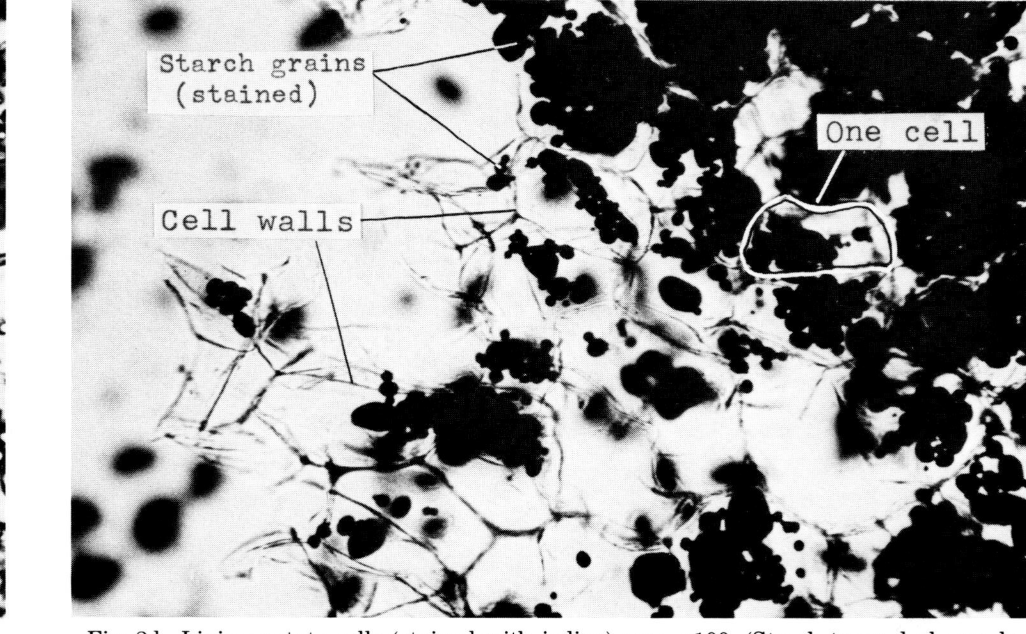

Fig. 2d Living potato cells (stained with iodine) w.m. x100. (Starch turns dark purple when stained with iodine.)

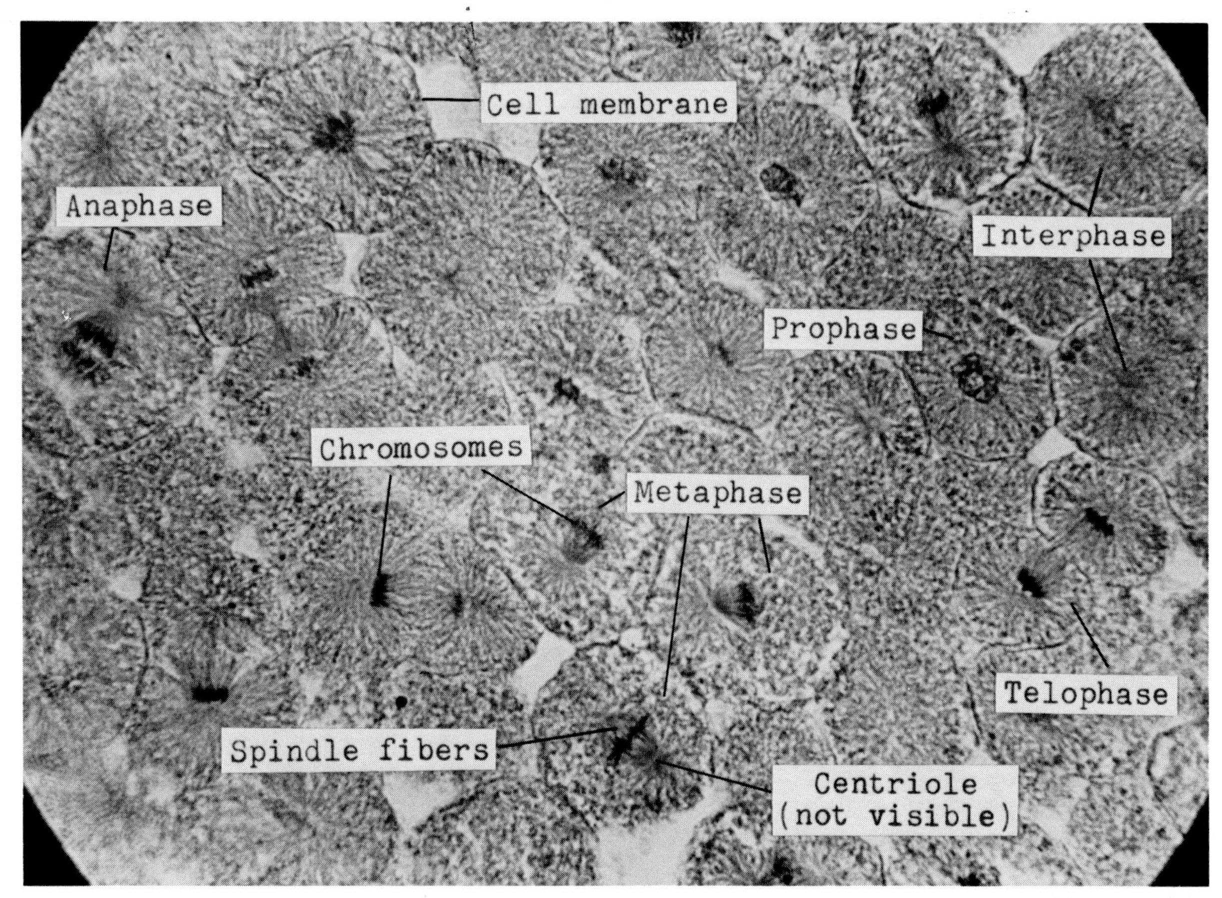

Fig. 3a Mitosis in cells of the Whitefish blastula x.s. x430. A blastula is a hollow ball of cells formed by successive mitotic divisions of a zygote (fertilized egg). (See Fig. 110g.)

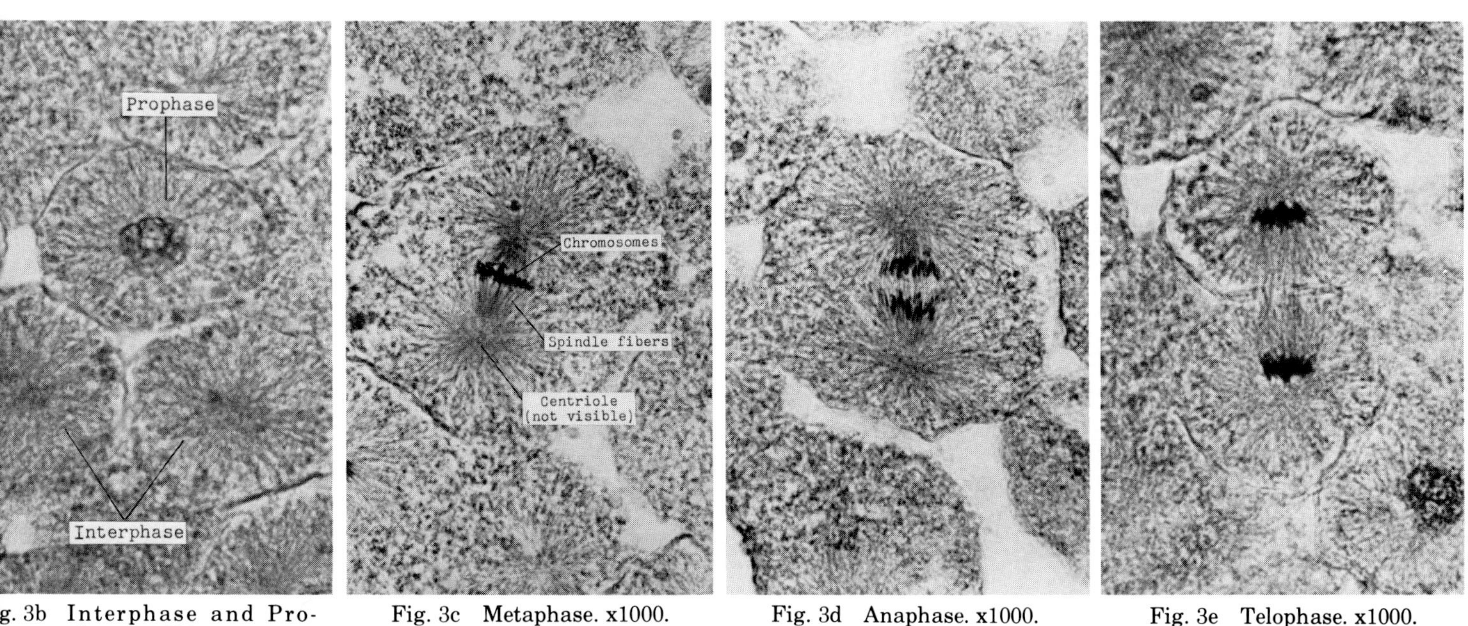

g. 3b Interphase and Prophase. x1000.

Fig. 3c Metaphase. x1000.

Fig. 3d Anaphase. x1000.

Fig. 3e Telophase. x1000.

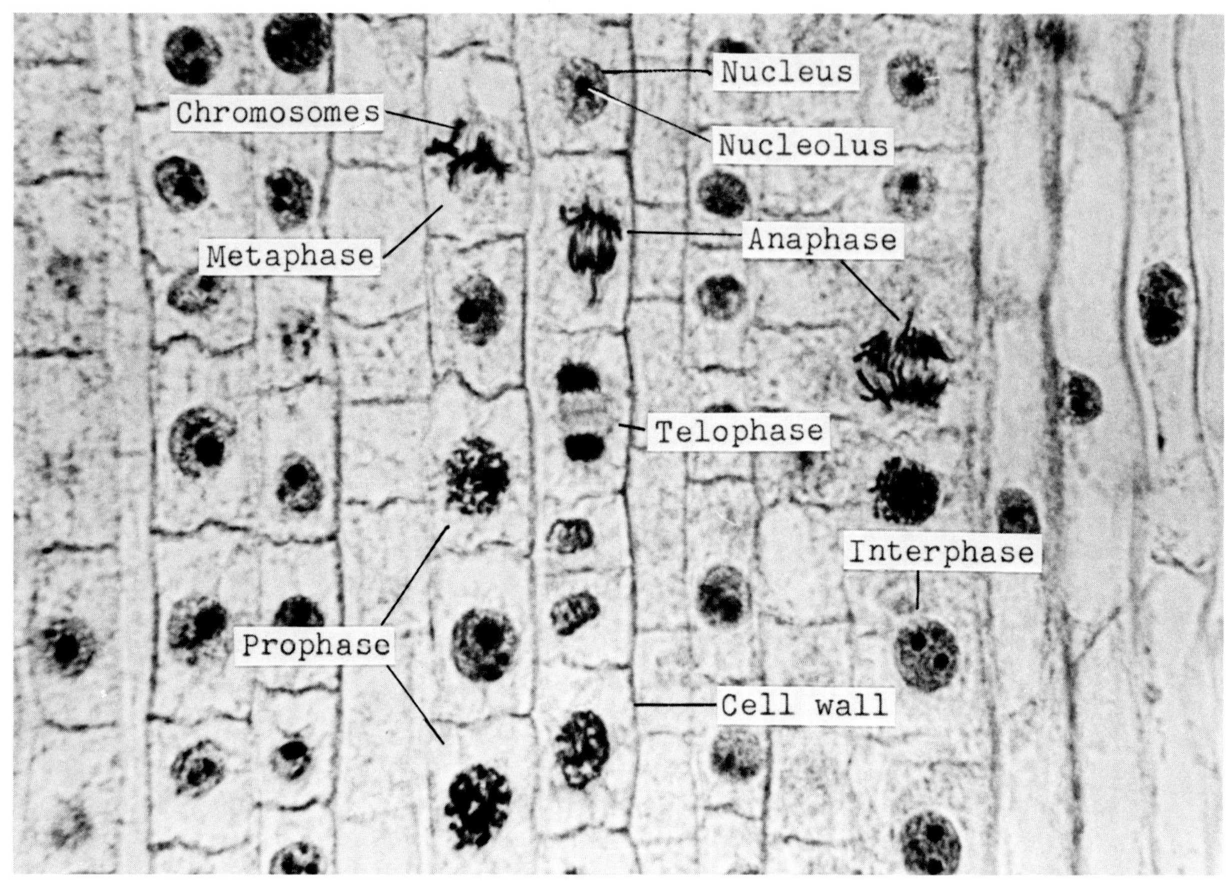

Fig. 4a Mitosis in onion (Allium) root tip l.s. x430. These cells are from the meristematic (embryonic) region of the root. (See Fig. 38d.)

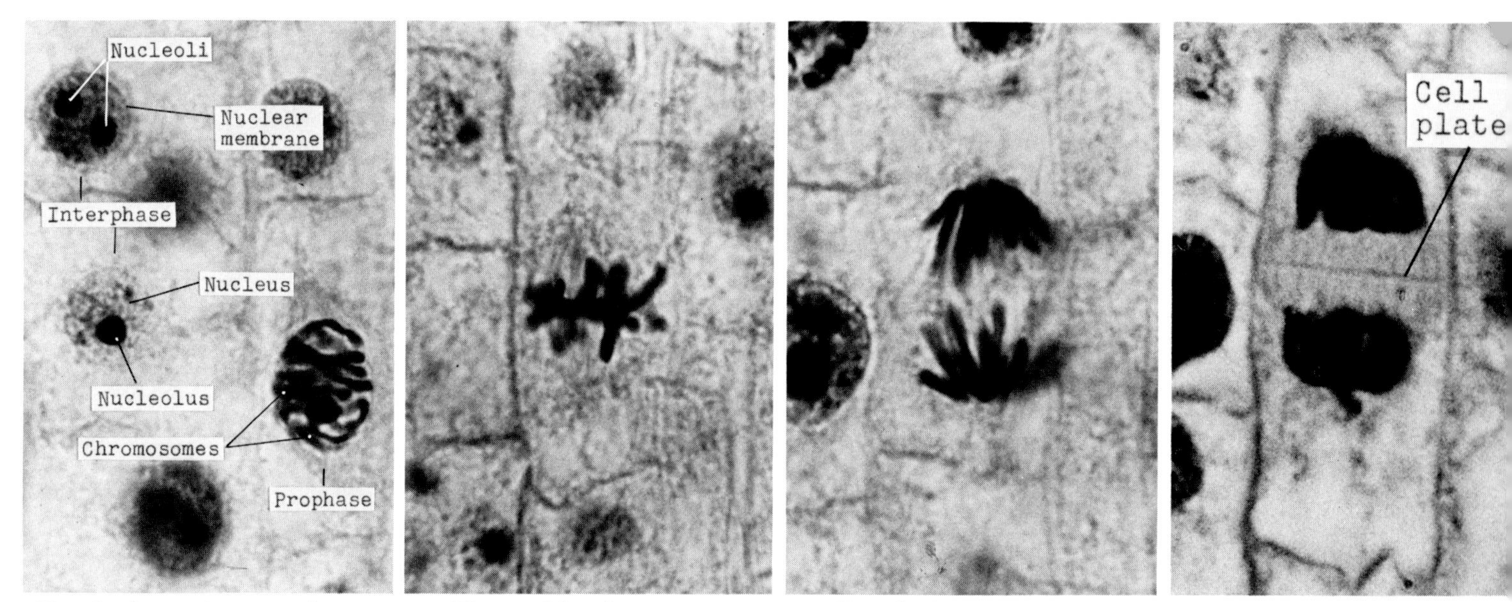

Fig. 4b Interphase and prophase. x1000.

Fig. 4c Metaphase. x1000.

Fig. 4d Anaphase. x1000.

Fig. 4e Telophase. x1000.

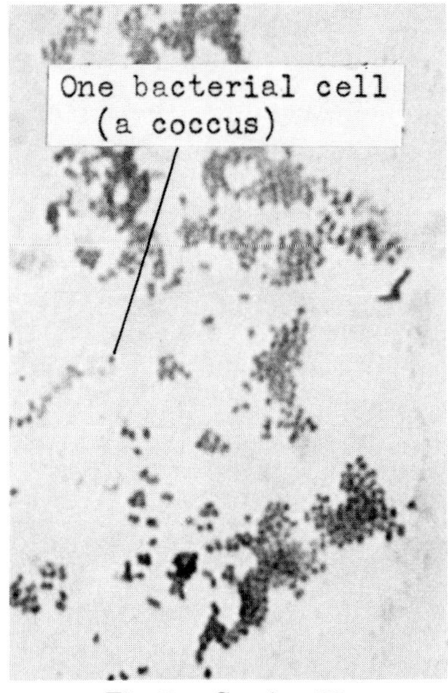

Fig. 5a Cocci. x430.

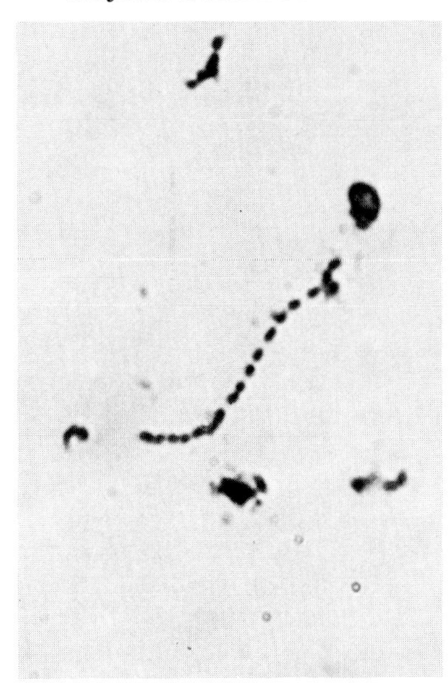

Fig. 5b Streptococci. x1000.

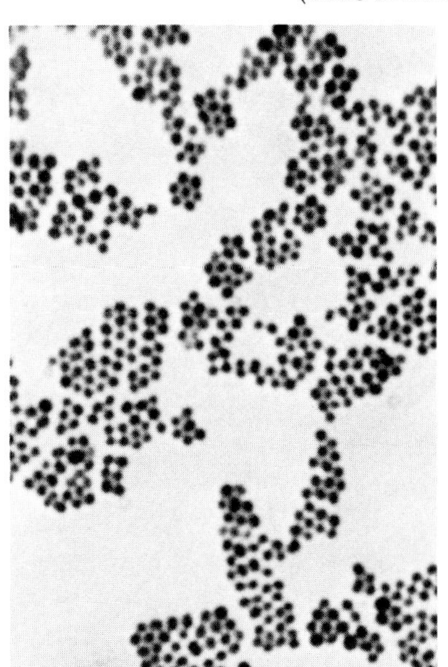

Fig. 5c Staphylococci. x1000.

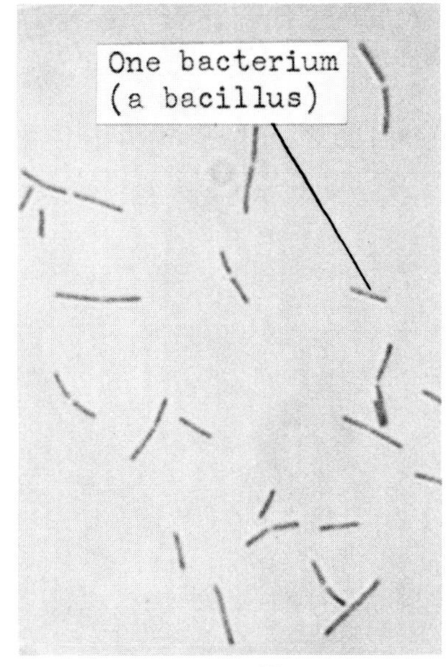

Fig. 5d Bacilli. x430.

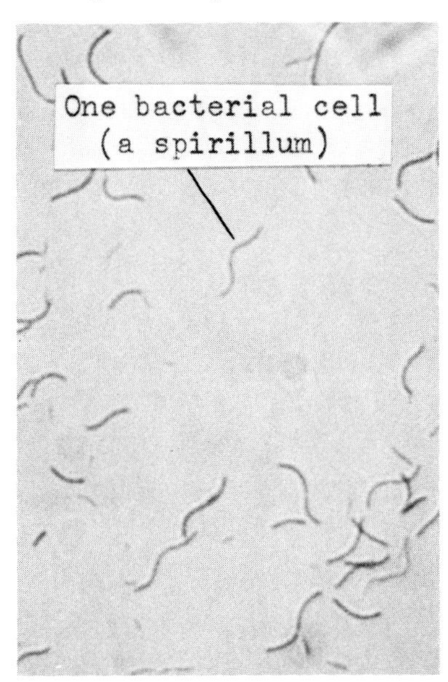

Fig. 5e Spirilla. x430.

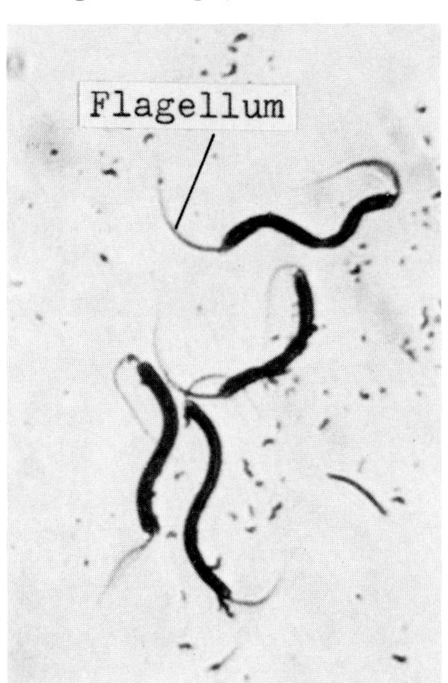

Fig. 5f Spirilla with flagella. x1000.

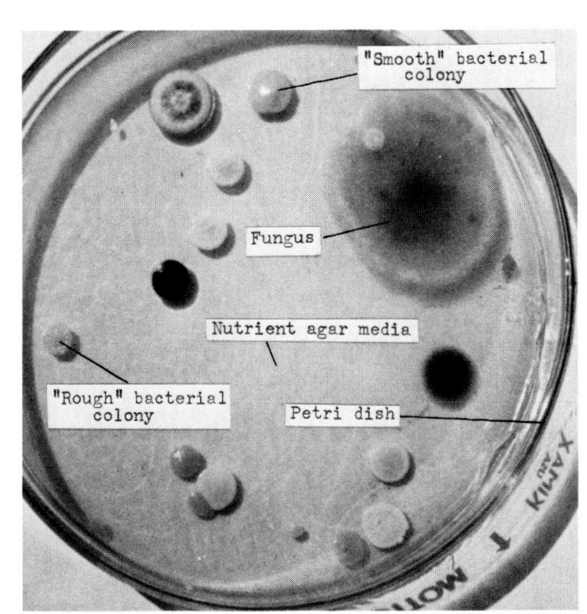

Fig. 5g Bacterial and fungal colonies on nutrient agar in a Petri dish. x1.

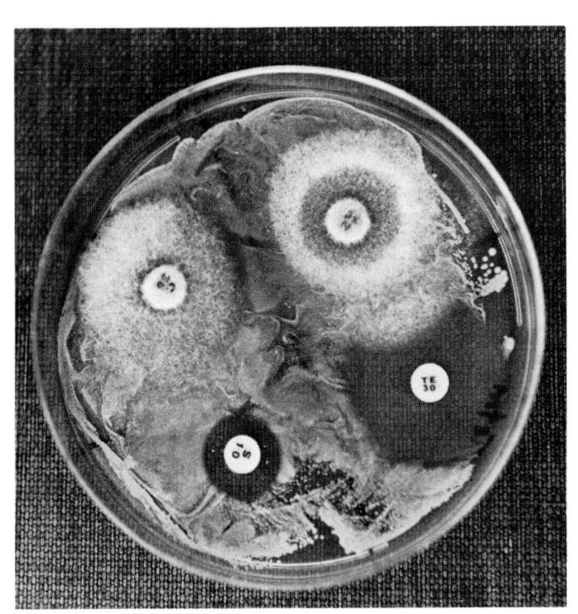

Fig. 5h Zones of inhibition of growth around antibiotic discs. The antibiotic in disc TE-30 obviously inhibited this particular species of bacteria best.

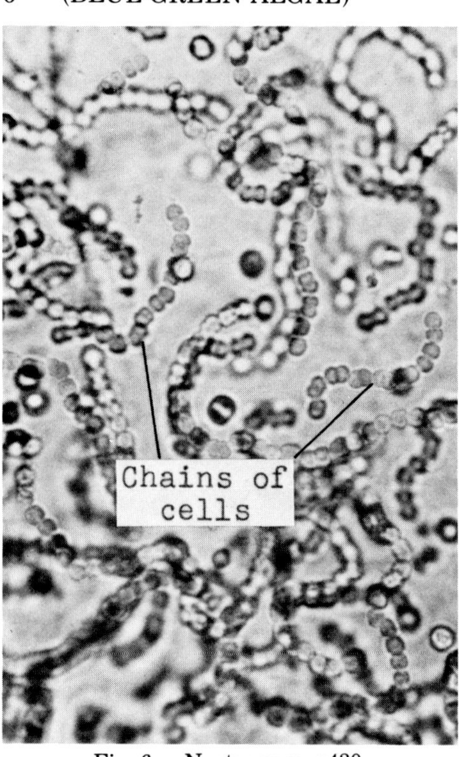

Fig. 6a Nostoc w.m. x430.

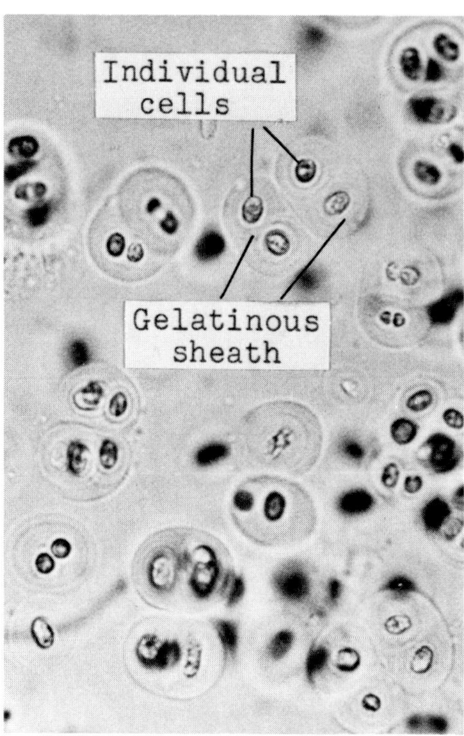

Fig. 6b Gleocapsa w.m. x430.

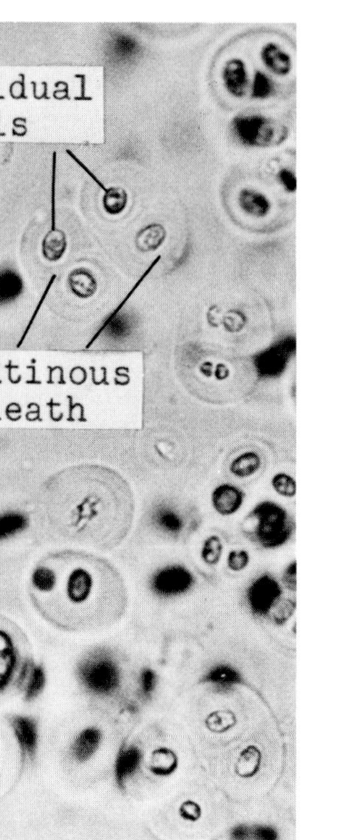

Fig. 6c Anabaena w.m. x430. The heterocysts are thought to function in nitrogen fixation. (Caddo Lake, Texas)

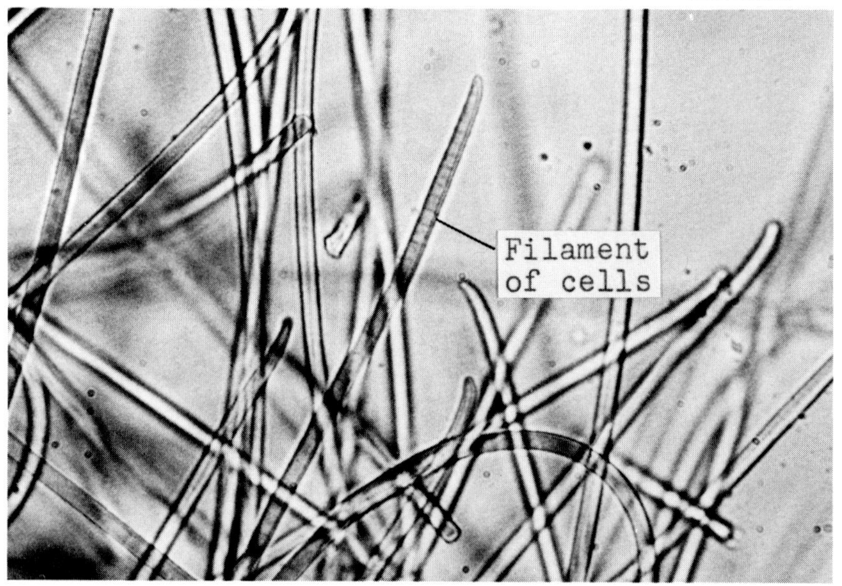

Fig. 6d Oscillatoria w.m. x430.

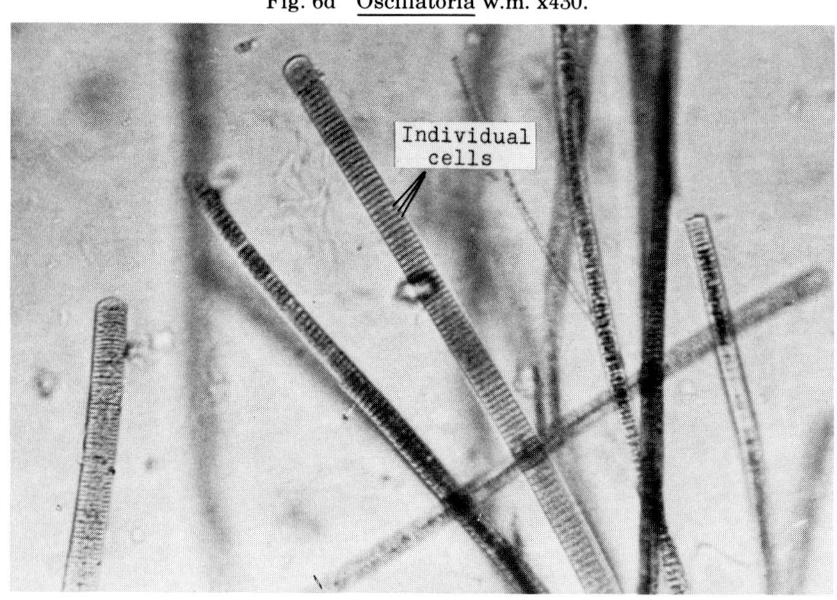

Fig. 6e Oscillatoria w.m. x430.

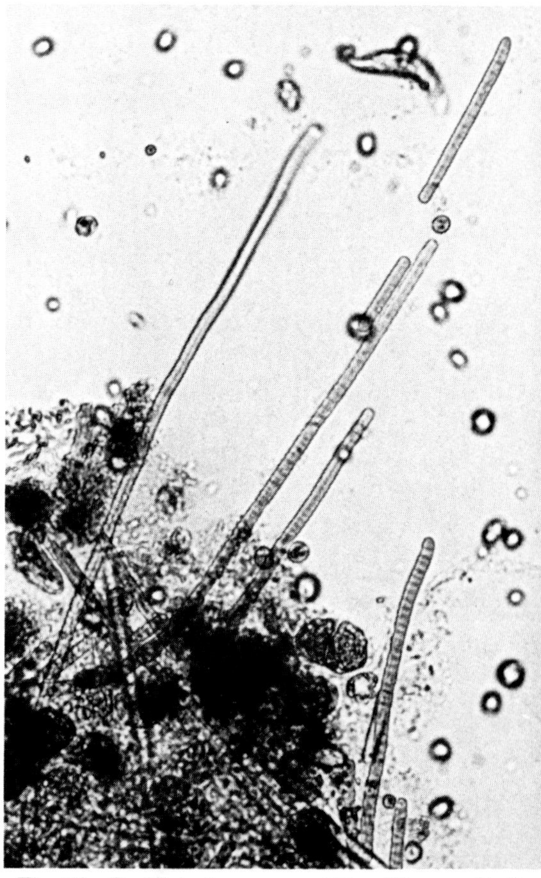

Fig. 6f Oscillatoria w.m. x430. The oscillating movements of the filaments gave this organism its name. The mechanism of movement is not understood.

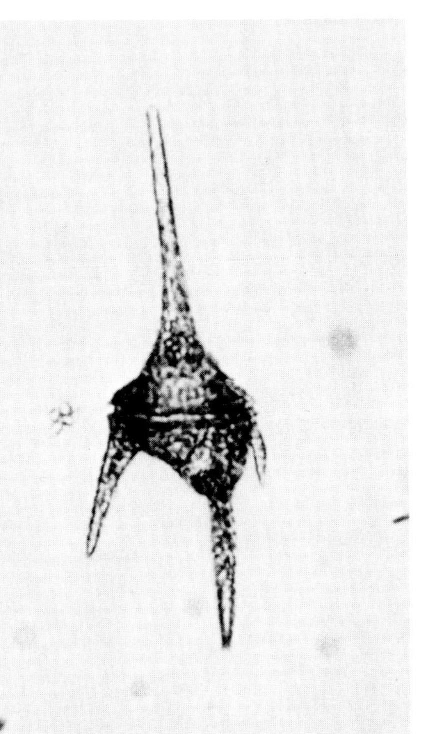

Fig. 7a Ceratium w.m. x430. (Medina Lake, Texas)

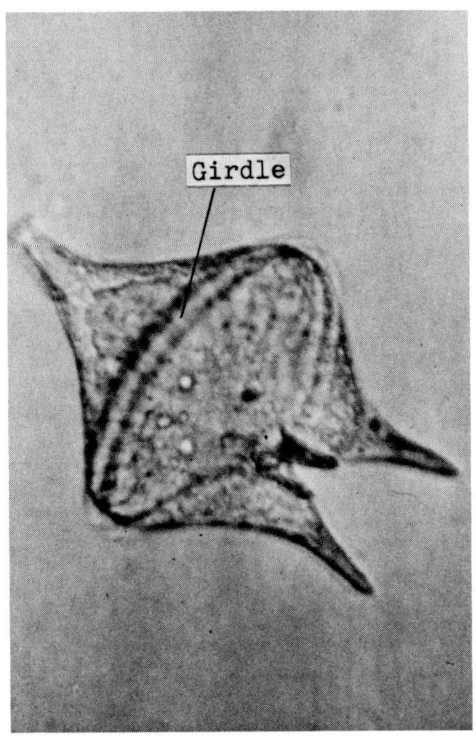

Fig. 7b Peridinium w.m. x430. One flagellum lies in the girdle and the other trails behind. Neither is visible in this photo. (Monterey Bay, Calif.)

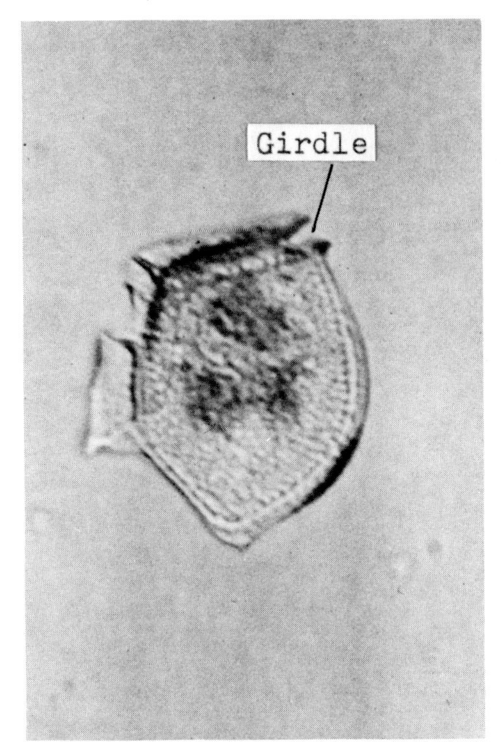

Fig. 7c Dinophysis w.m. x430. (Pacific Grove, Calif.)

Phylum CHRYSOPHYTA Class Bacillariophyceae (DIATOMS)

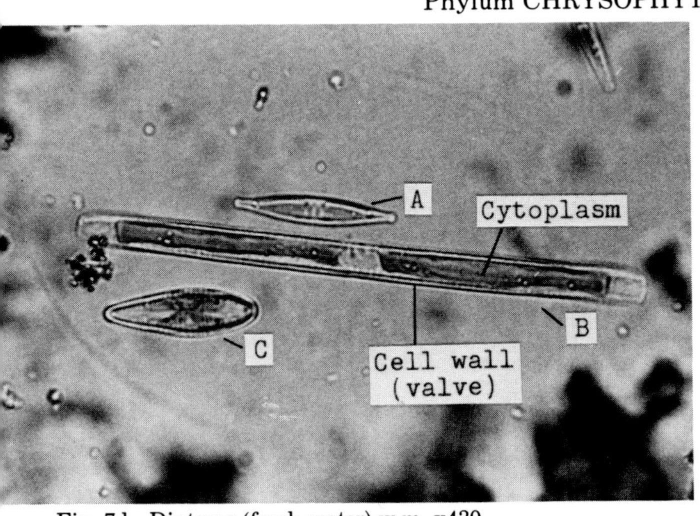

Fig. 7d Diatoms (fresh water) w.m. x430.
A - Navicula, B - Pinnularia, C - Surirella.

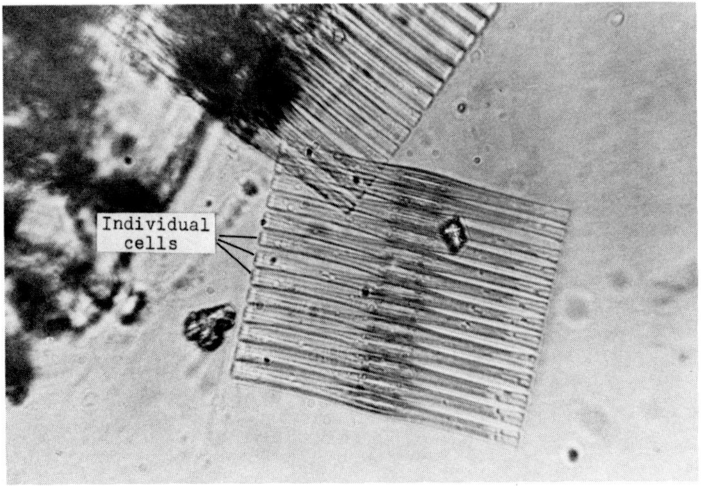

Fig. 7e Fragilaria w.m. x430. A cluster of 13 fresh water diatoms.

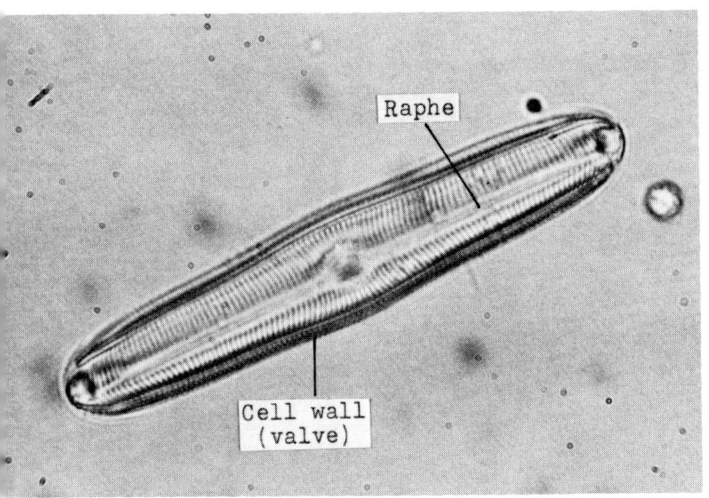

Fig. 7f Pinnularia (empty cell walls) (valve view) w.m. x430. Oily protoplasm moving along the raphe is thought to aid in locomotion. The cell walls are made of glass (SiO_2).

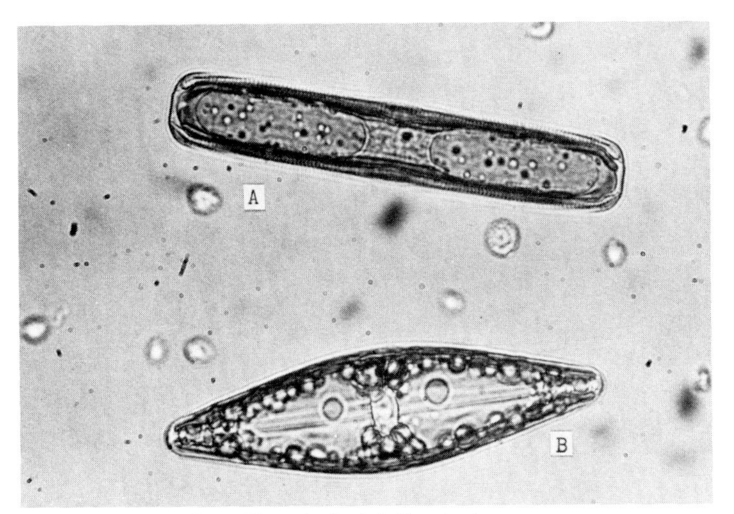

Fig. 7g A - Live Pinnularia (girdle view)
B - Live Frustulia w.m. x430. (San Antonio, Texas)

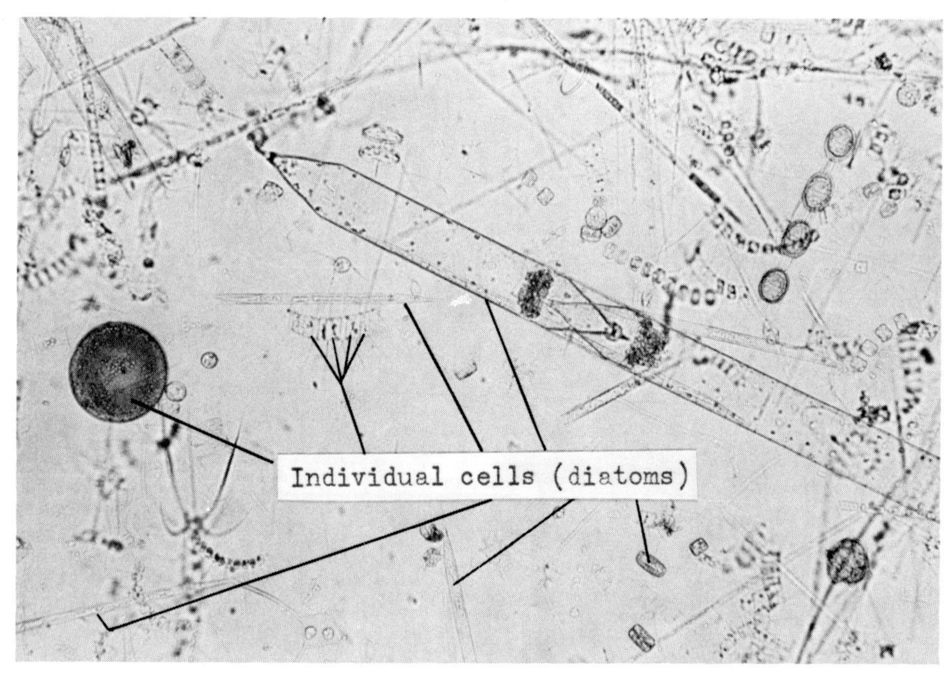

Fig. 8a Marine diatoms w.m. x100. Diatoms are the primary producers of food in the sea. The oxygen they release during photosynthesis provides the air with about 80% of its oxygen. There are about 5 tons of diatoms per surface acre of ocean.

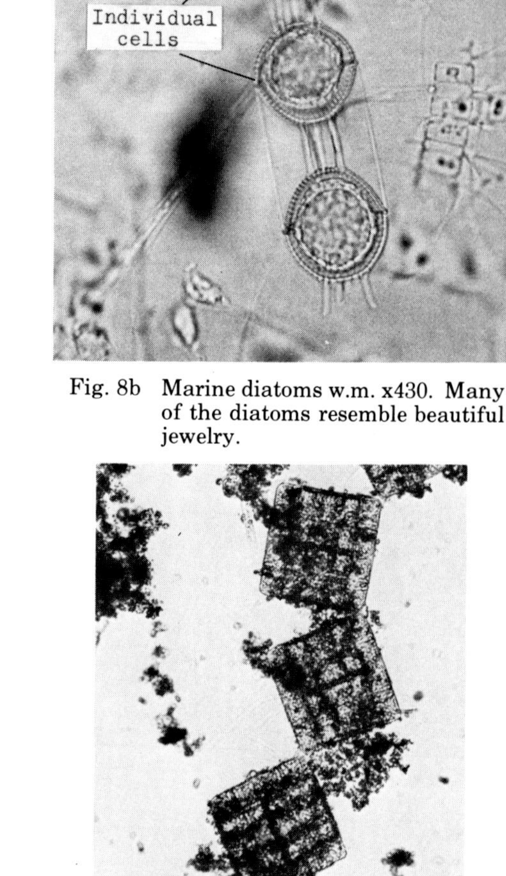

Fig. 8b Marine diatoms w.m. x430. Many of the diatoms resemble beautiful jewelry.

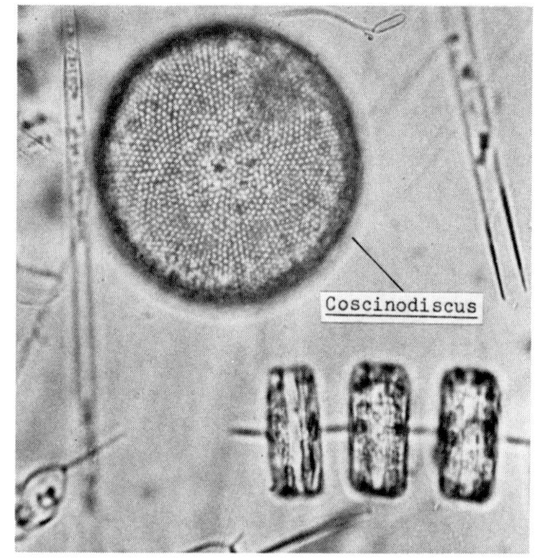

Fig. 8c Coscinodiscus, (valve view w.m.) x430. (Monterey Bay, Calif.)

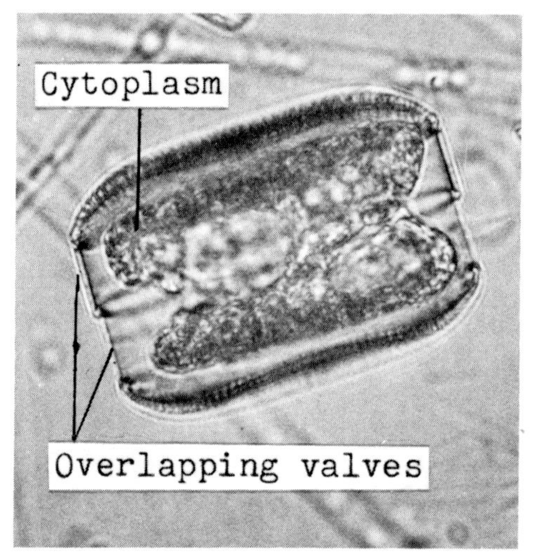

Fig. 8d Coscinodiscus in division w.m. x430. (Girdle view.) This diatom is shaped like a Petri dish. The valves (glass cell walls) separate during cell division and new ones form.

Fig. 8e Square diatoms w.m. x100. (San Pedro Park, Texas.)

(YELLOW-GREEN ALGAE) Class Xanthophyceae

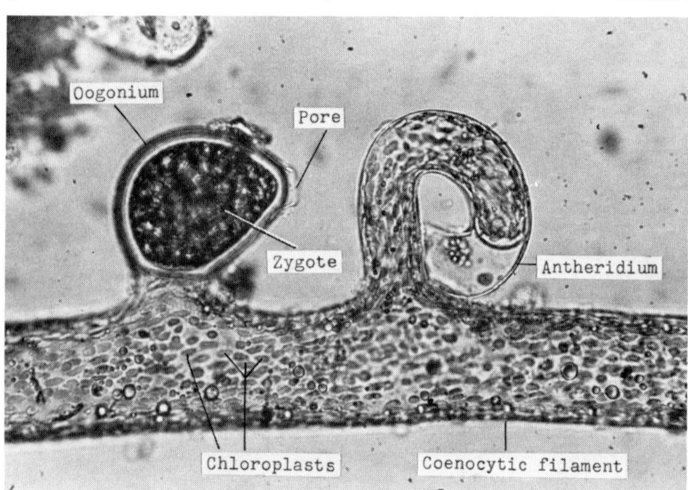

Fig. 8f Vaucheria sessilis (live) w.m. x430. Sperm from the antheridium swim to and enter the pore of the oogonium. The filament is coenocytic. (San Antonio, Texas)

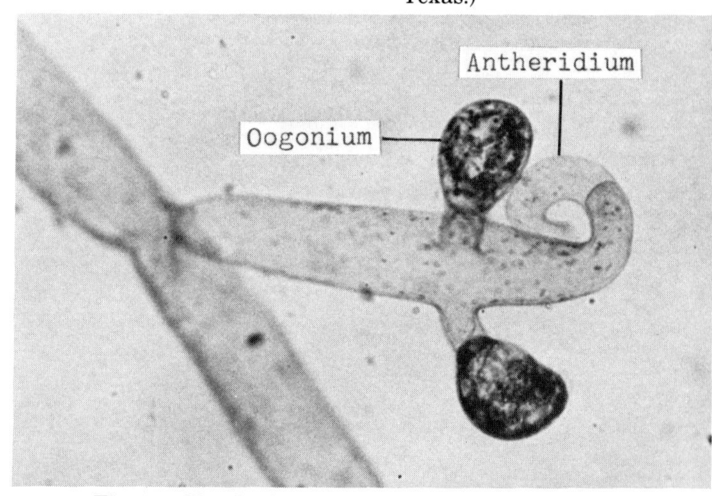

Fig. 8g Vaucheria geminata (prepared slide) w.m. x430.

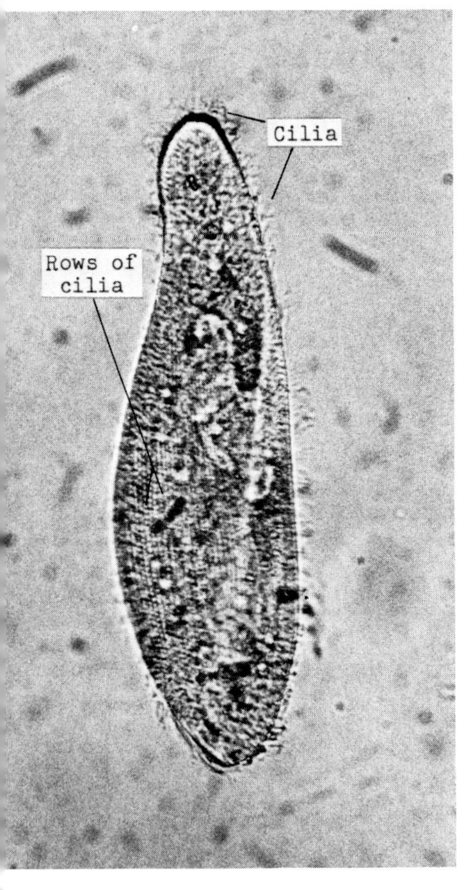

Fig. 9a Paramecium (live) w.m. x430. (Photo courtesy of Stephen Davenport.)

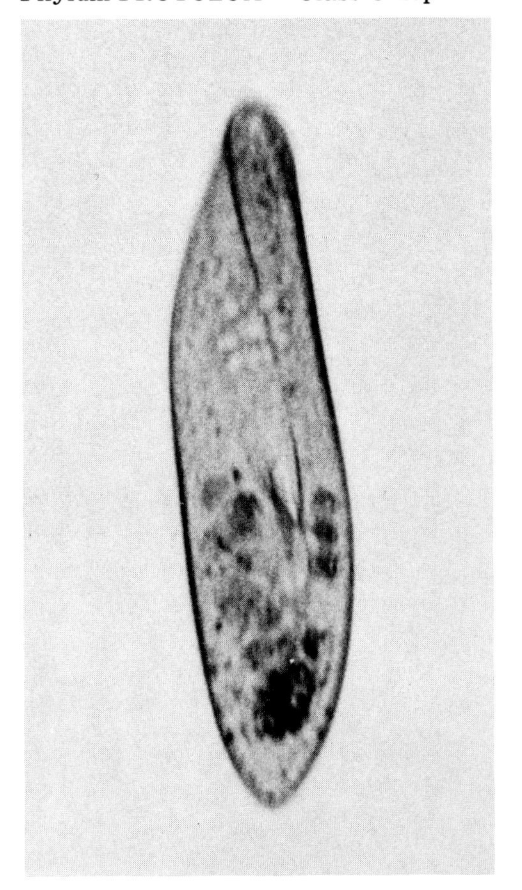

Fig. 9b Paramecium (live) w.m. x430.

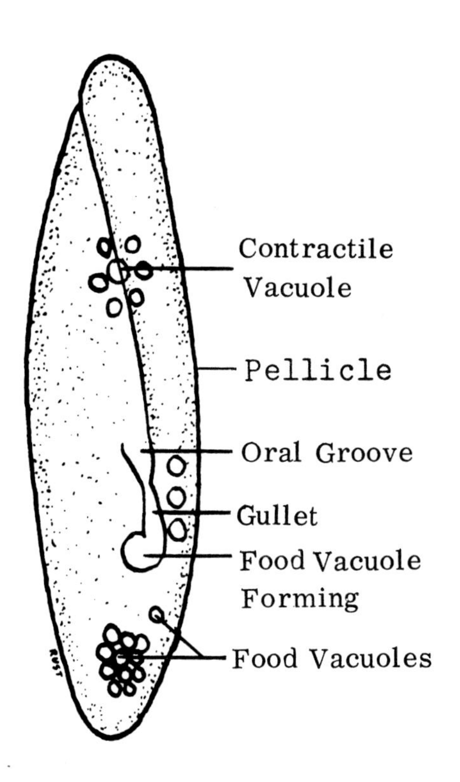

Fig. 9c Paramecium diagram. Compare with Fig. 9b.

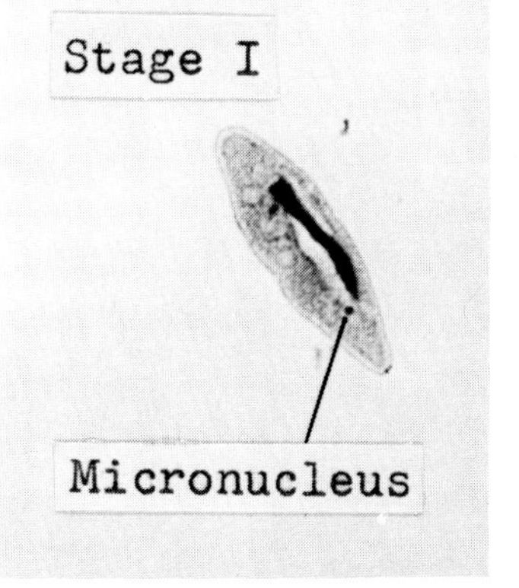

Fig. 9d Paramecium undergoing fission (early stage) w.m. x100.

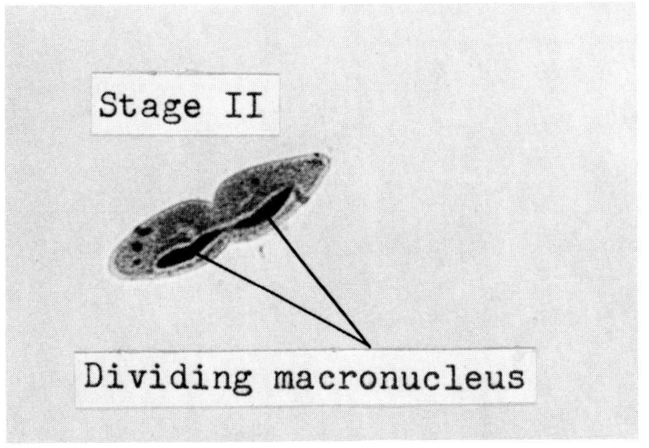

Fig. 9e Paramecium undergoing fission (later stage) w.m. x100.

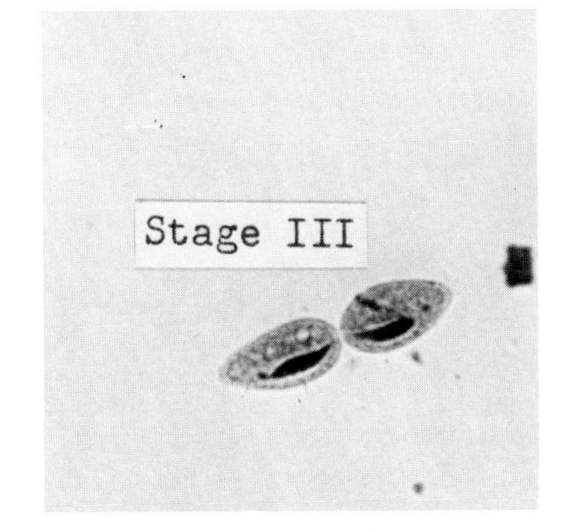

Fig. 9f Paramecium undergoing fission (final stage) w.m. x100.

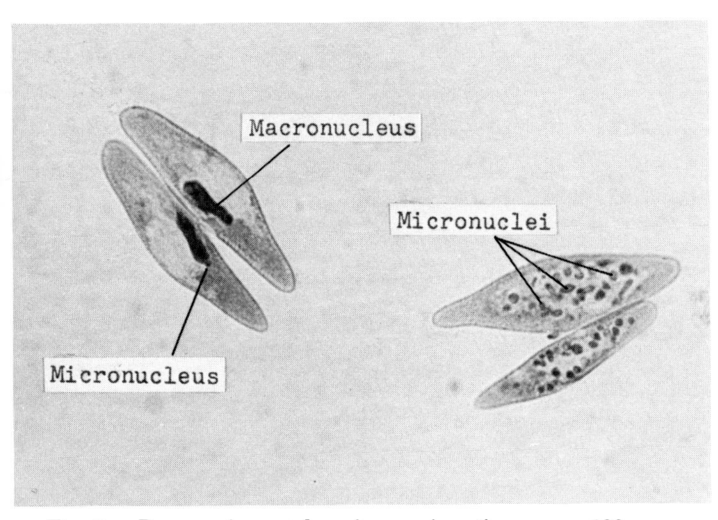

Fig. 9g Paramecium undergoing conjugation w.m. x100.

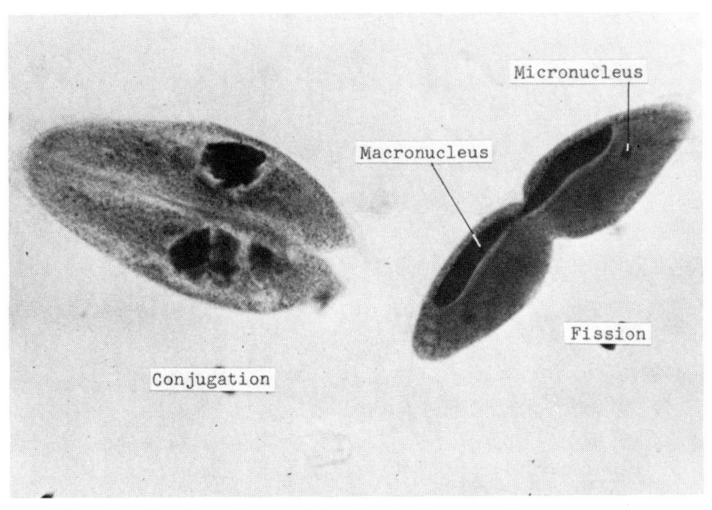

Fig. 9h Conjugation and fission in Paramecium w.m. x430.

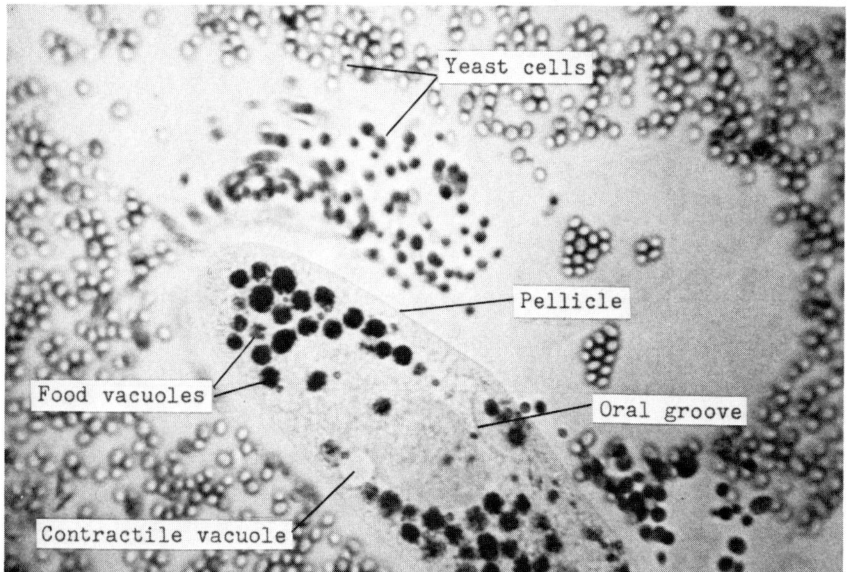

Fig. 10a Paramecium eating yeast cells stained with Congo Red w.m. x430. As the food vacuoles begin to digest the yeast, their acidic secretions turn the indicator dark blue. (See those in upper end.) Ciliary action is moving some of the yeast cells.

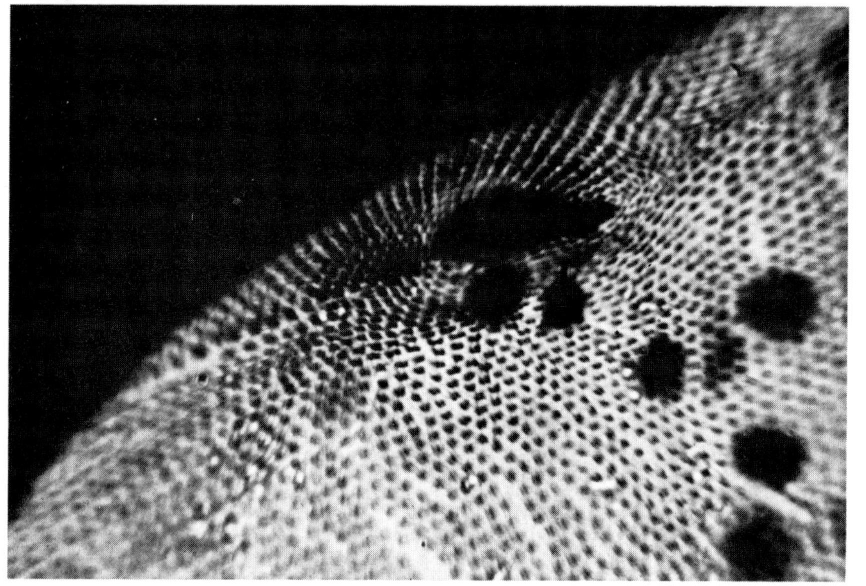

Fig. 10b Paramecium stained with Nigrosin w.m. x1000. This stain demonstrates surface features of the pellicle. Notice the pattern of cilia insertions in the oral groove region.

Fig. 10c Paramecium bursaria (live) w.m. x430. The algal cells (zoochlorellae) live symbiotically within the Paramecium. The Paramecium provides a place to live for the algae and they provide food by photosynthesis. Both benefit from the relationship.

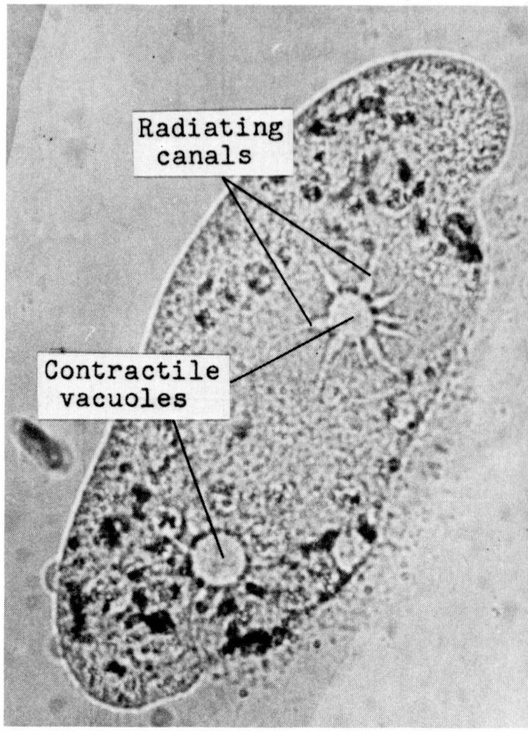

Fig. 10d Contractile vacuoles in Paramecium (liv w.m. x430. The radiating canals a thought to be part of the cell's endoplasmic reticulun

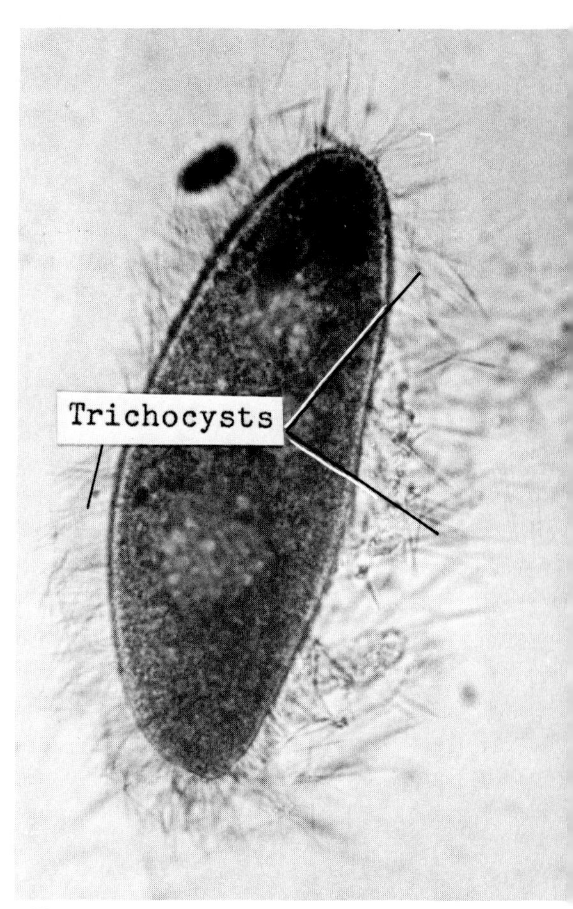

Fig. 10e Trichocysts discharged by Paramecium in response to methylene blue w.m. x430.

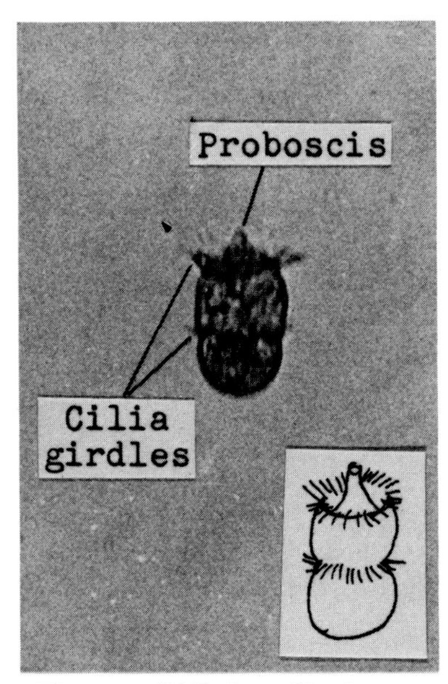

Fig. 11a Didinium (live) w.m. x100.

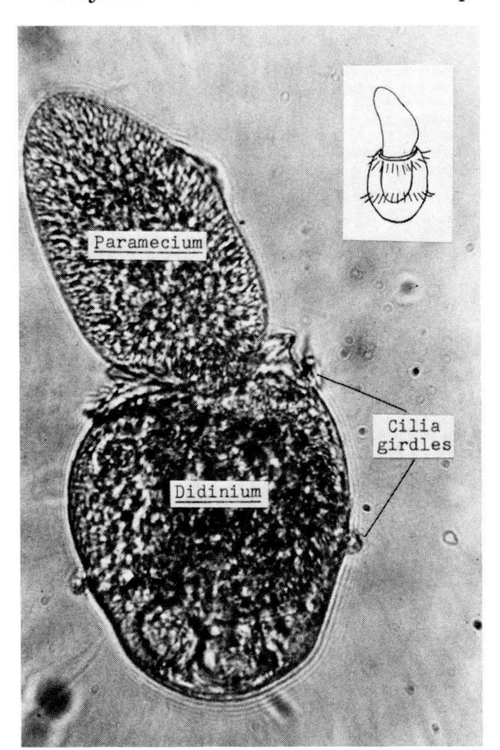

Fig. 11b Didinium eating a Paramecium (live) w.m. x430. Didinium eats Paramecium almost exclusively, consuming up to a dozen or more per day.

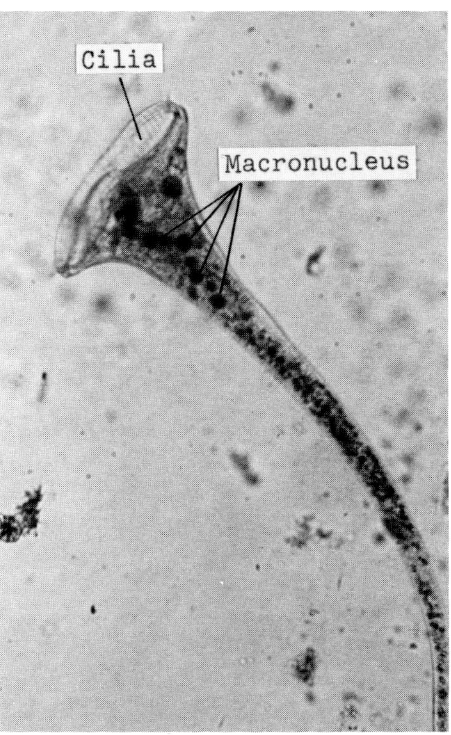

Fig. 11c Stentor (live) w.m. x100. The macronucleus resembles a chain of beads.

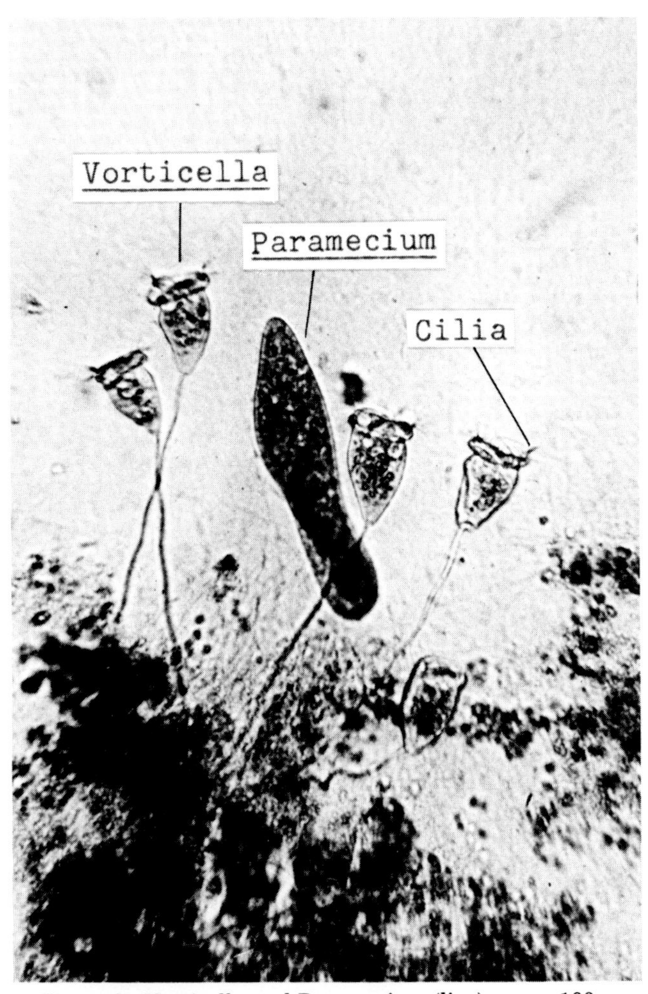

Fig. 11d Vorticella and Paramecium (live) w.m. x100.

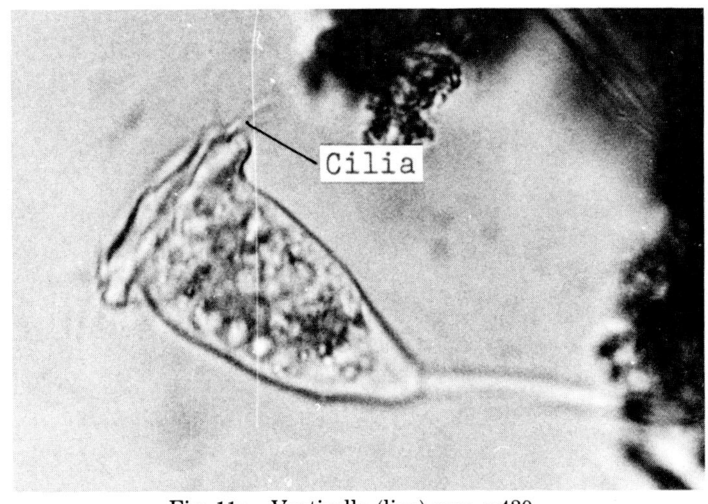

Fig. 11e Vorticella (live) w.m. x430.

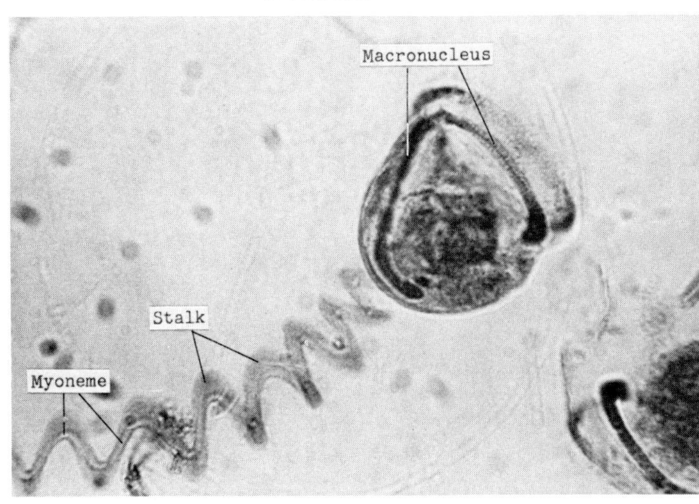

Fig. 11f Vorticella (prepared slide) w.m. x430.

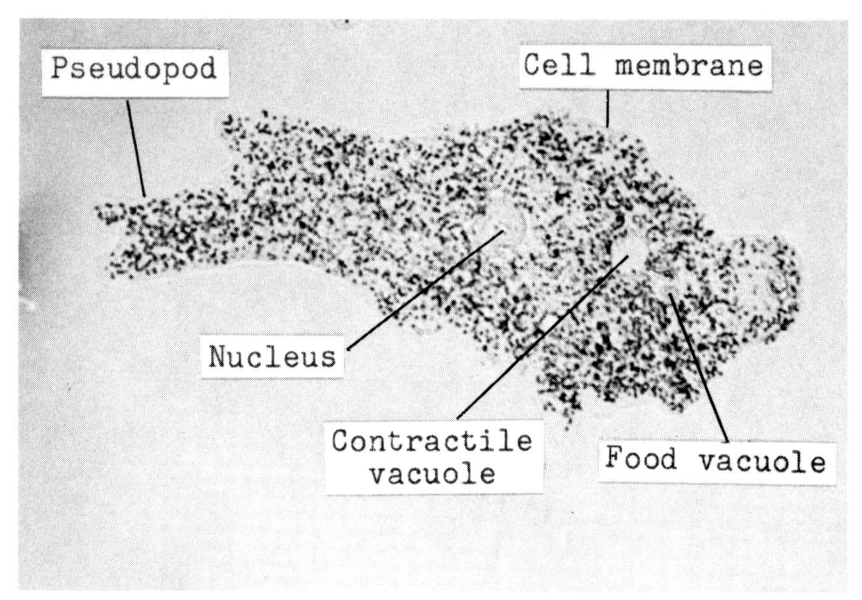

Fig. 12a Amoeba (live) w.m. x100.

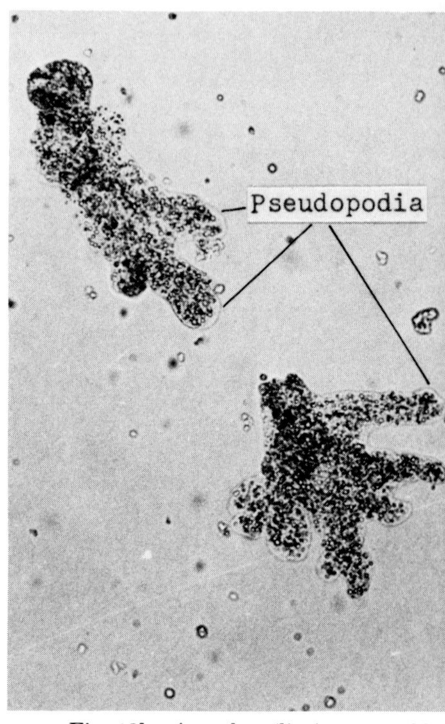

Fig. 12b Amoebae (live) w.m. x100.

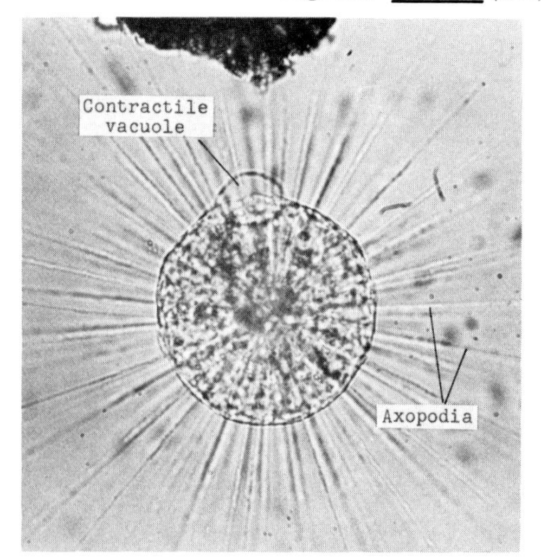

Fig. 12c Actinophrys sol (live) w.m. x430. Commonly known as the "sun animal." Axopodia are thin unbranched pseudopodia around an axial filament. (Caddo Lake, Texas)

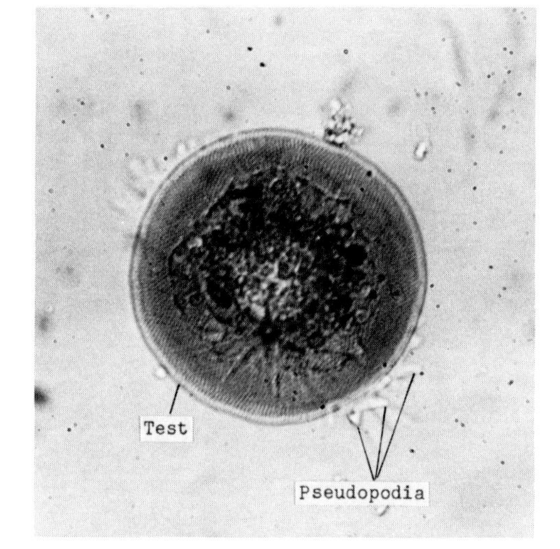

Fig. 12d Arcella (live) w.m. x430. The test (shell) is hemispherical with a hole in the bottom (visible as light circle). The amoeba's cell can be seen in the test with pseudopodia extended. (Caddo Lake, Texas)

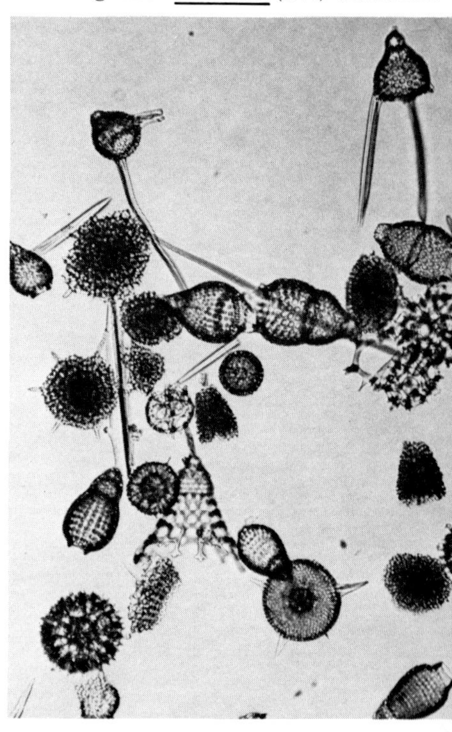

Fig. 12e Glass shells of Radiolaria w.m. x40. Marine amoebae produce and live in these, extending needle-like pseudopodia through tiny holes in the shells.

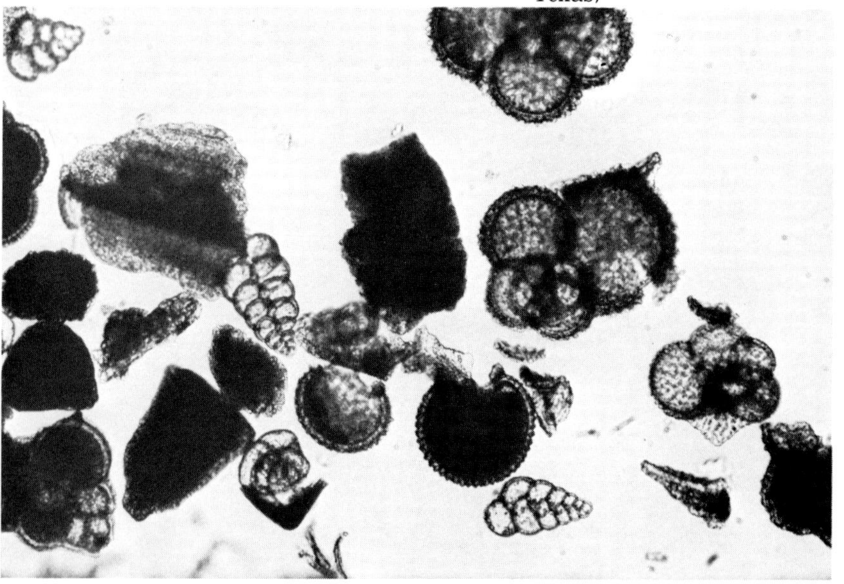

Fig. 12f Foraminifera shells w.m. x40. These are the $CaCO_3$ shells of some marine amoebae. Each shell contained one cell. Needle-like pseudopodia extended through tiny holes in the shells (foramen means hole). At death, these shells sink and form a sediment on the ocean bottom ultimately becoming a part of limestone.

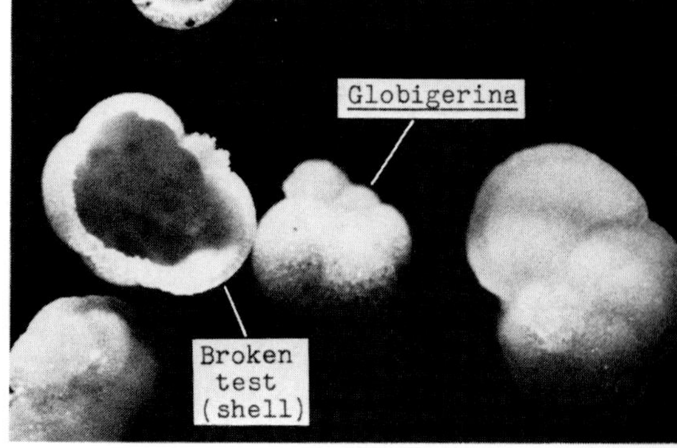

Fig. 12g Foraminifera shells w.m. x40. (Pacific Grove, Calif.)

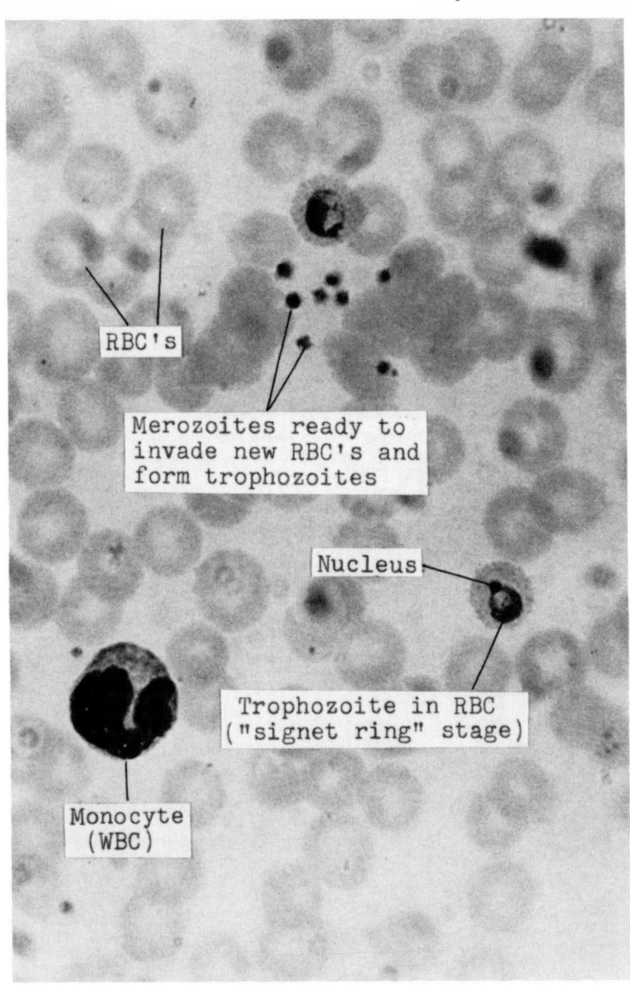

Fig. 13a Plasmodium vivax w.m. x1000. This protozoan (sporozoan) causes malaria. Notice the characteristic "ring" stage formed in the infected red blood cell (RBC) by the protozoan. P. vivax is transmitted by the female Anopheles mosquito.

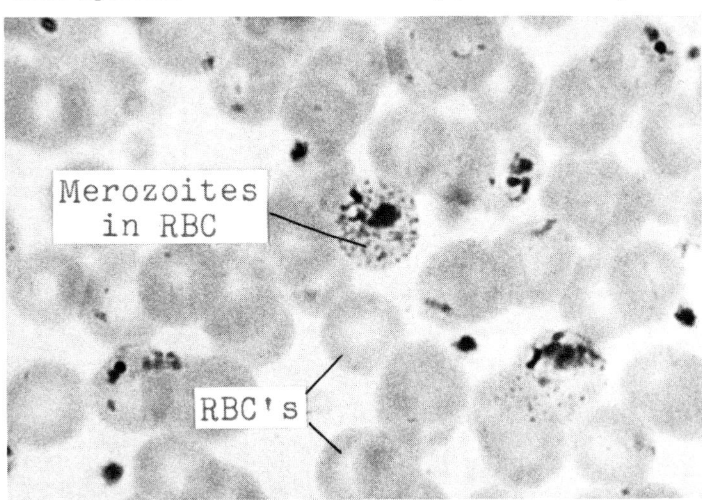

Fig. 13b Plasmodium vivax w.m. x1000. Merozoites in an RBC.

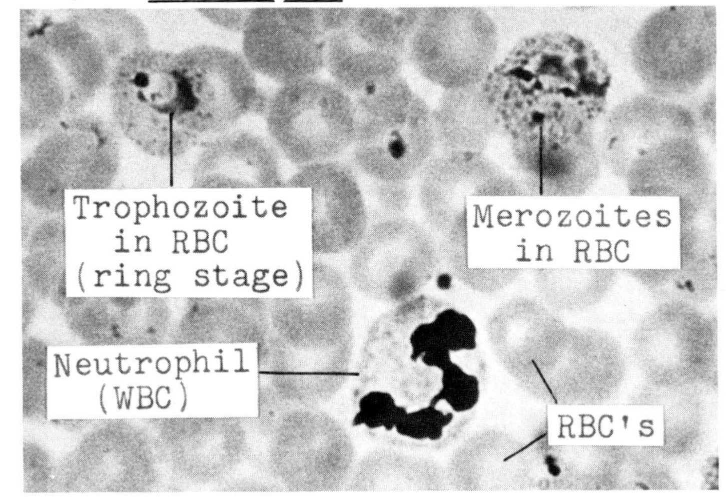

Fig. 13c Plasmodium vivax w.m. x1000. Trophozoites and merozoites in RBC's.

Class Mastigophora (NON-PHOTOSYNTHETIC FLAGELLATES)

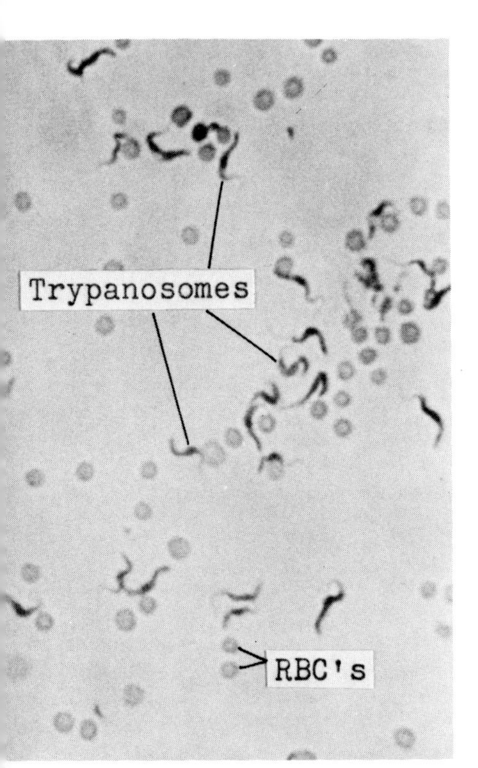

Fig. 13d Trypanosoma w.m. x100. These flagellated protozoans cause African sleeping sickness.

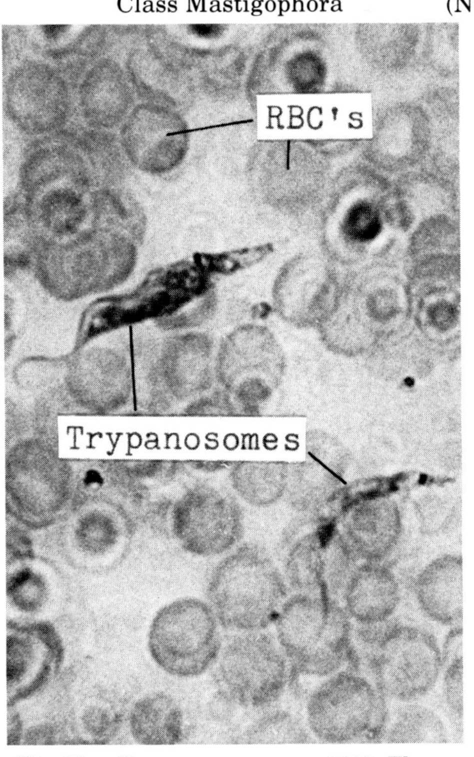

Fig. 13e Trypanosoma w.m. x1000. These organisms leave the blood and invade the fluids of the brain and spinal cord causing unconsciousness. It is transmitted by the Tsetse fly.

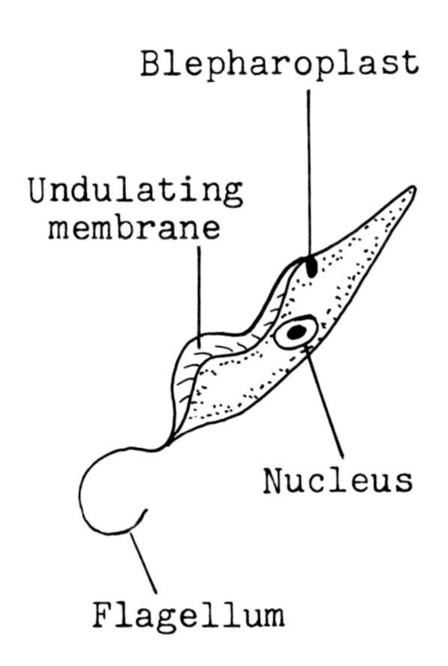

Fig. 13f Trypanosoma diagram. Compare with Fig. 13e. The flagellum is thought to originate from the blepharoplast.

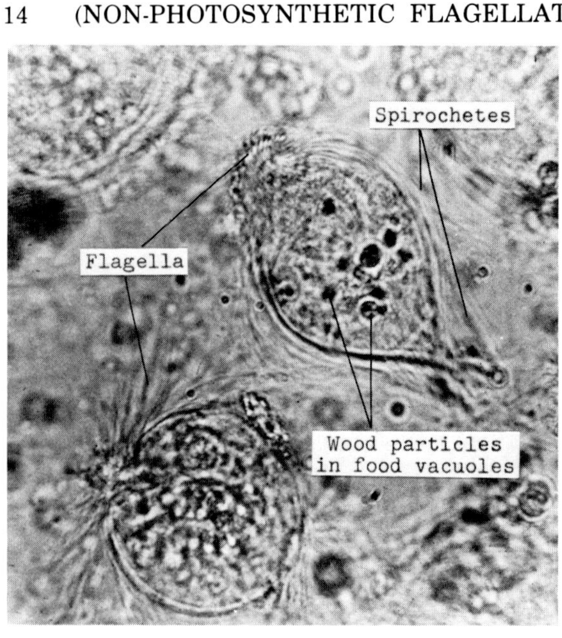

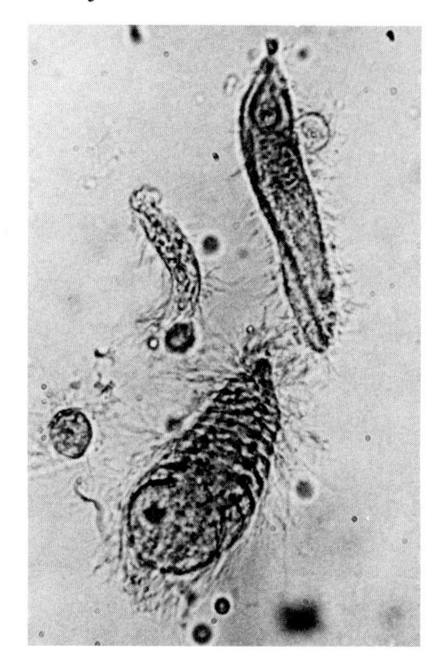

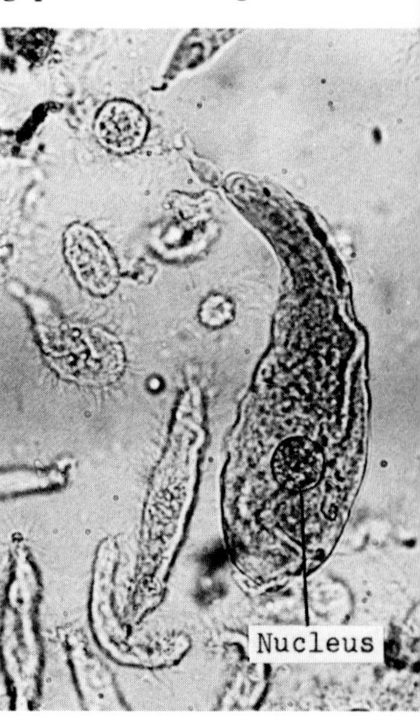

Fig. 14a, 14b, 14c Live protozoans (mutualistic flagellates) from the gut of a termite w.m. x430. These protozoans have the enzymes necessary to digest wood (cellulose) that the termite does not. They take their share of the digested cellulose and the termite gets the rest.

(PHOTOSYNTHETIC FLAGELLATES) Phylum EUGLENOPHYTA

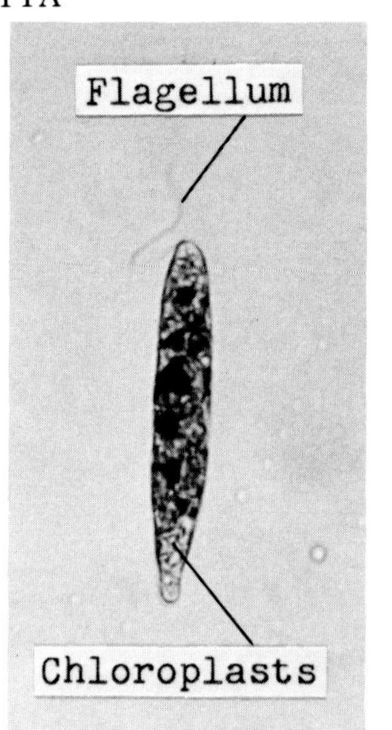

Fig. 14e Euglena w.m. x430.

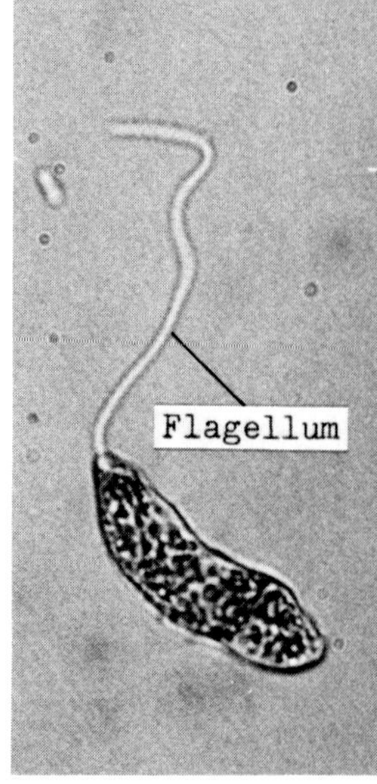

Fig. 14f Peranema w.m. x430. Peranema has no chloroplasts & is non-photosynthetic.

Fig. 14d Euglena and Paramecium (live) w.m. x430. The Paramecium is beginning to divide (notice the furrow around its middle). The bacteria are Paramecium's principal food. Euglena is photosynthetic. (Cedar Falls, Iowa)

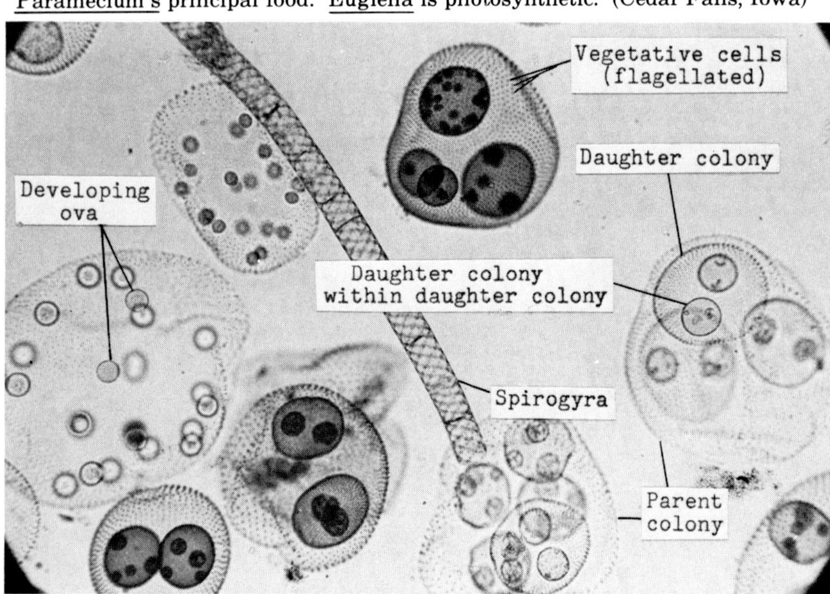

Fig. 14g Volvox colonies (prepared slide) w.m. x100.

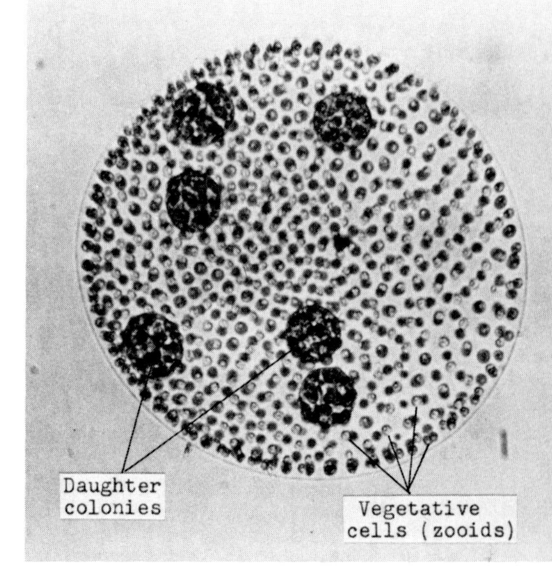

Fig. 14h Volvox colony (live) w.m. x430. (Caddo Lake, Texas) (Sometimes classified with the Chlorophyta.)

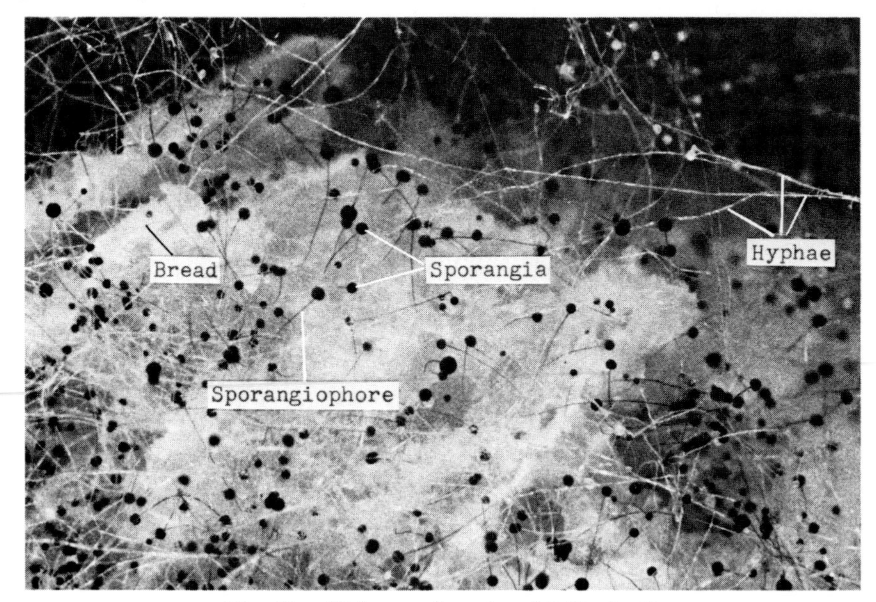

Fig. 15a Rhizopus (bread mold) (live) x10.

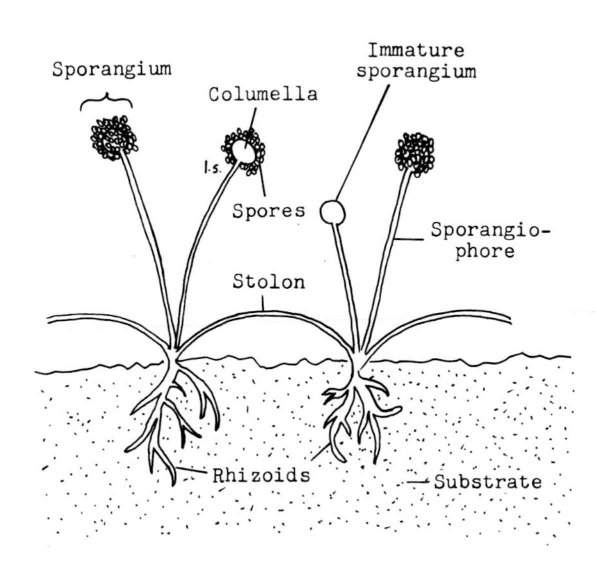

Fig. 15b Rhizopus anatomy.

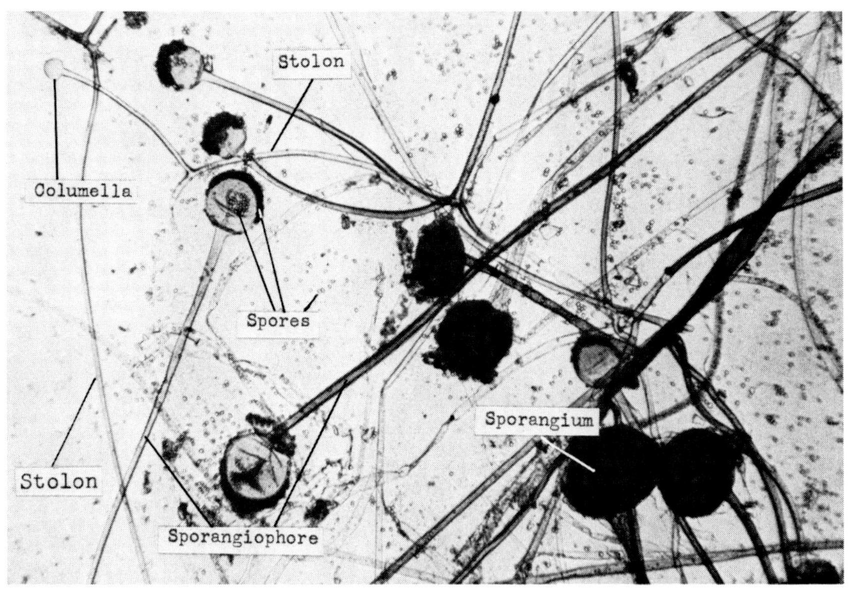

Fig. 15c Rhizopus (live) w.m. x100.

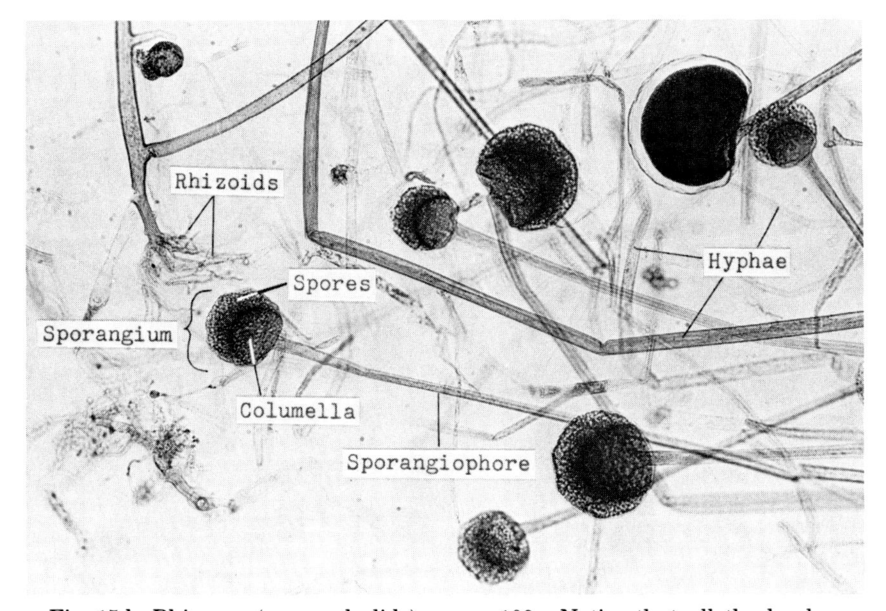

Fig. 15d Rhizopus (prepared slide) w.m. x100. Notice that all the hyphae (filaments) are coencytic.

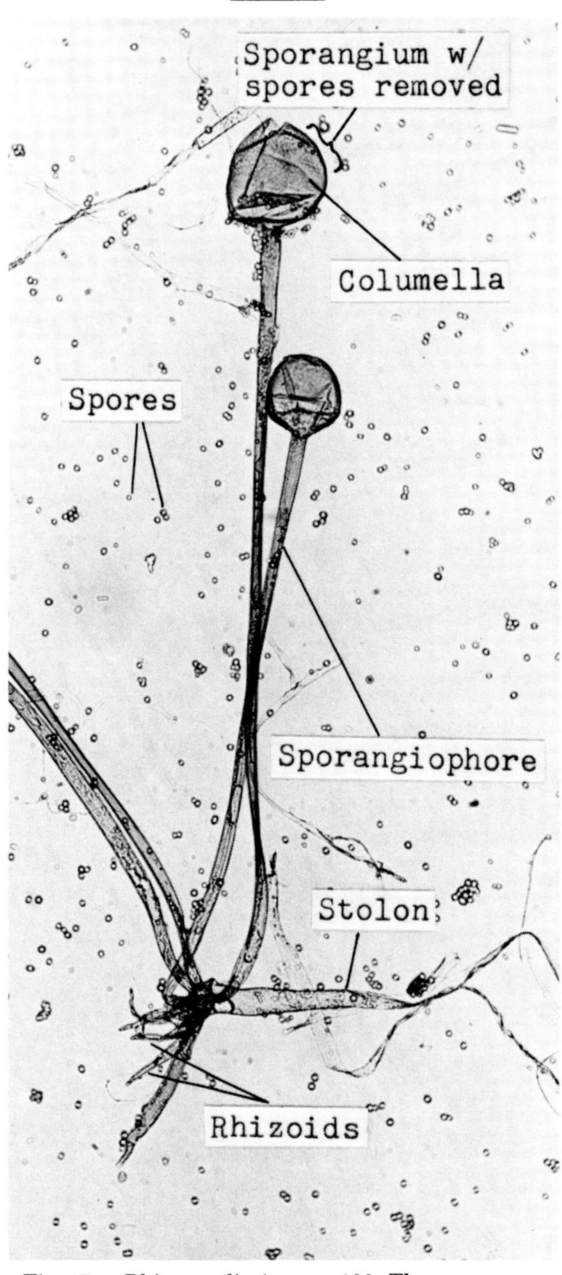

Fig. 15e Rhizopus (live) w.m. x100. The spores were accidentally knocked off the sporangia in the slide-making process.

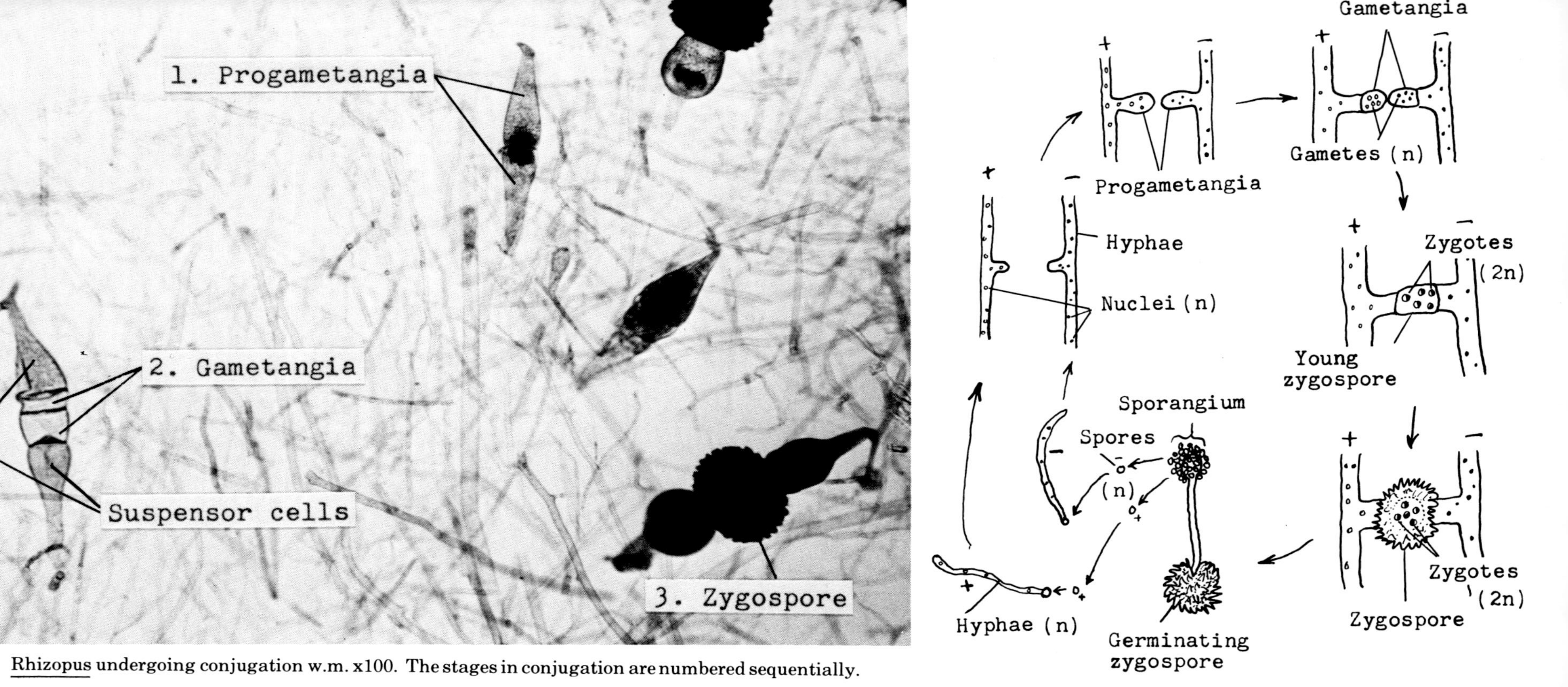

Fig. 16a Rhizopus undergoing conjugation w.m. x100. The stages in conjugation are numbered sequentially.

Fig. 16b Conjugation in Rhizopus. The zygospore contains many zygotes. Meiosis occurs during zygospore germination to produce haploid spores.

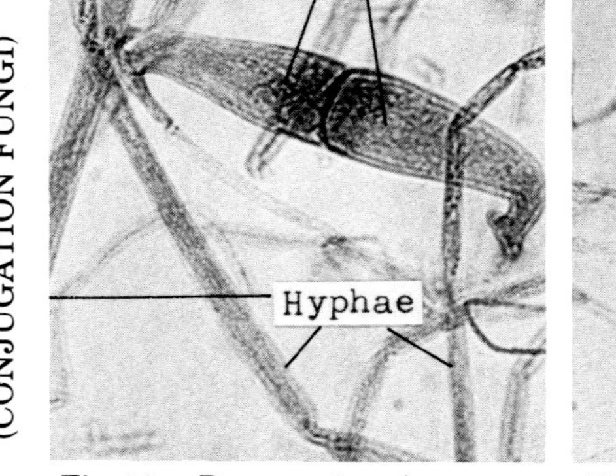

Fig. 16c Progametangia w.m. x430.

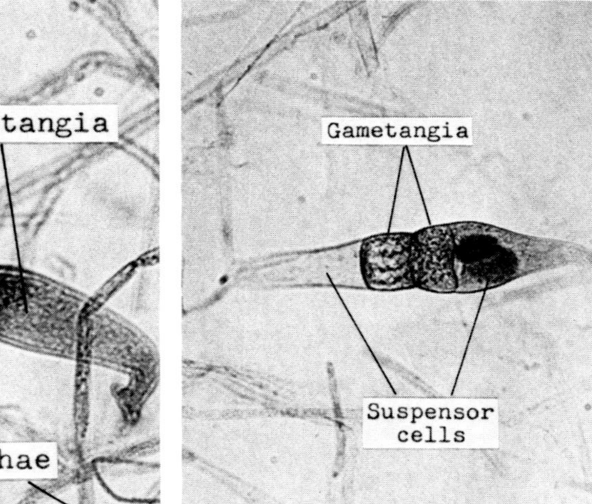

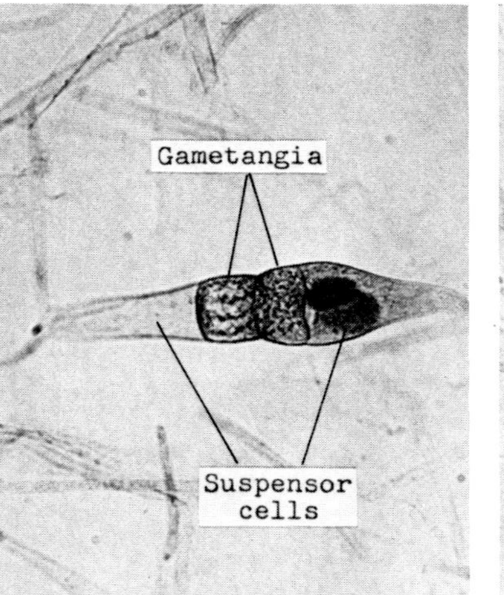

Fig. 16d Gametangia (containing gametes) w.m. x430.

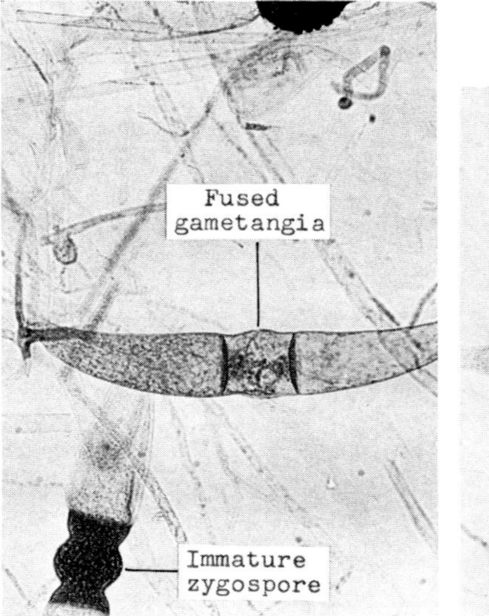

Fig. 16e Fused gametangia w.m. x430. (Gametes fusing to form zygotes.)

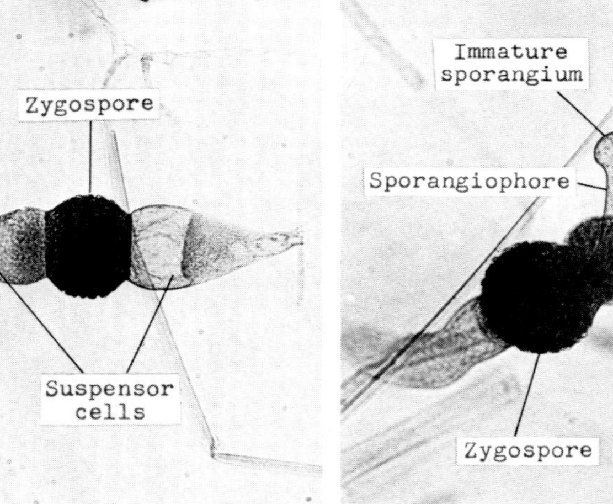

Fig. 16f Zygospore (containing zygotes) w.m. x430.

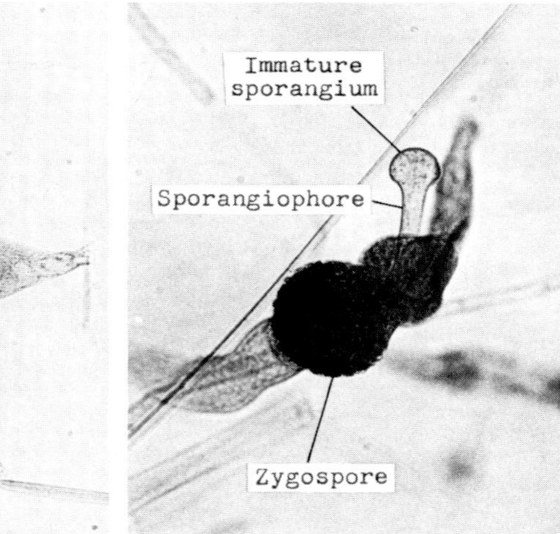

Fig. 16g Germinating zygospore w.m. x430.

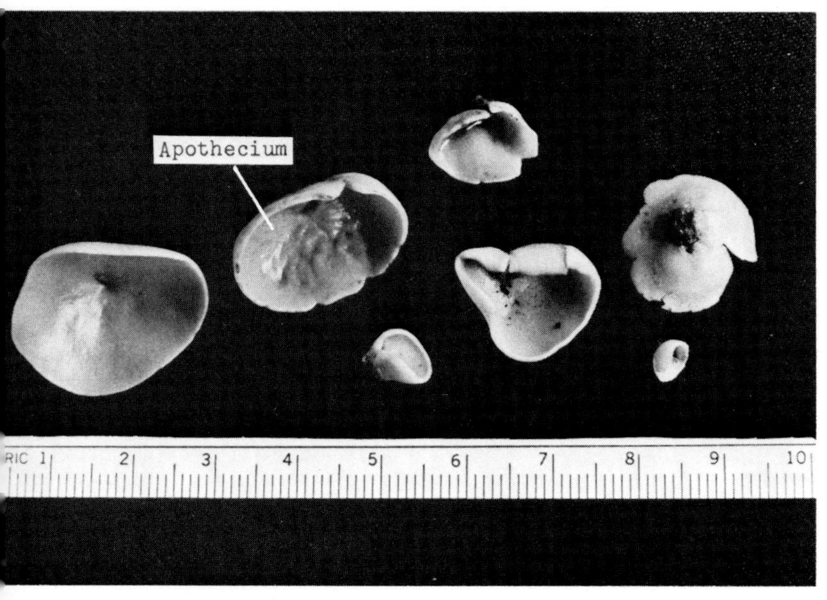

Fig. 17a Whole Peziza x1. The apothecium (fruiting body) is cup-shaped.

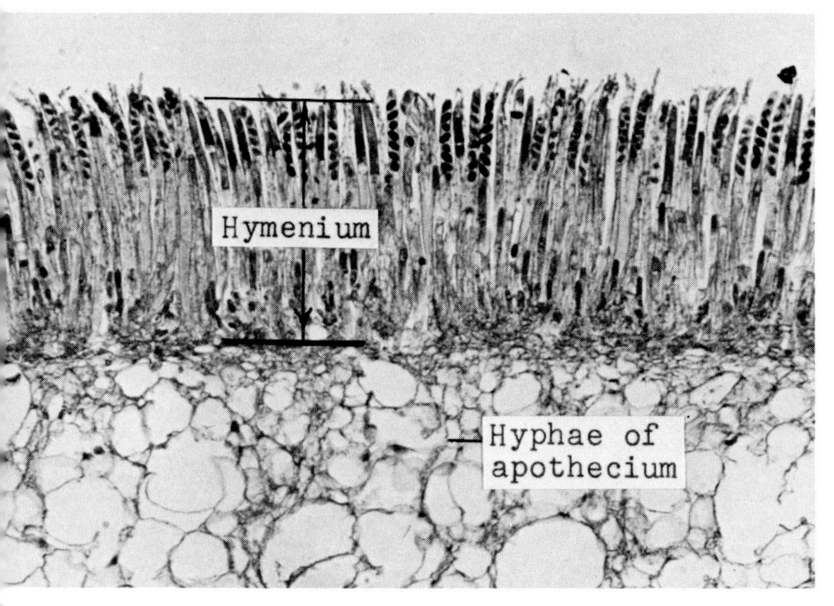

Fig. 17b Peziza apothecium l.s. x100. The hymenium is the layer of spore-producing hyphae.

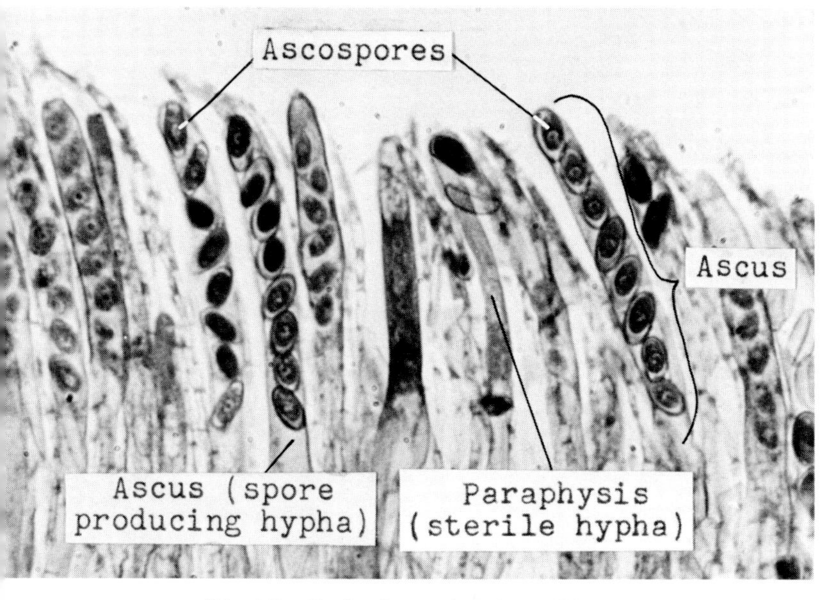

Fig. 17c Peziza hymenium l.s. x430.

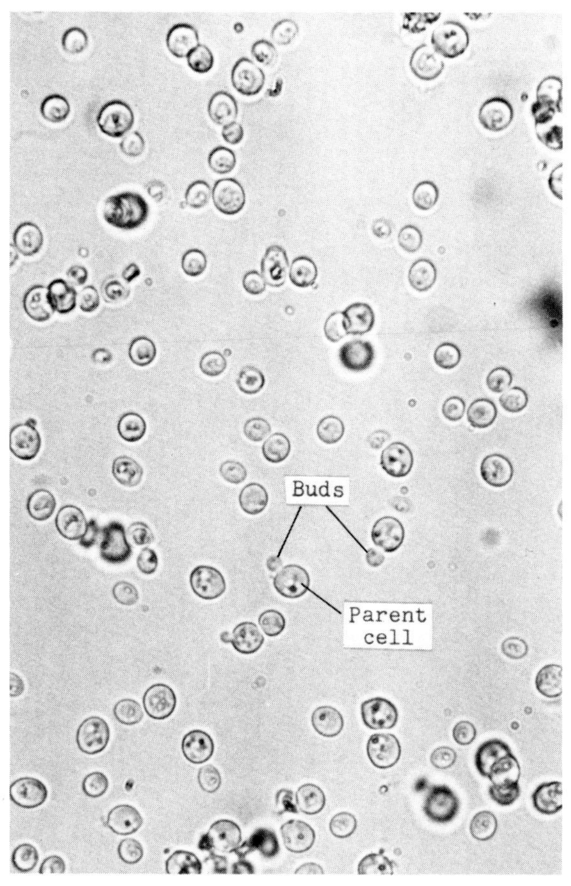

Fig. 17d Budding yeast cells (live) w.m. x1000.

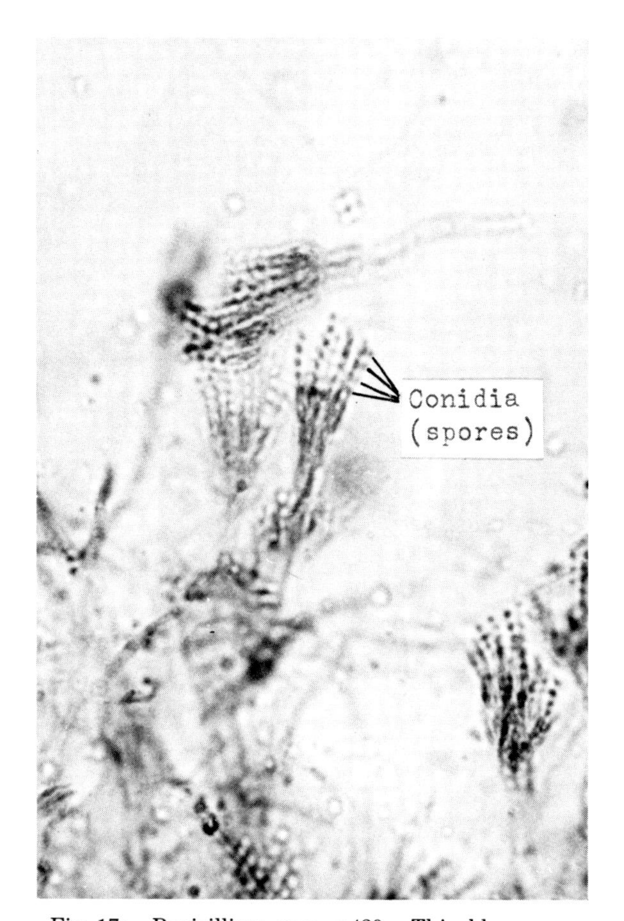

Fig. 17e Penicillium w.m. x430. This blue-green mold is commonly found in refrigerators on cheese, bacon and long forgotten left-overs. One species is used to produce penicillin. Another species of Penicillium produces the flavor in roquefort cheese.

Fig. 18a Wild mushroom (Grand Teton Nat'l Park, Wyo.) x½.

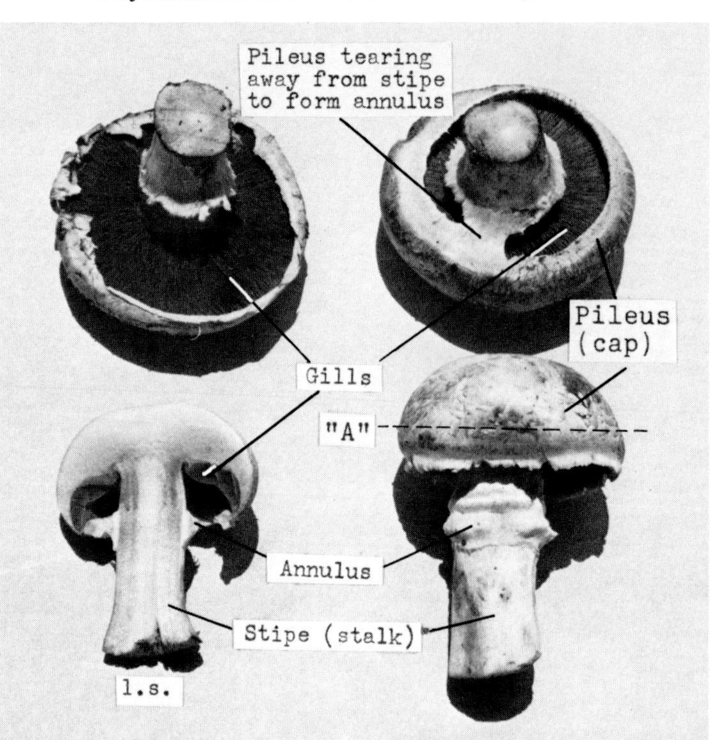

Fig. 18b Mushroom anatomy x1.

Fig. 18c Bracket (shelf) fungi. (Forked Lake, N.Y.)

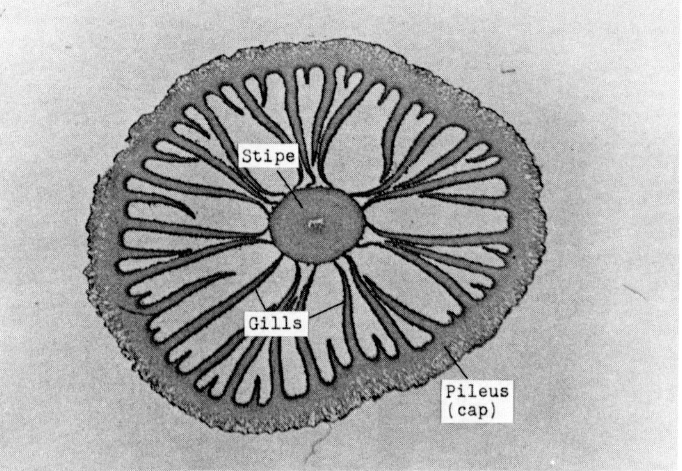

Fig. 18d Mushroom (Coprinus) pileus x.s. x40. Section taken at location "A" in Fig. 18b.

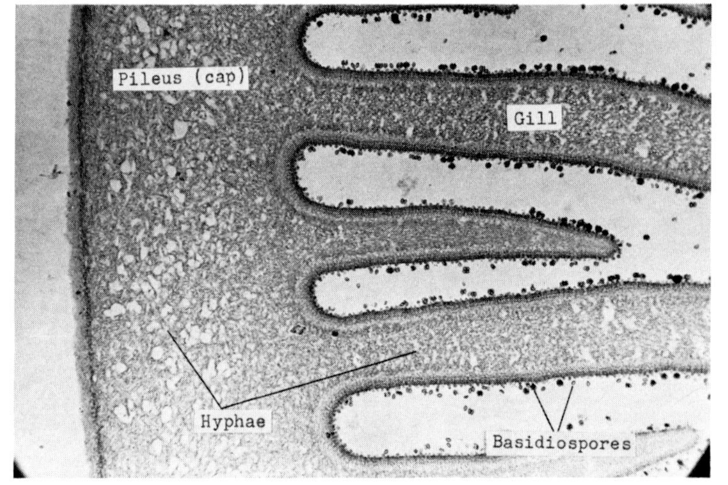

Fig. 18e Mushroom (Coprinus) pileus x.s. x100.

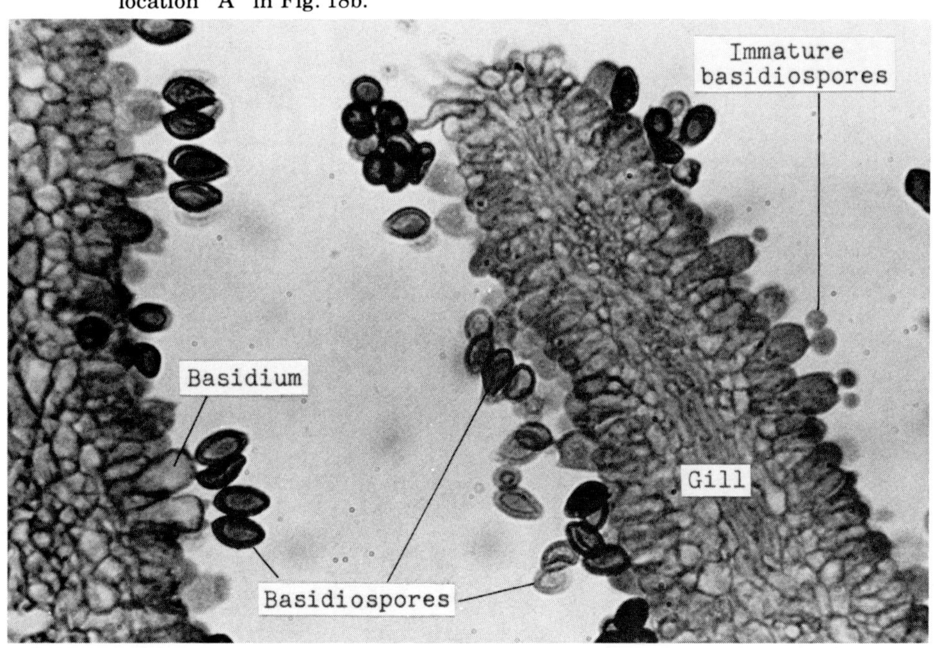

Fig. 18f Coprinus gill x.s. x430. Usually four basidiospores form on each basidium.

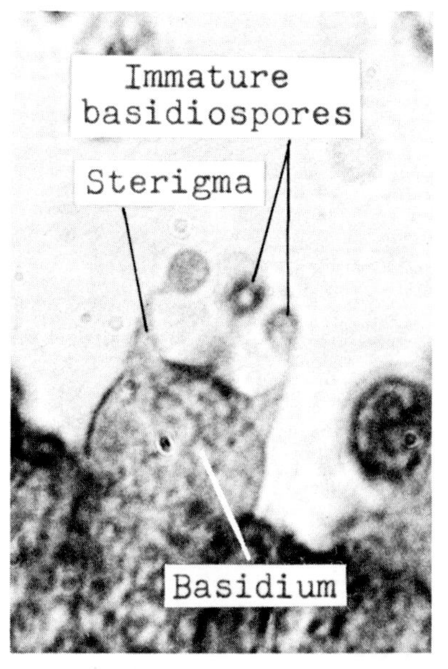

Fig. 18g Coprinus gill x.s. x1000. Basidium with immature basidiospores.

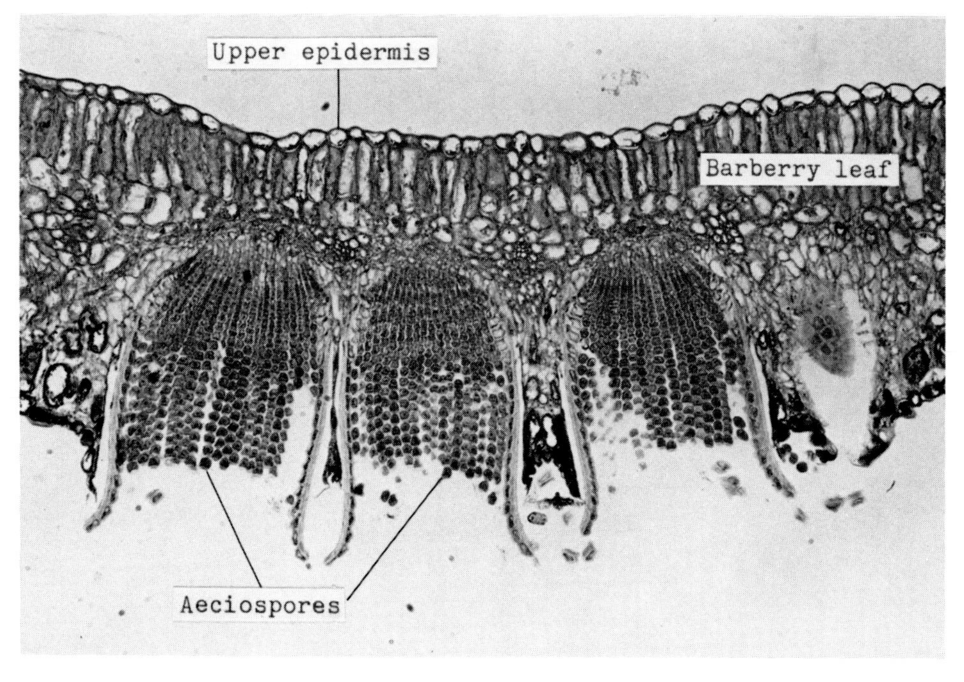

Fig. 19a Puccinia graminis (wheat rust) aeciospores on Barberry leaf x.s. x100.

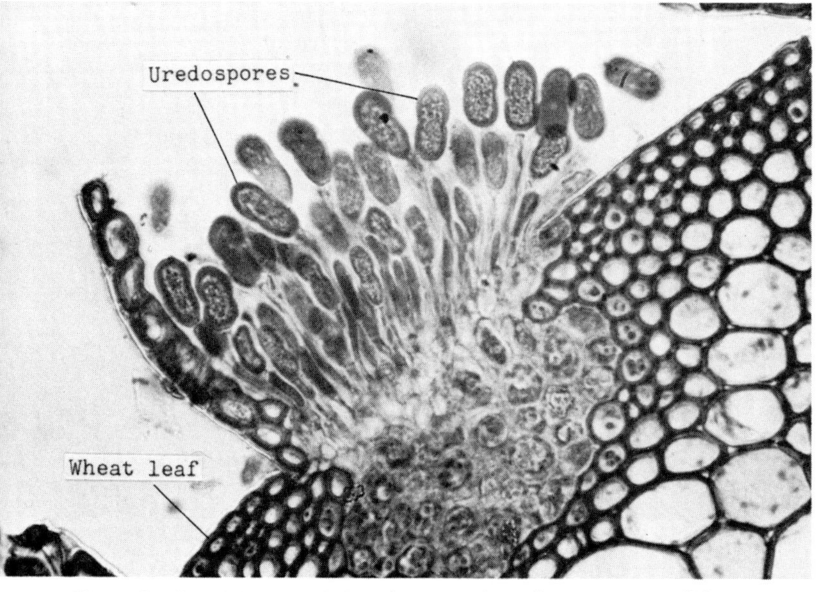

Fig. 19b Puccinia graminis (wheat rust) uredospores x.s. x430.

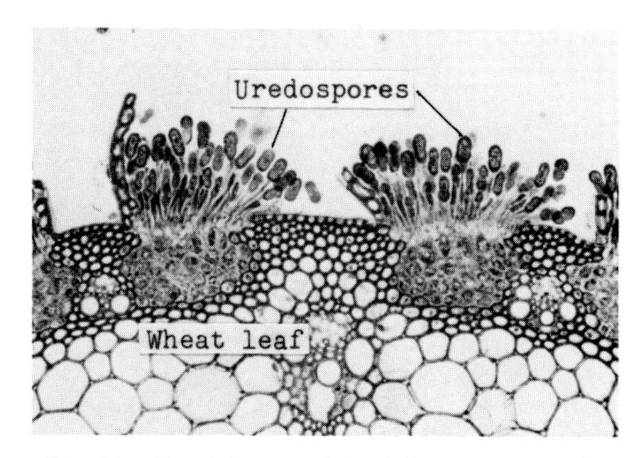

Fig. 19c Puccinia graminis (wheat rust) uredospores on wheat leaf x.s. x100.

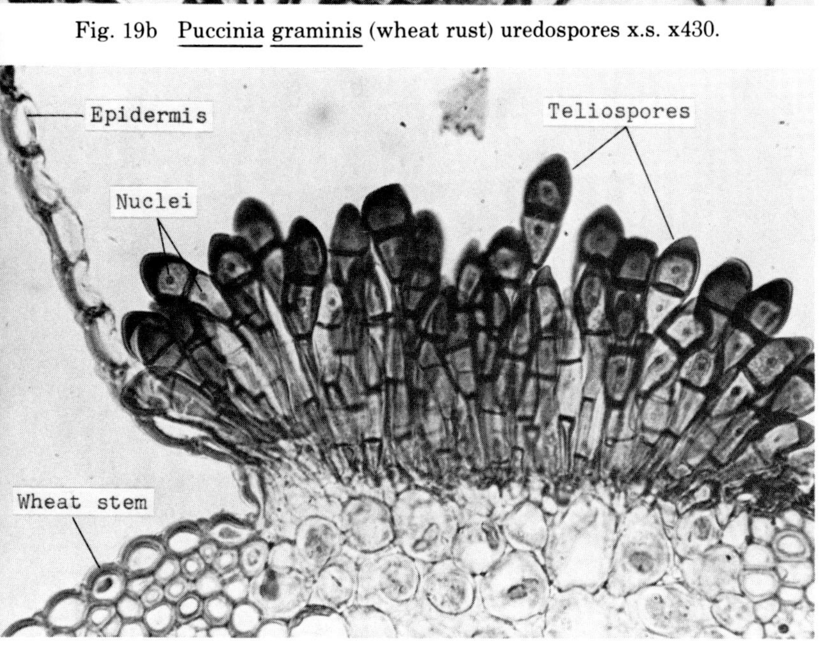

Fig. 19d Puccinia graminis (wheat rust) teliospores x.s. x430.

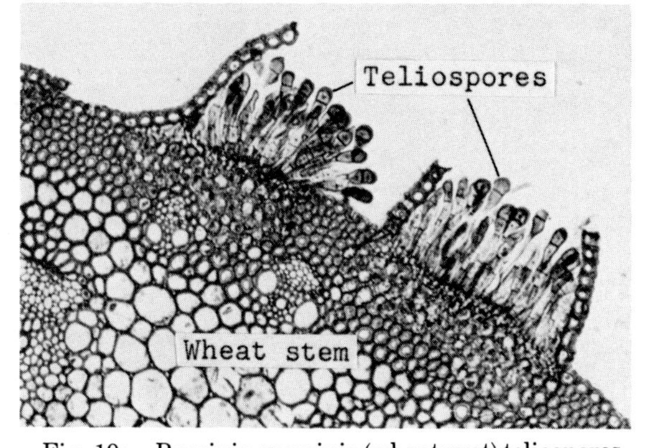

Fig. 19e Puccinia graminis (wheat rust) teliospores on wheat stem x.s. x100.

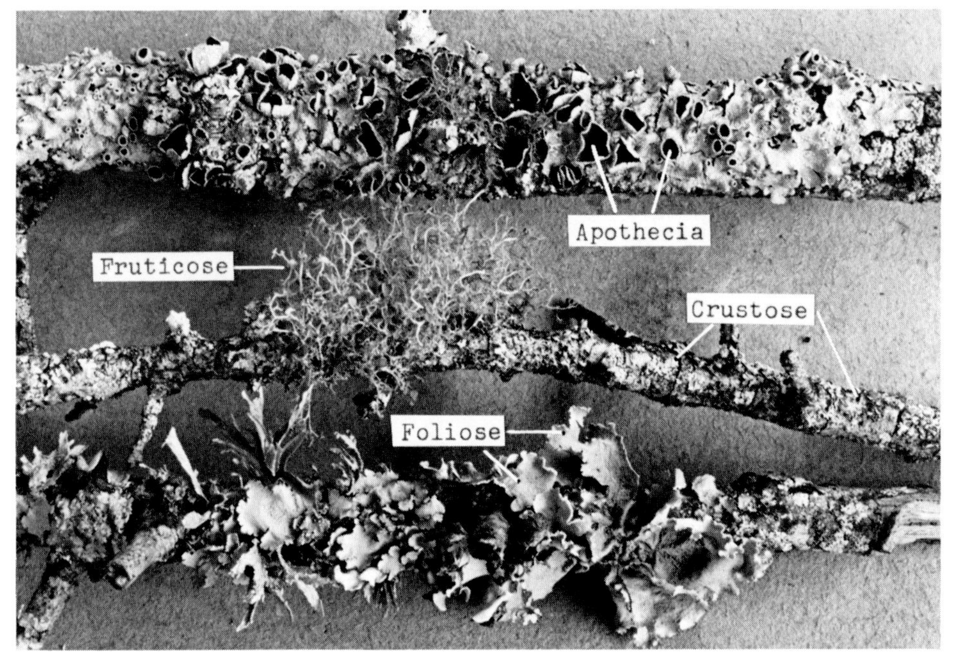

Fig. 20a Lichens (crustose, foliose and fruticose) x1. The various colors and shapes within these three groups are determined by which species of fungus has become associated with which species of algae.

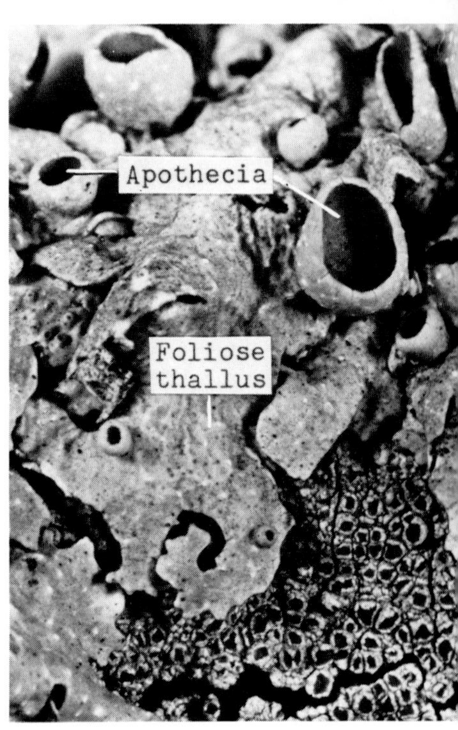

Fig. 20b Apothecia on foliose lichen x3. Fungus spores are produced in the apothecia.

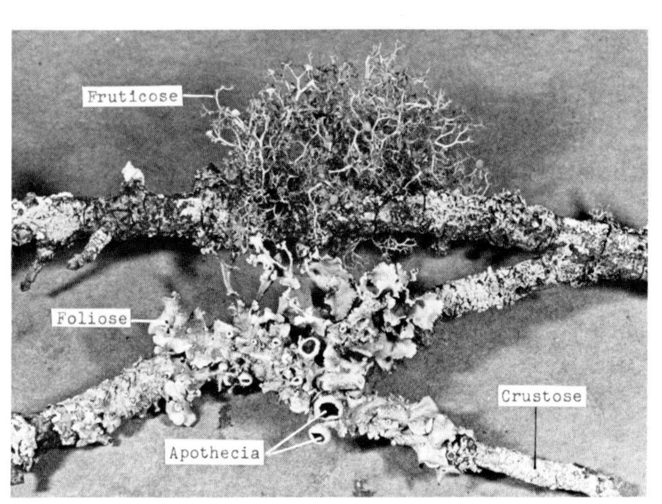

Fig. 20c Various lichen types x1.

Fig. 20d Crustose lichens on rocks excrete acidic wastes which erode the rock, helping to form the mineral portion of soil.

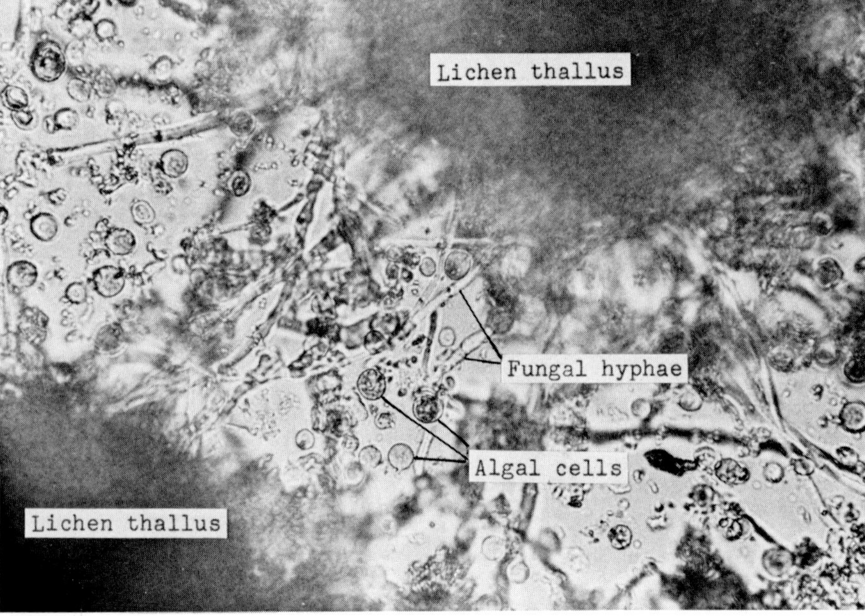

Fig. 20e Living lichen thallus (body) teased apart to show the fungal hyphae and algal cells that comprise it w.m. x430.

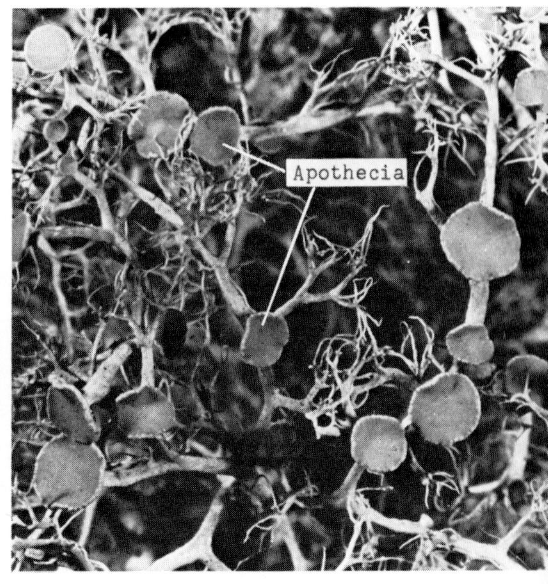

Fig. 20f Fruticose lichen with apothecia x5. Apothecia will be formed only if the fungus involved is an ascomycete.

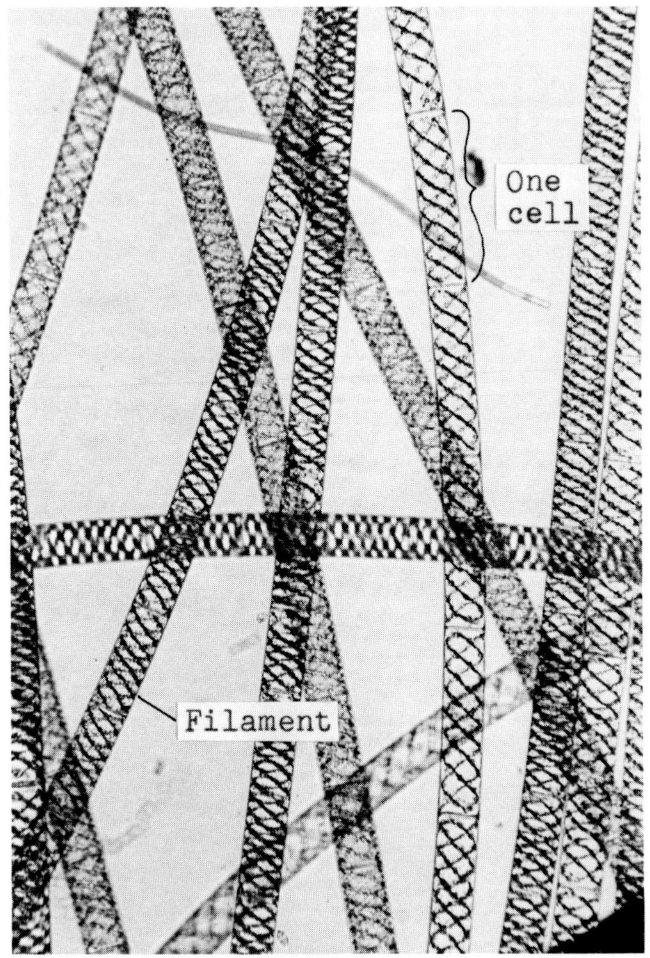

Fig. 21a Spirogyra filaments (live) w.m. x100. The filaments are made of cells connected end to end.

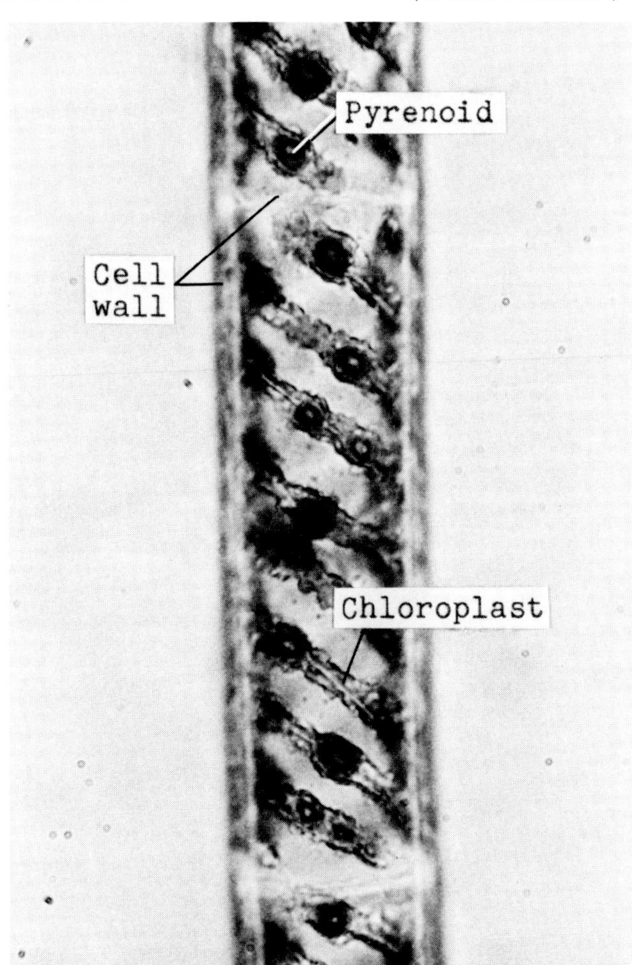

Fig. 21b Spirogyra cell (live) w.m. x430.

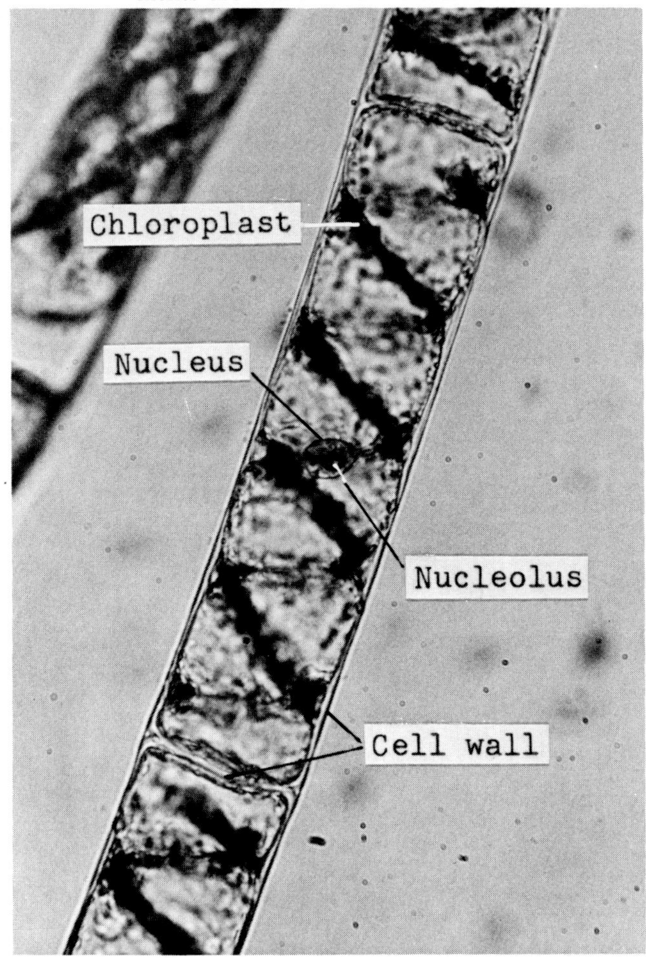

Fig. 21c Spirogyra cell (live) w.m. x430.

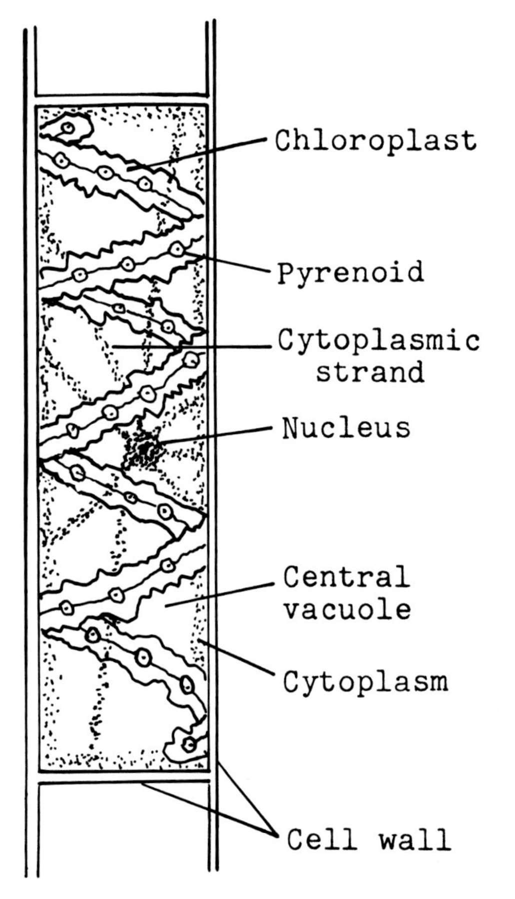

Fig. 21d Spirogyra anatomy.

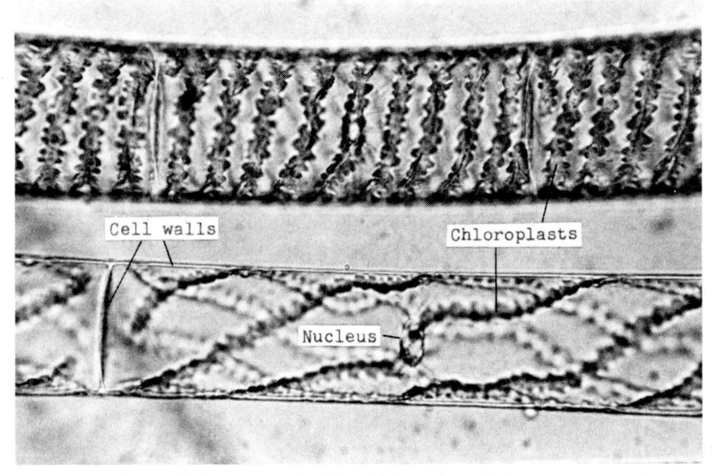

Fig. 22a Spirogyra (live) 2 different varieties w.m. x430.

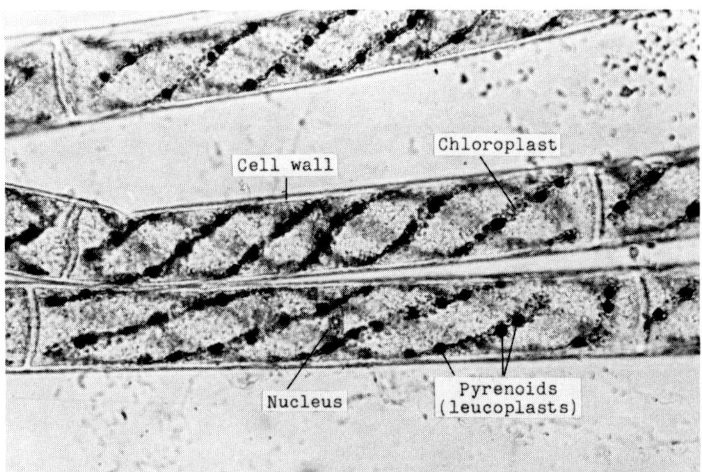

Fig. 22b Spirogyra (live) stained with iodine w.m. x100. Notice the pyrenoids have turned dark indicating the presence of starch.

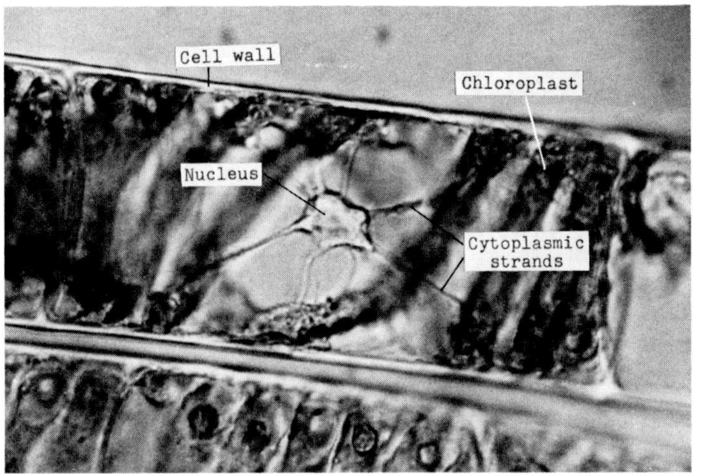

Fig. 22c Spirogyra (live) showing nucleus and supporting cytoplasmic strands w.m. x430. Photo courtesy of Stephen Davenport.

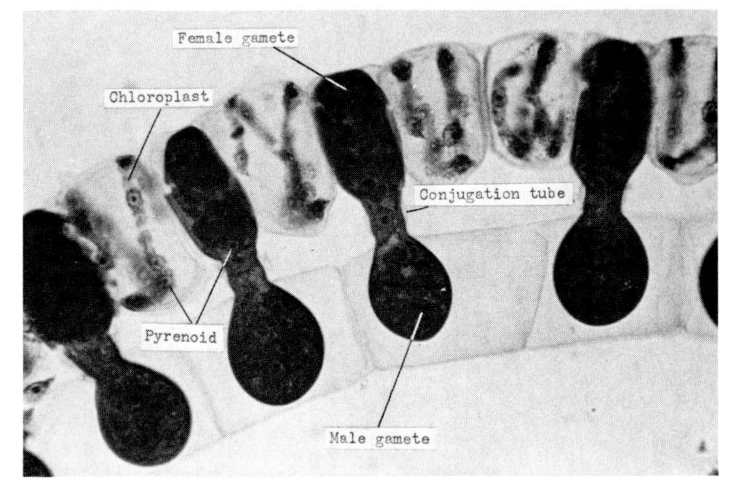

Fig. 22d Spirogyra undergoing conjugation w.m. x430.

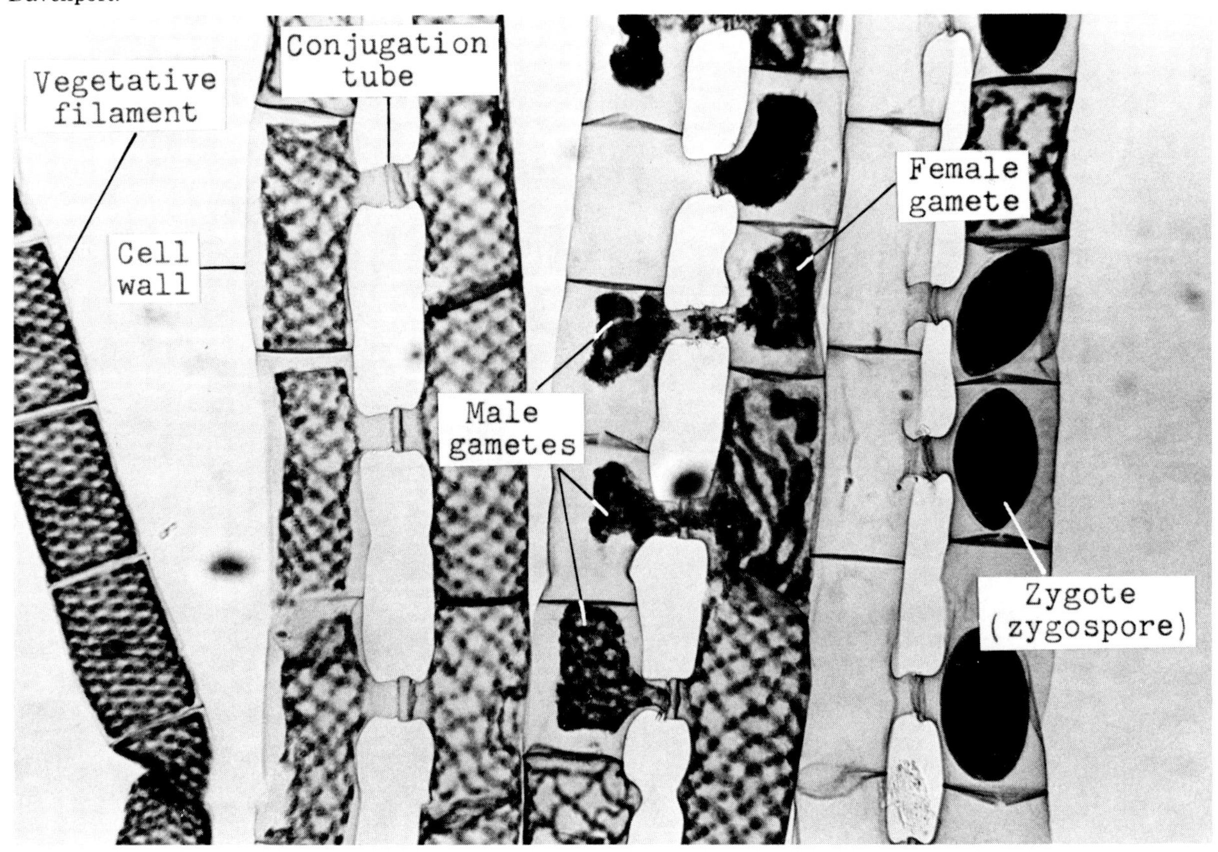

Fig. 22e Spirogyra undergoing conjugation w.m. x100. Early stages are on the left, progressing to later stages on the right.

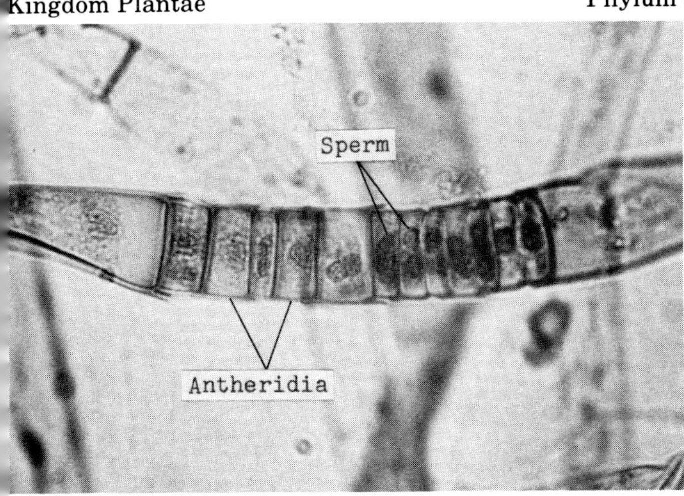

Fig. 23a Oedogonium antheridium (prepared slide) w.m. x430.

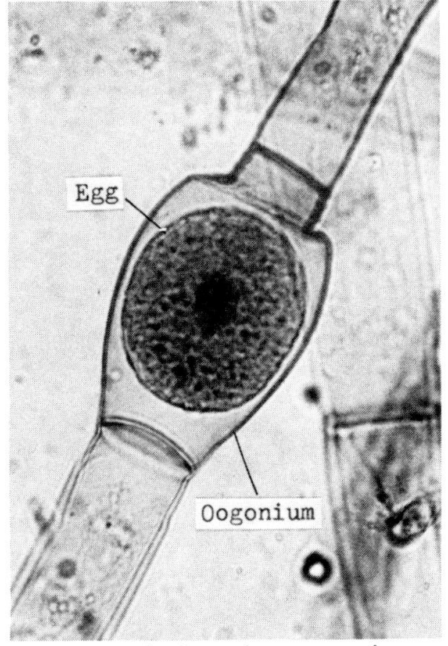

Fig. 23b Oedogonium oogonium (prepared slide) w.m. x430.

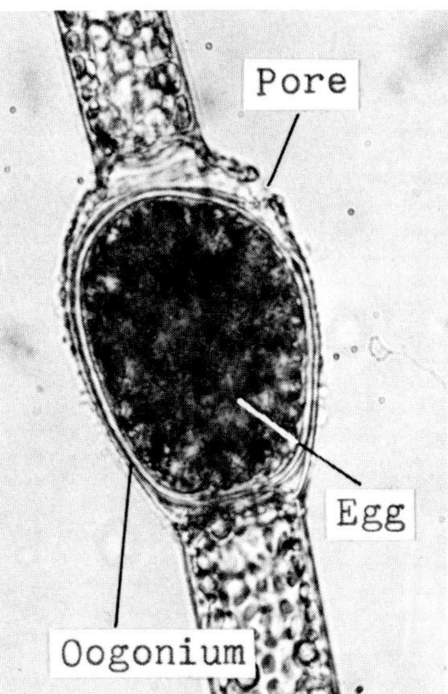

Fig. 23c Oedogonium oogonium (live) w.m. x430.

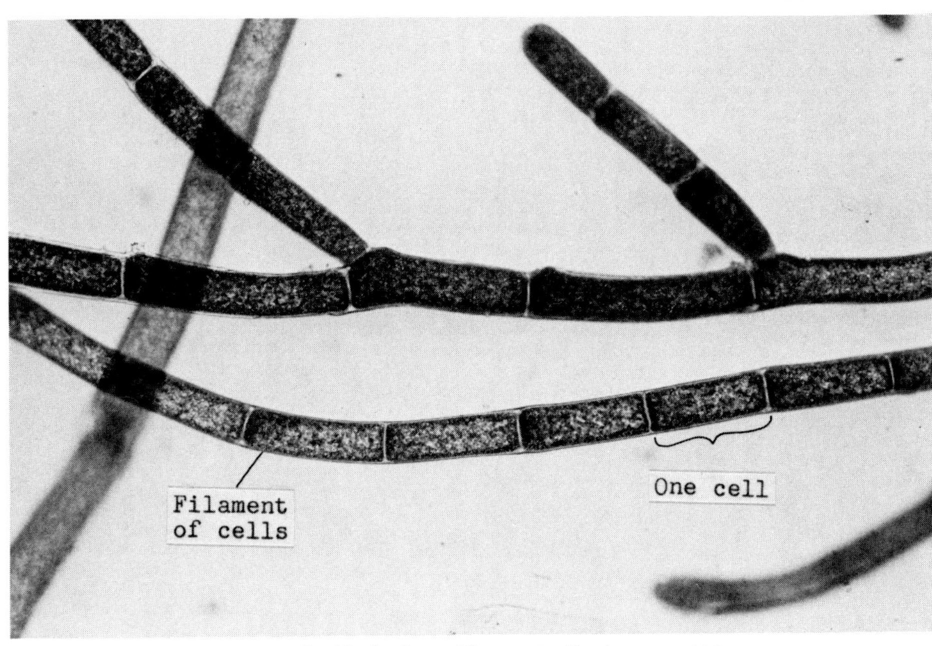

Fig. 23d Cladophora filaments (live) w.m. x100.

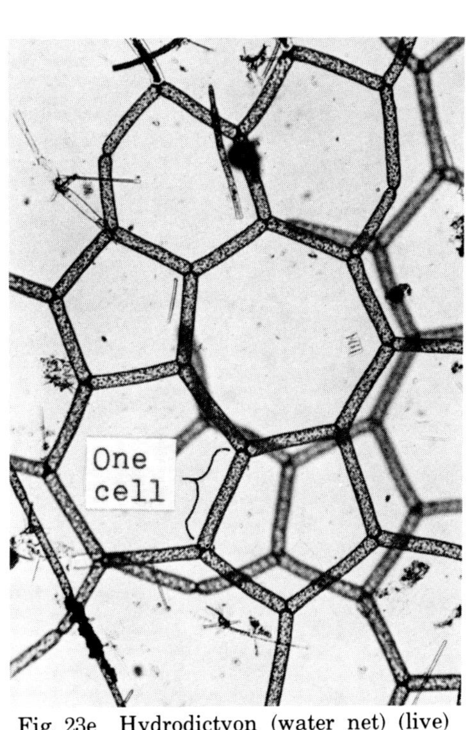

Fig. 23e Hydrodictyon (water net) (live) w.m. x40.

(DESMIDS)

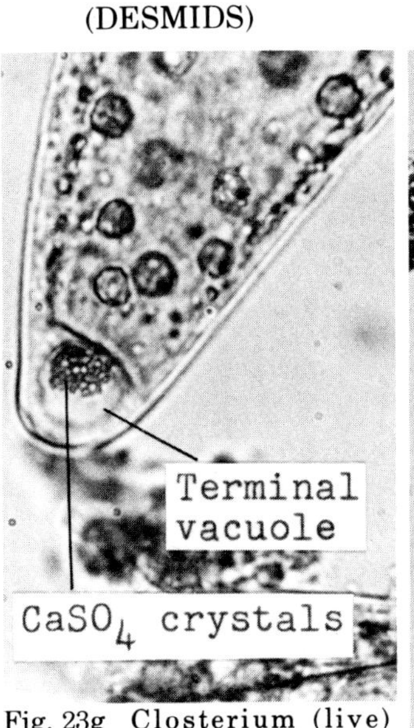

Fig. 23f Closterium (live) w.m. x100.

Fig. 23g Closterium (live) w.m. x430. The $CaSO_4$ crystals exhibit Brownian movement.

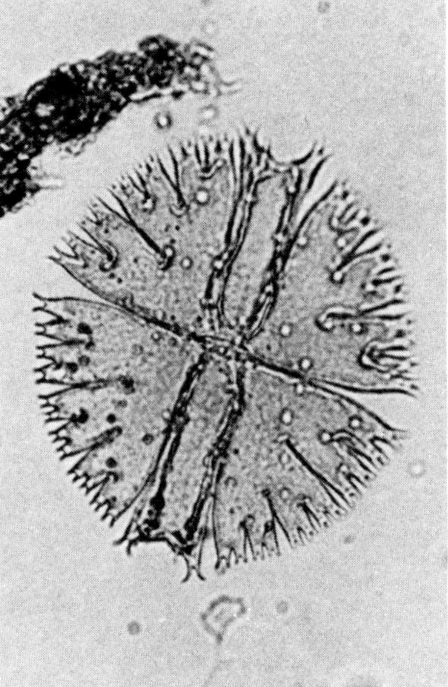

Fig. 23h Micrasterias (live) w.m. x430. (Caddo Lake, TX)

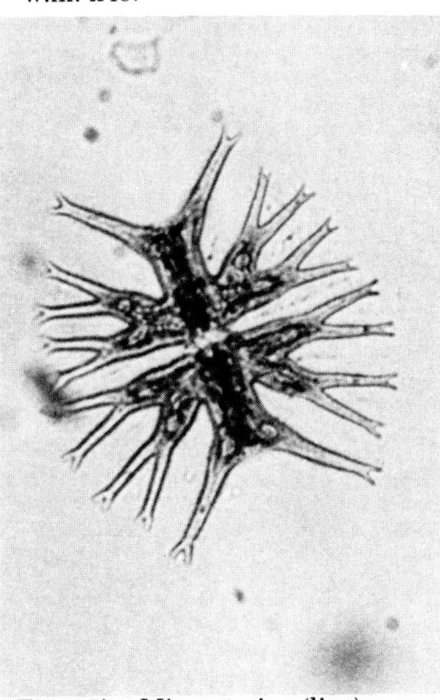

Fig. 23i Micrasterias (live) w.m. x100. (Caddo Lake, TX)

Fig. 24a Fucus (rockweed) x½. A common seaweed on rocky coasts.

Fig. 24b Nereocystis (live). This seaweed grows to 300 feet in length. (Pacific Grove, Calif.)

Fig. 24c Macrocystis (live). This seaweed contains algin and alginic acid which are added to foods to thicken and "velvetize" their texture. (Pacific Grove, Calif.)

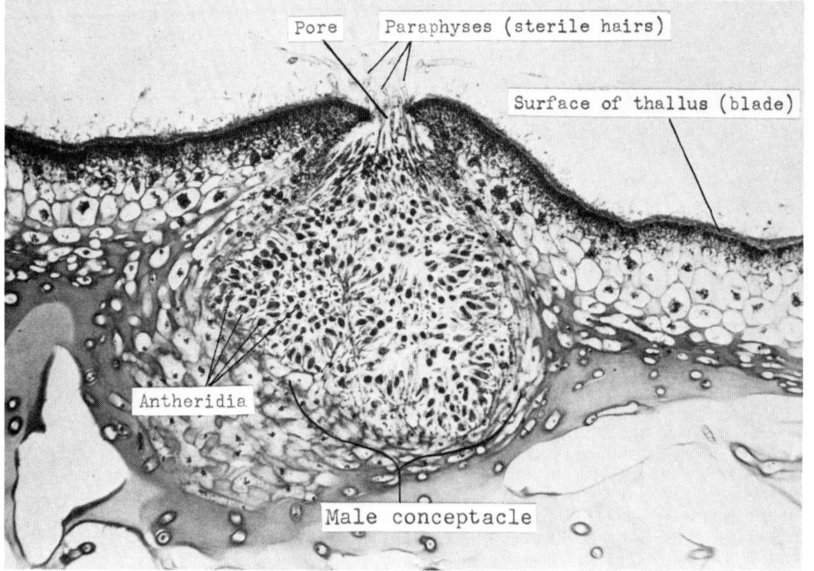

Fig. 24d Fucus, male conceptacle l.s. x100.

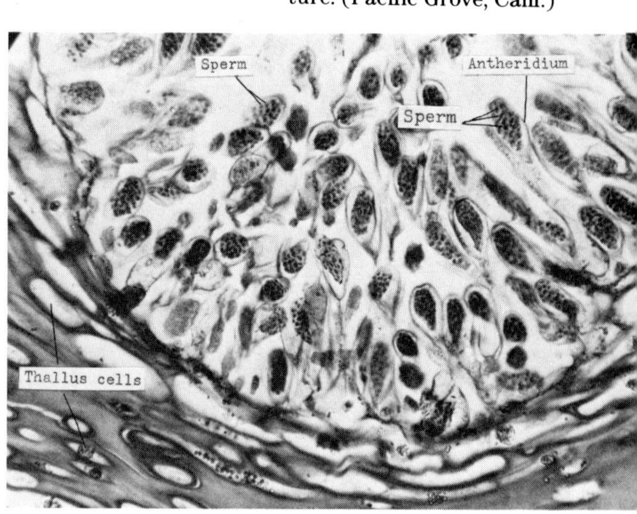

Fig. 24e Fucus, male conceptacle l.s. x430.

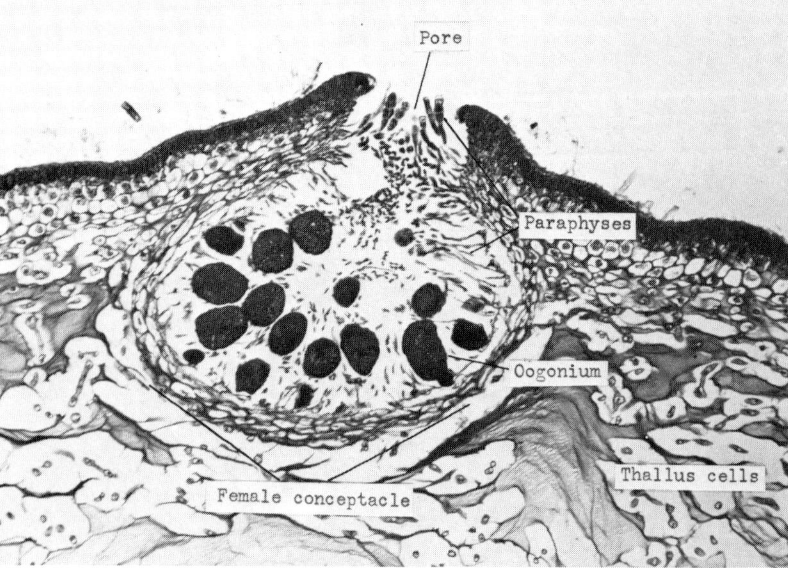

Fig. 24f Fucus, female conceptacle l.s. x100.

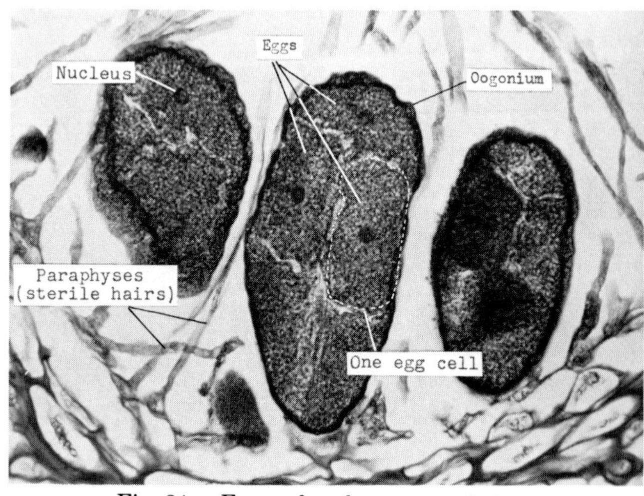

Fig. 24g Fucus, female conceptacle l.s. x430.

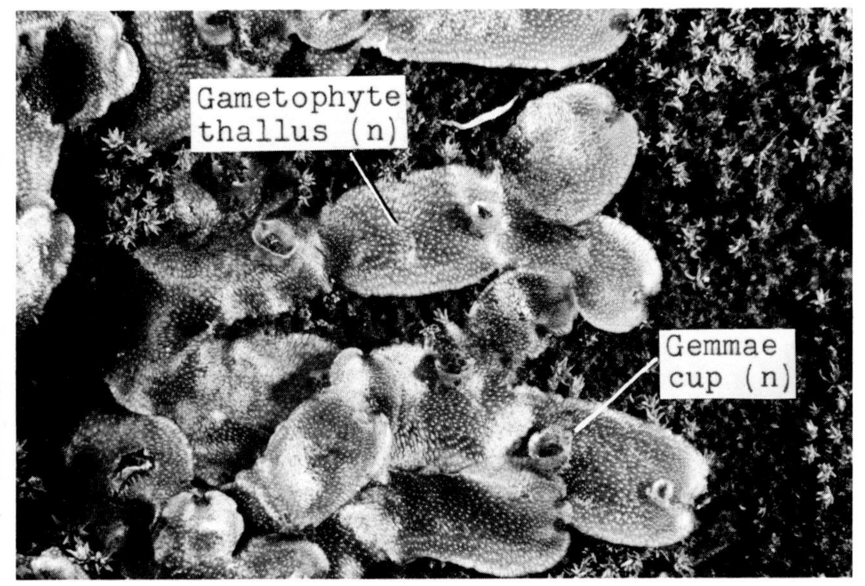

Fig. 25a Liverwort gametophyte colony (live) x½. Found growing 12″ above waterline on bank of Guadalupe River, Texas.

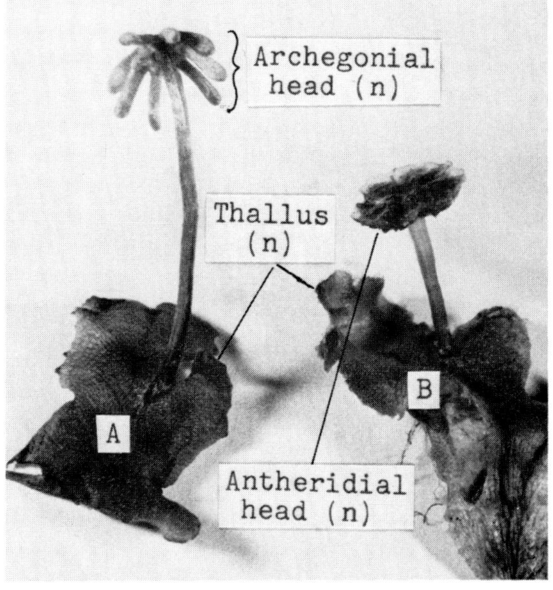

Fig. 25b Liverwort gametophyte plants x1. A - female, B - male.

Gametophyte thallus (n)
Gemmae cup (n)

Fig. 25c Liverwort thalli with gemmae cups x2.

Fig. 25d Liverwort gemmae cups x10. Each gemma can float away and grow into a new thallus.

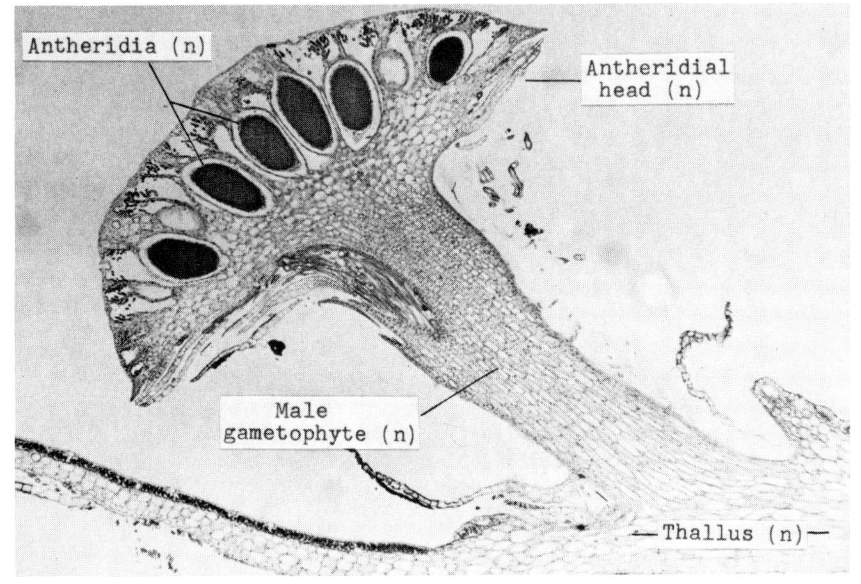

Fig. 25e Liverwort (Marchantia) male gametophyte l.s. x40.

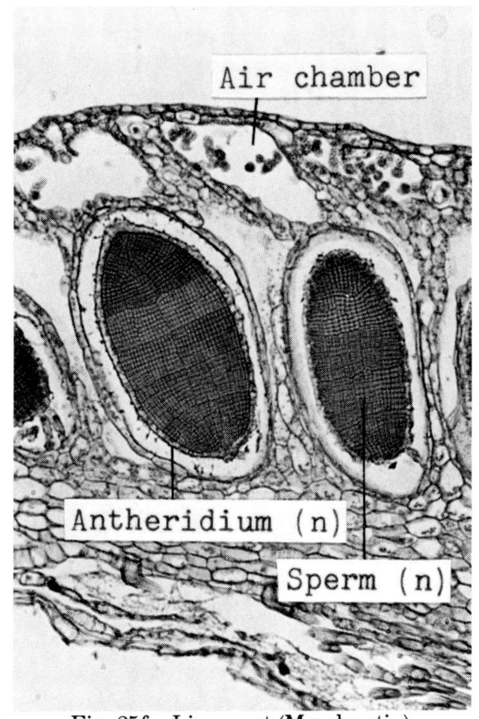

Fig. 25f Liverwort (Marchantia) antheridial head l.s. x100.

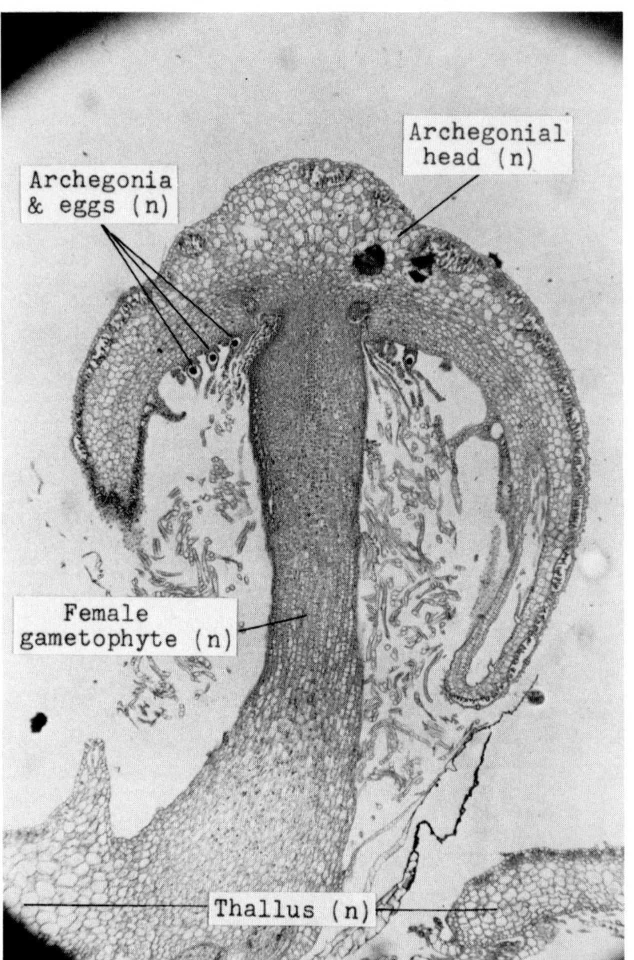

Fig. 26a Liverwort (Marchantia) female gametophyte l.s. x40.

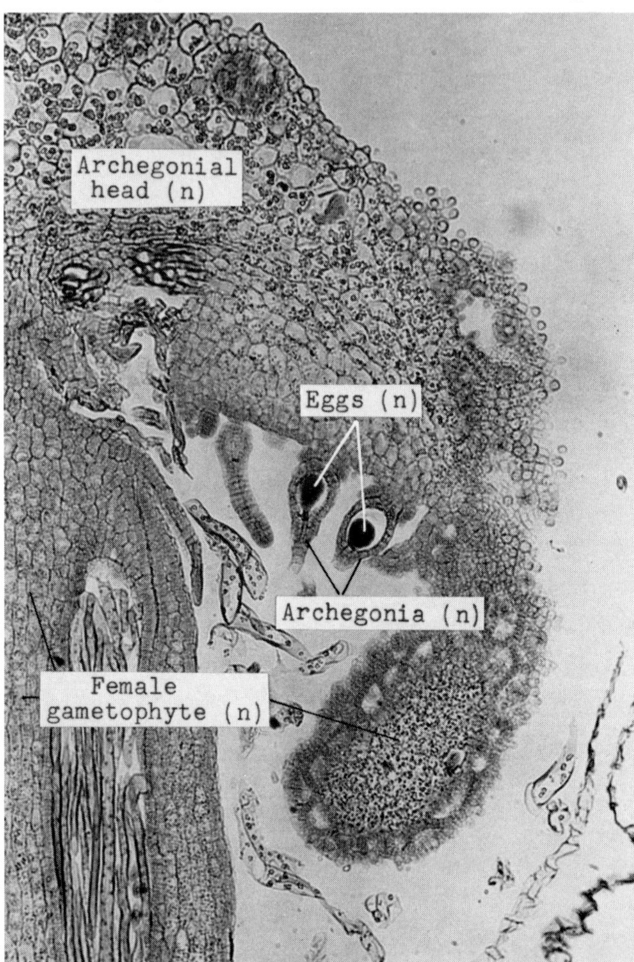

Fig. 26b Liverwort (Marchantia) archegonial head l.s. x100.

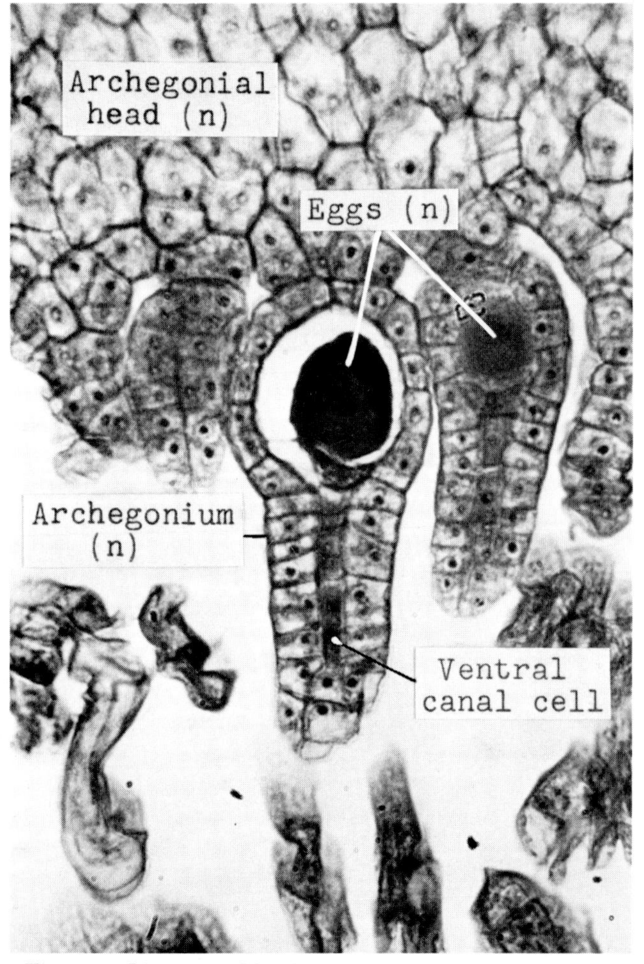

Fig. 26c Liverwort (Marchantia) archegonium l.s. x430.

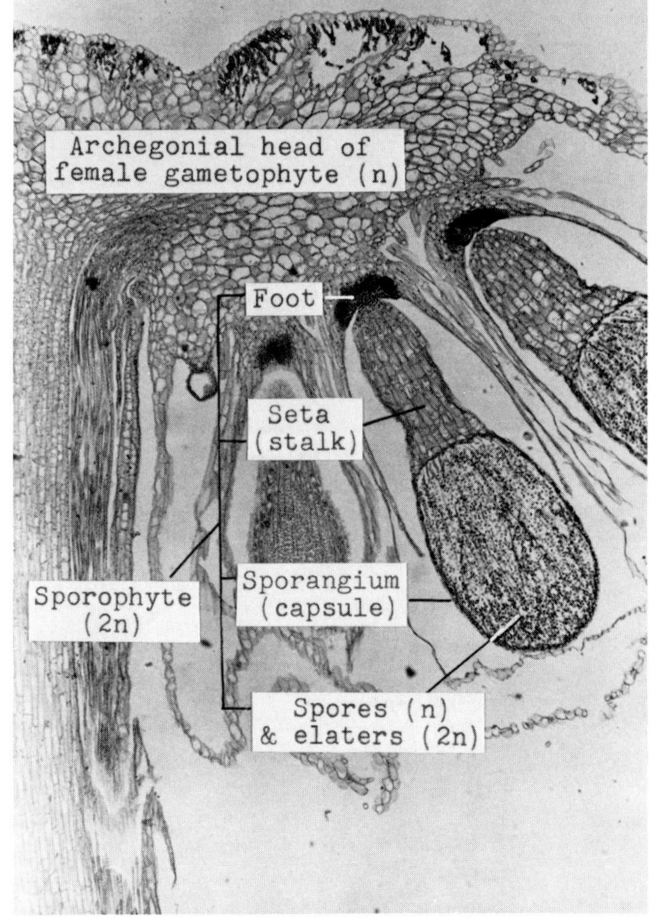

Fig. 26d Liverwort (Marchantia) sporophyte (in archegonial head of a female gametophyte) l.s. x100.

Fig. 27a Moss colony x2.

Fig. 27b Moss gametophyte and sporophyte plants x5. The calyptera is part of the old archegonium.

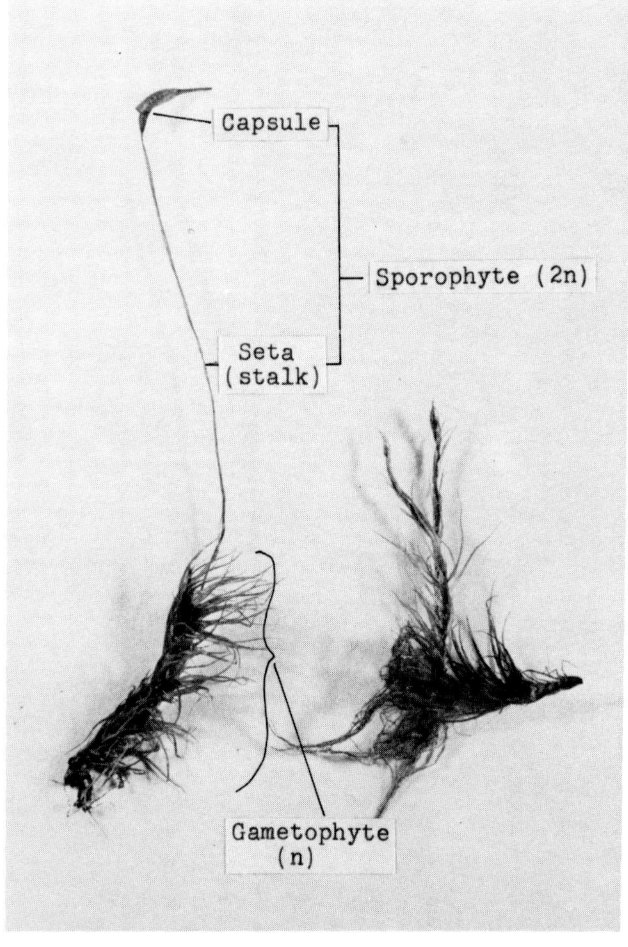

Fig. 27c Moss sporophyte growing out of the female gametophyte x3.

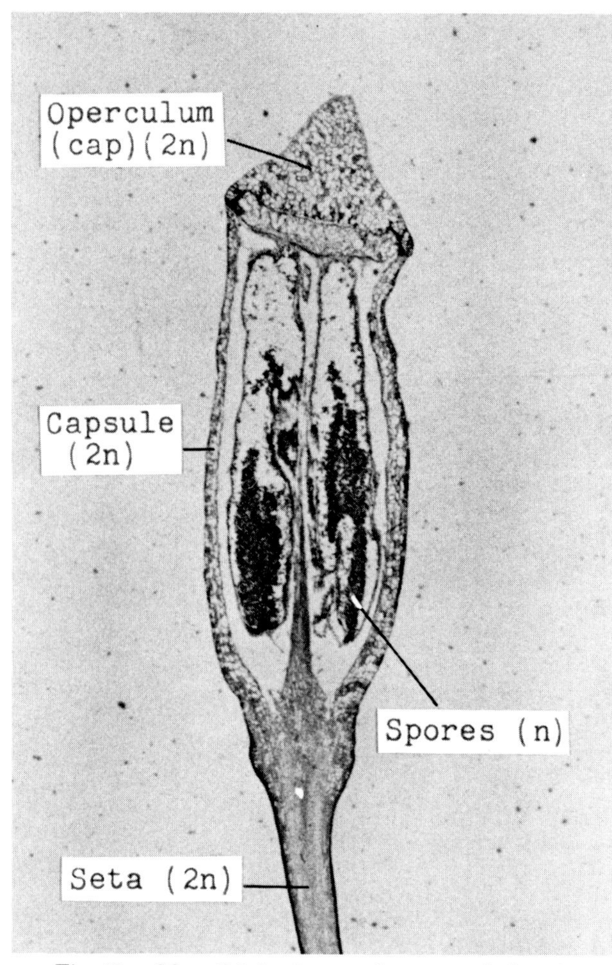

Fig. 28a Moss (Mnium) sporophyte capsule l.s. x20.

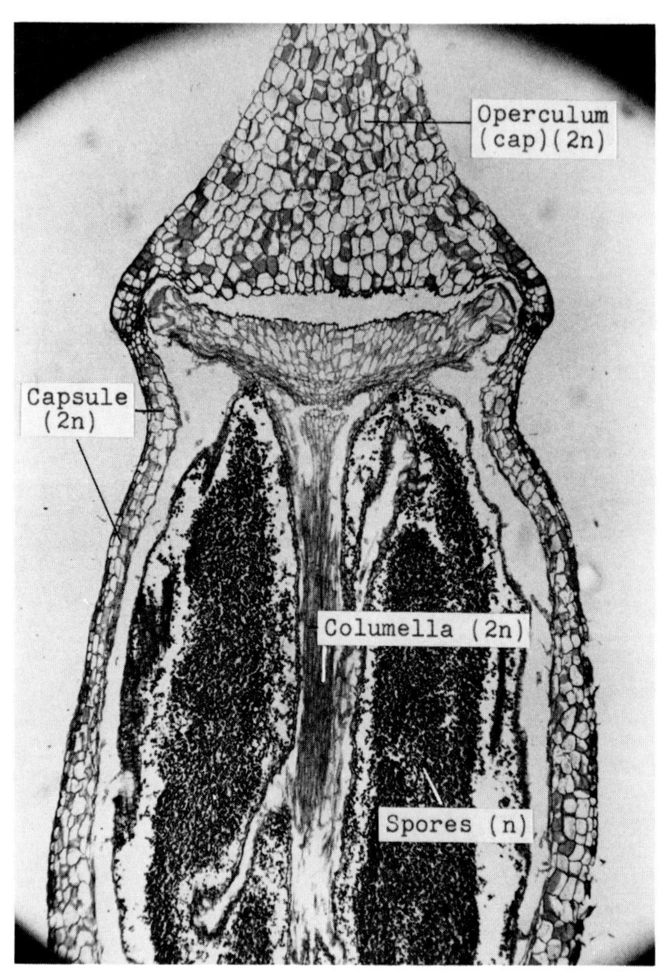

Fig. 28b Moss (Mnium) sporophyte capsule l.s. x40.

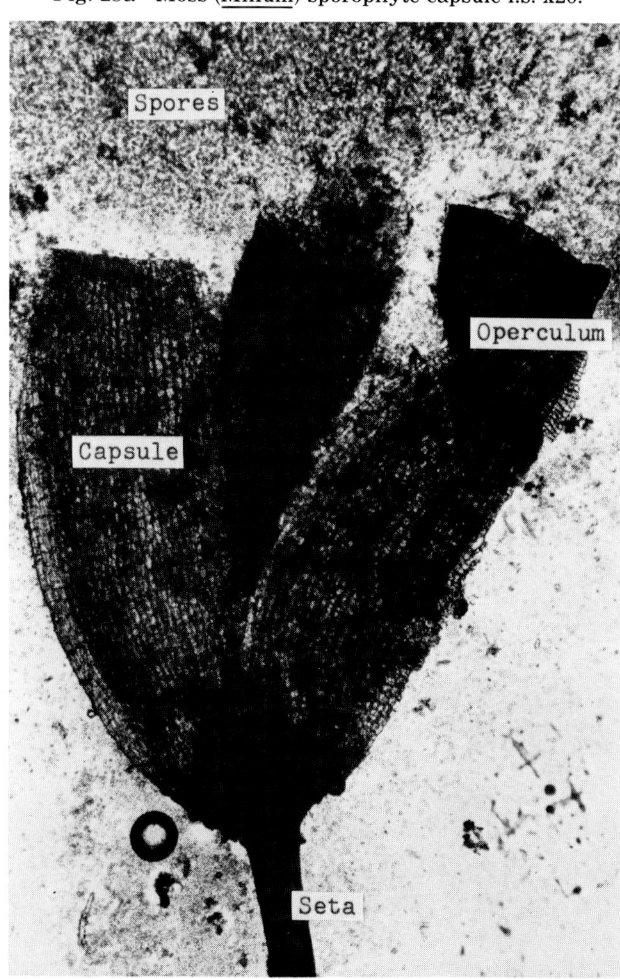

Fig. 28c Living moss capsule, crushed to show spores w.m. x40.

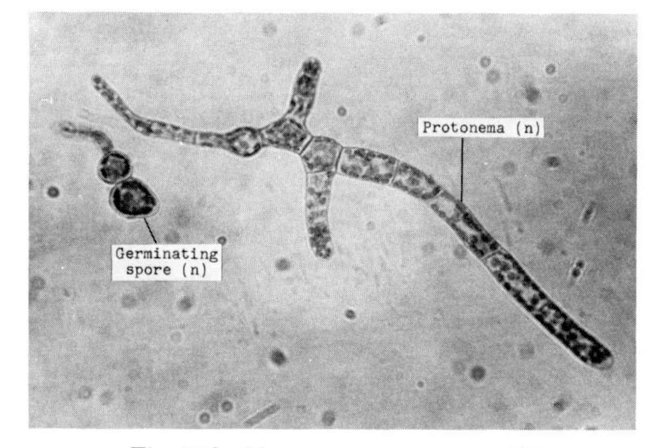

Fig. 28d Moss protonema w.m. x430.

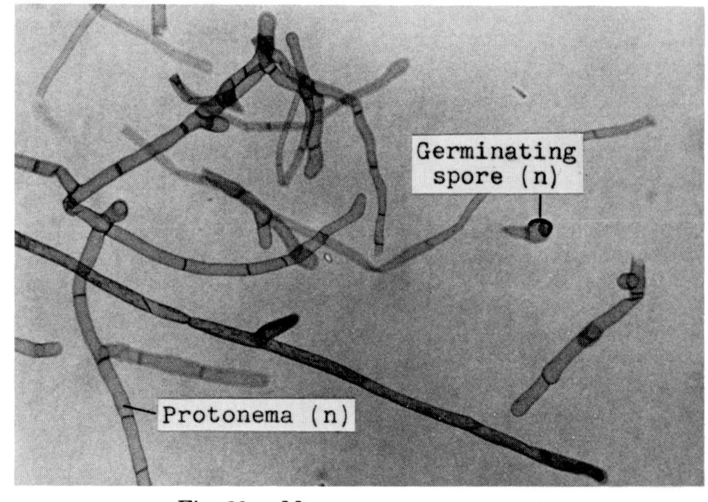

Fig. 28e Moss protonema w.m. x100.

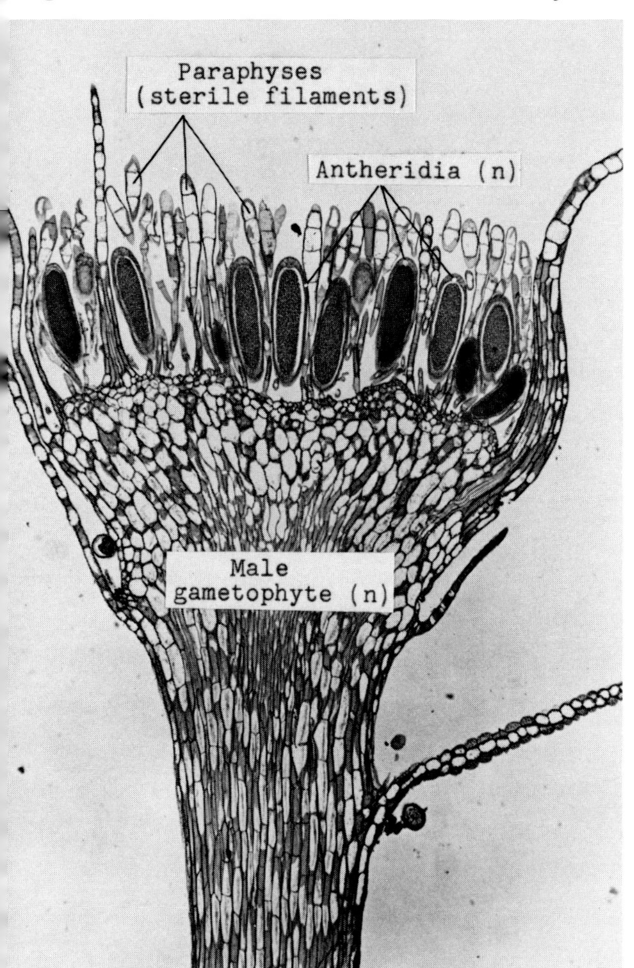

Fig. 29a Moss (Mnium) antheridial head l.s. x40.

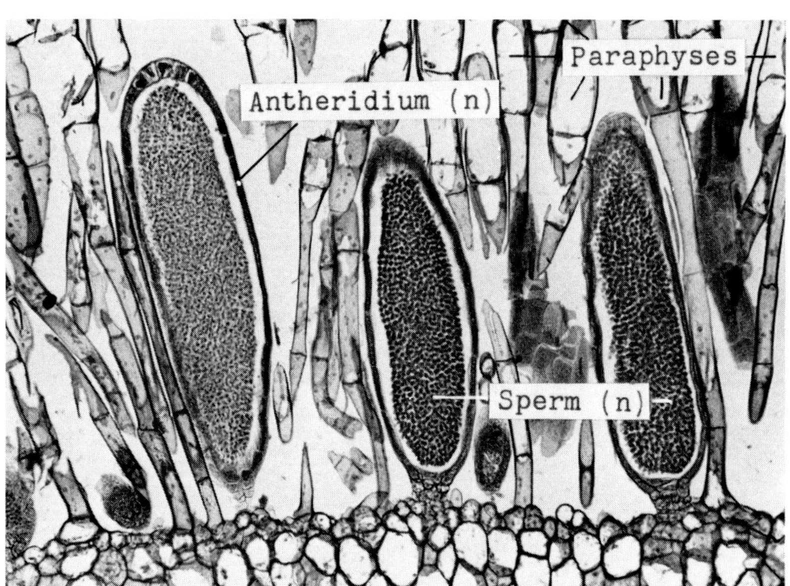

Fig. 29b Moss (Mnium) antheridia l.s. x100.

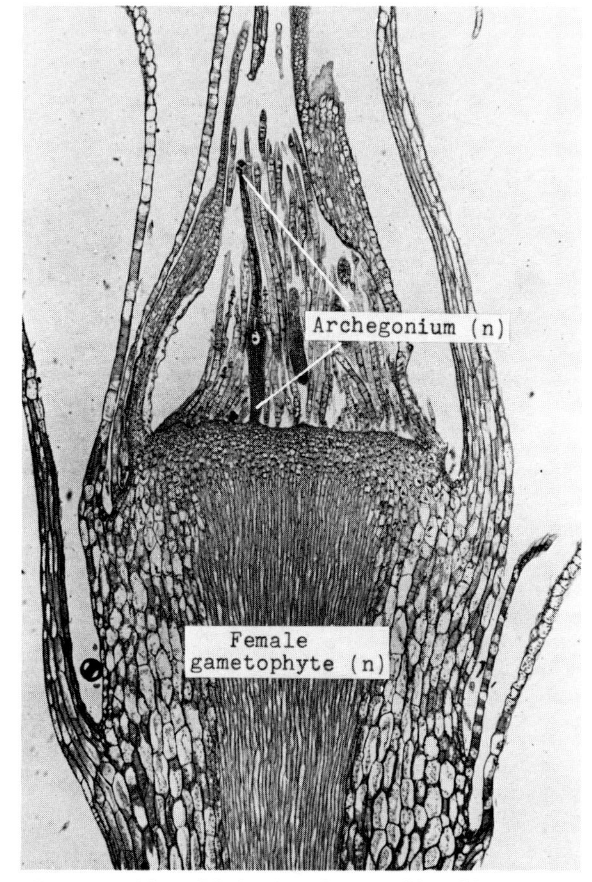

Fig. 29c Moss (Mnium) archegonial head l.s. x40.

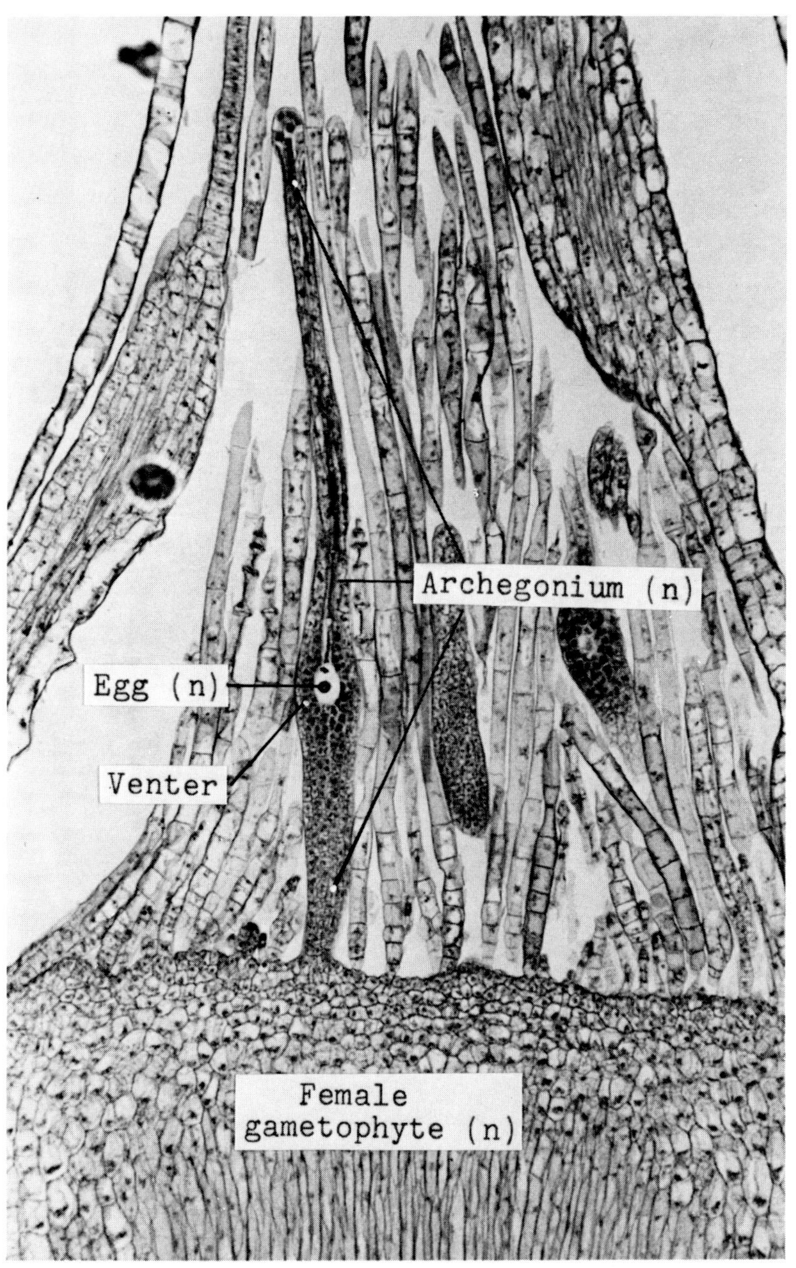

Fig. 29d Moss (Mnium) archegonium l.s. x100. The venter is the enlarged part of the archegonium that produces the egg.

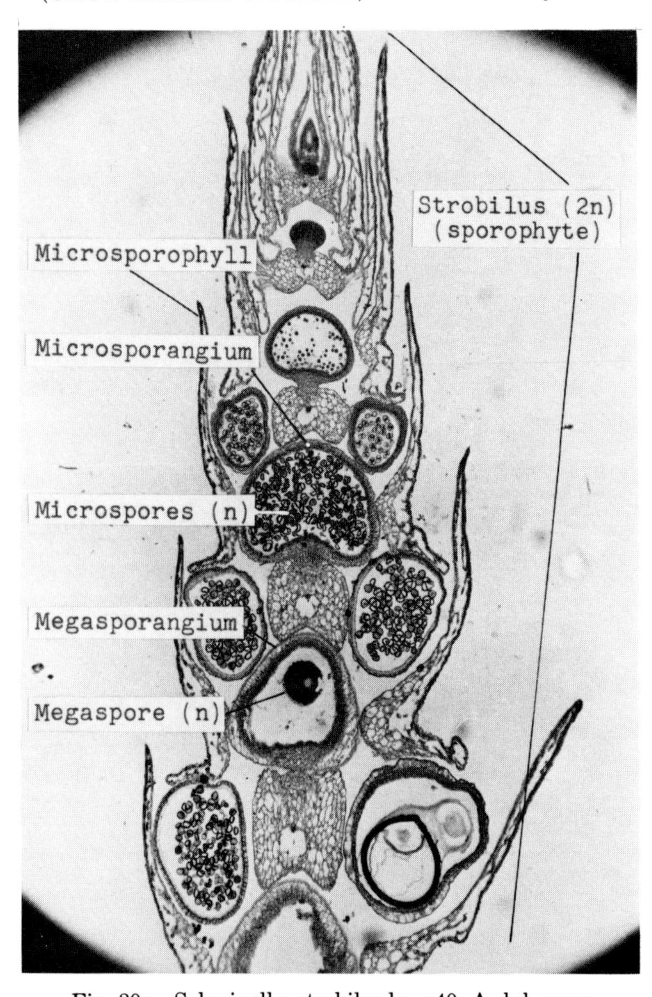

Fig. 30a Selaginella, strobilus l.s. x40. A club moss.

Fig. 30b Ferns, fiddleheads (immature fronds) x½.

Fig. 30c Fern; mature frond x½.

Fig. 30d Fern spore print x½. A frond with mature sori was left lying on a white paper overnight. The spores that discharged from the sori made this pattern.

Fig. 31a Sori on underside of a fern pinna x10. Each sorus contains a cluster of sporangia.

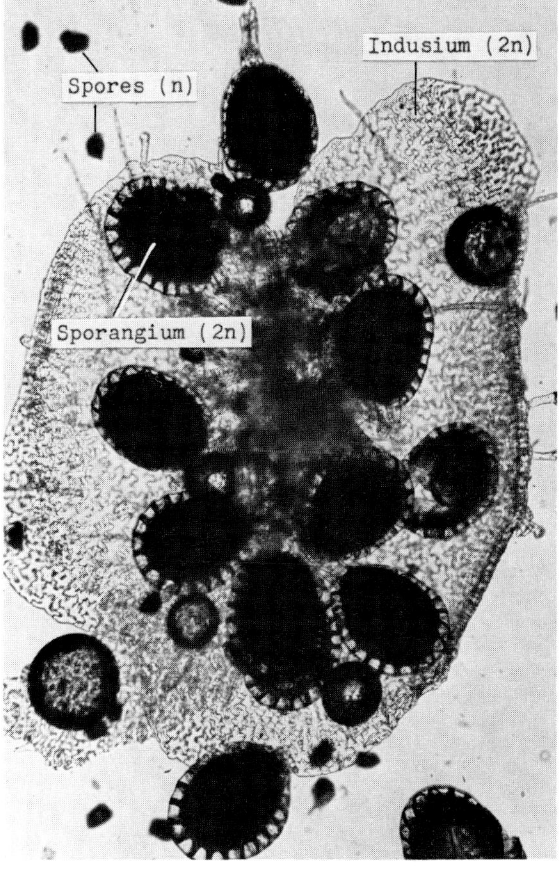

Fig. 31b Sorus w.m. x100.

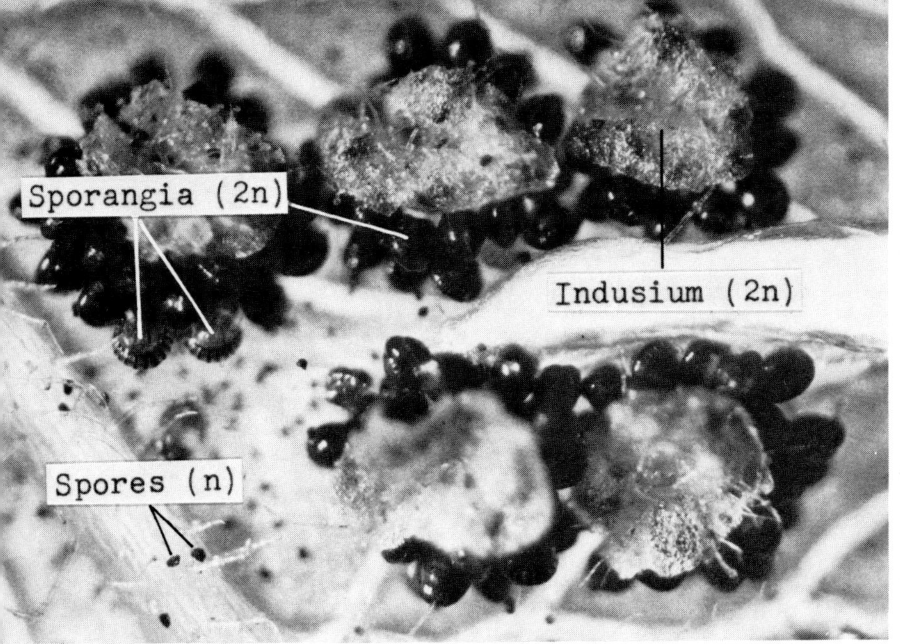

Fig. 31c Five sori w.m. x40.

Fig. 31d Sporangia discharging their spores w.m. x40.

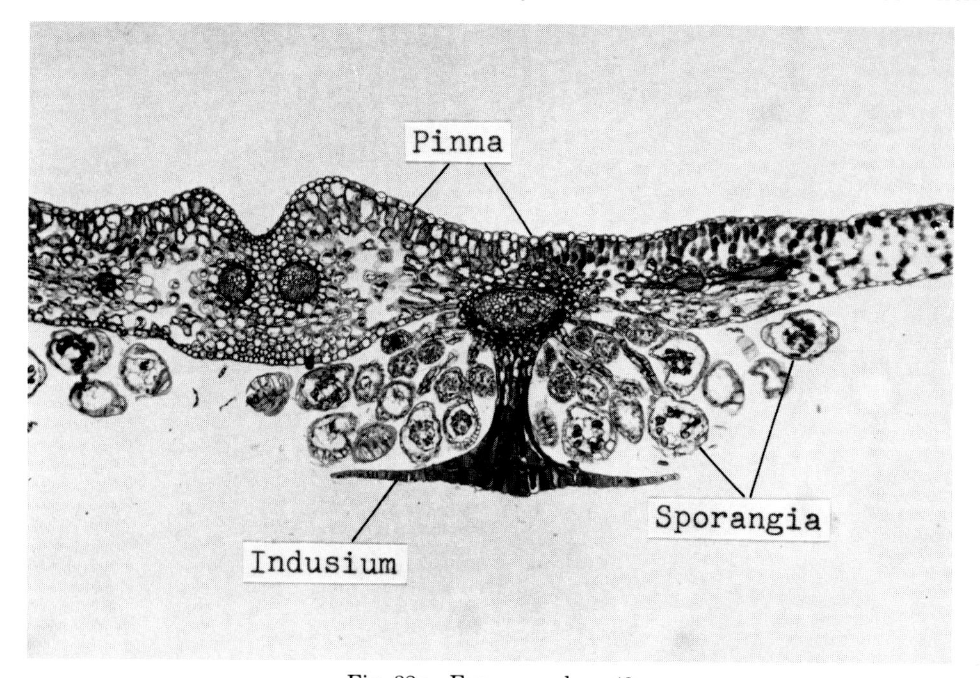

Fig. 32a Fern sorus l.s. x40.

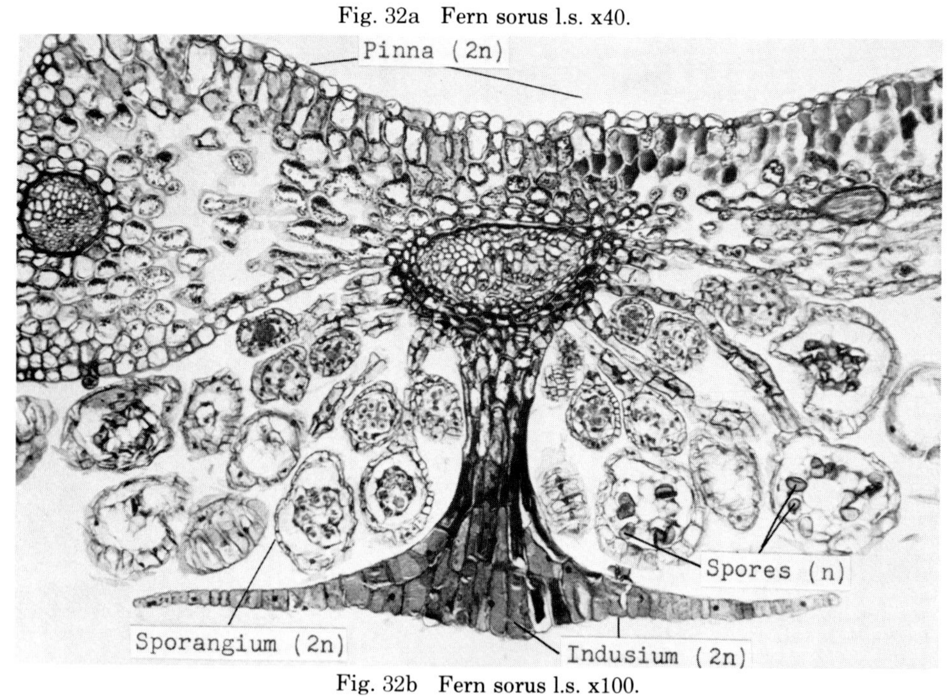

Fig. 32b Fern sorus l.s. x100.

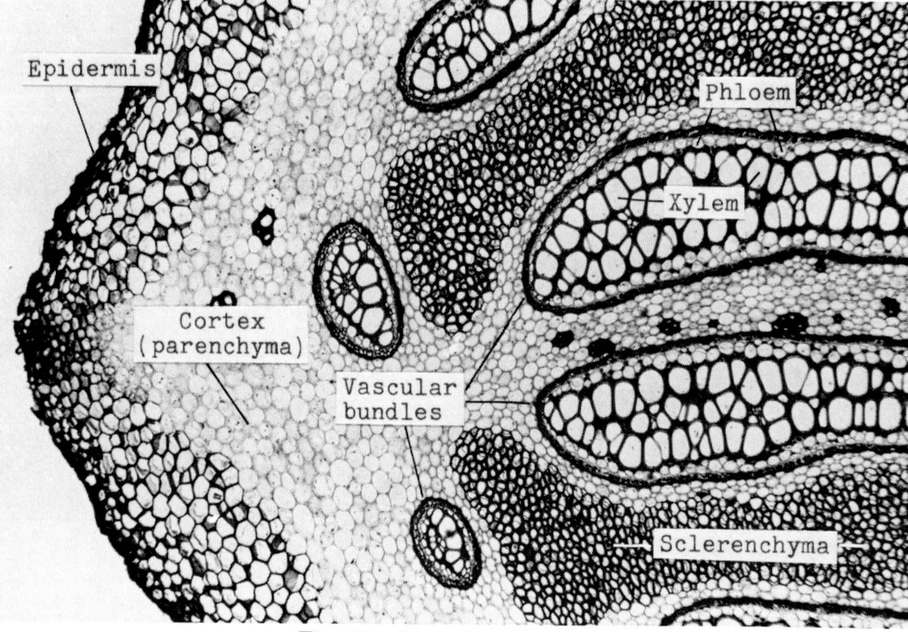

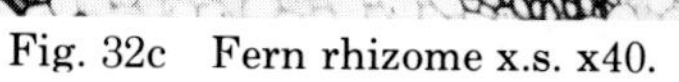

Fig. 32c Fern rhizome x.s. x40.

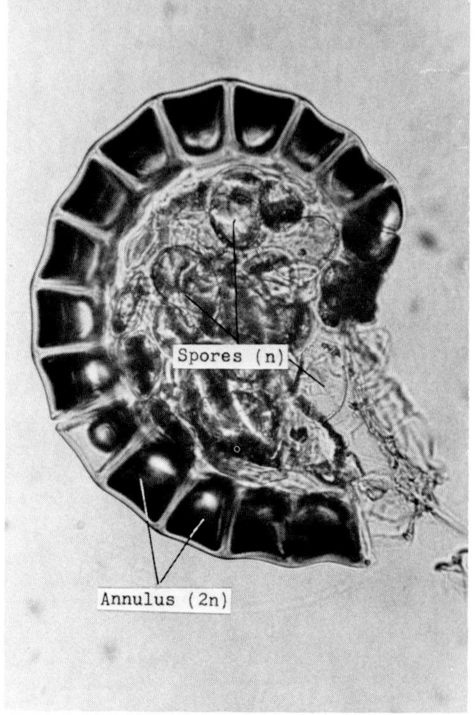

Fig. 32d Sporangium containing spores w.m. x430.

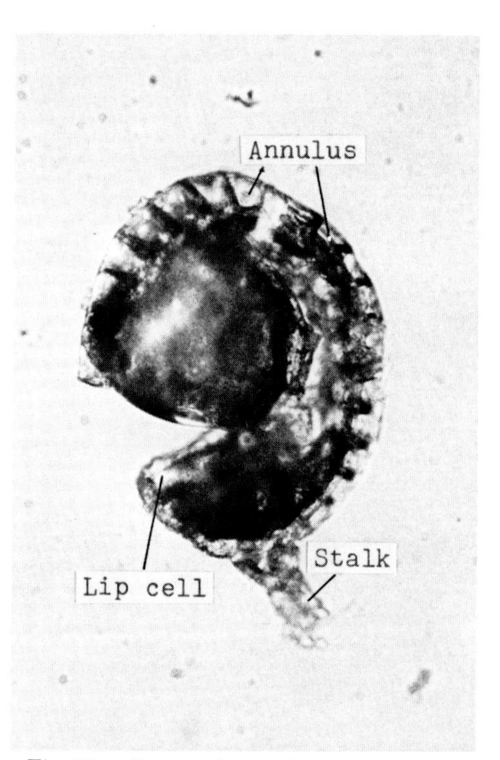

Fig. 32e Sporangium after discharging spores w.m. x430.

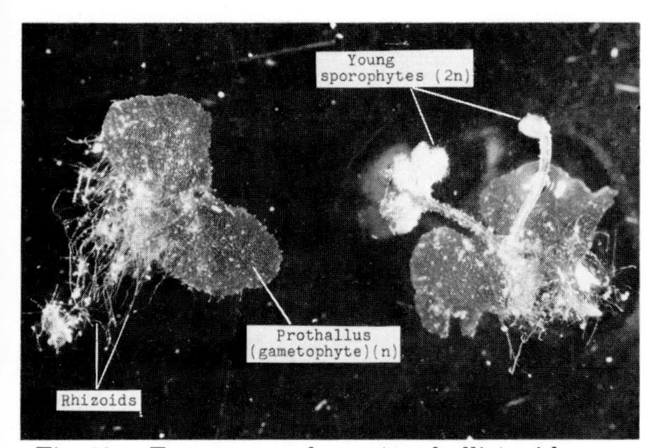

Fig. 33a Fern gametophytes (prothallia) with new sporophytes growing out of a prothallus (live) w.m. x5.

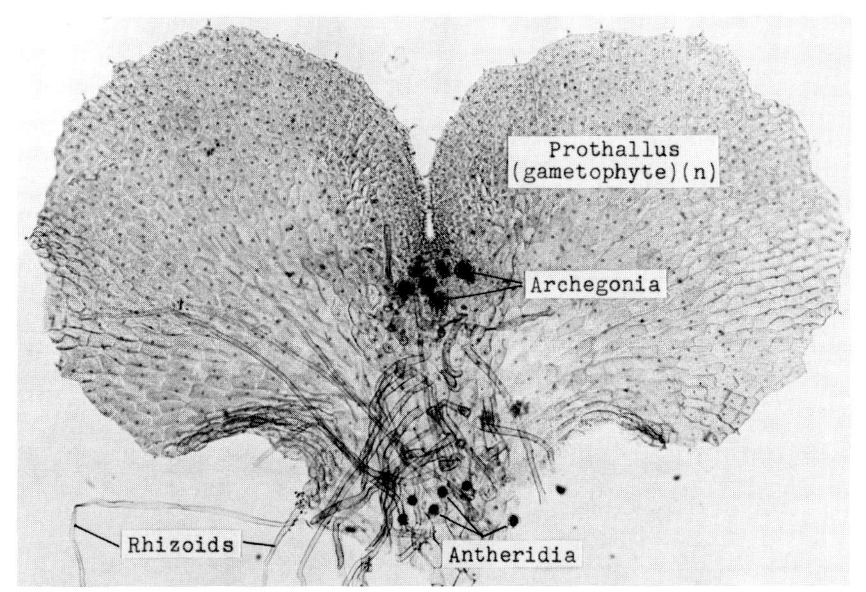

Fig. 33b Fern gametophyte (prothallus) w.m. x40.

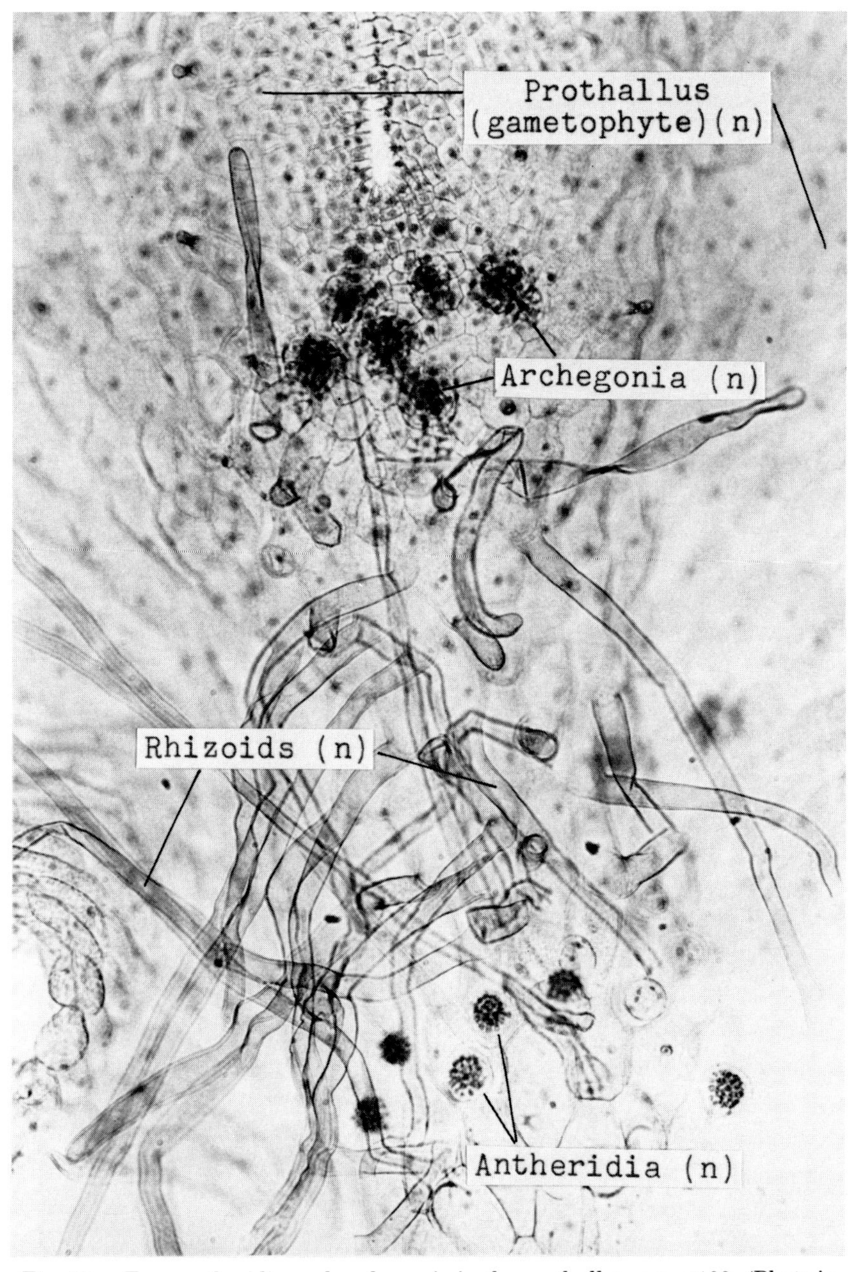

Fig. 33c Fern antheridia and archegonia in the prothallus w.m. x100. (Photo is enlarged central region of Fig. 33b.)

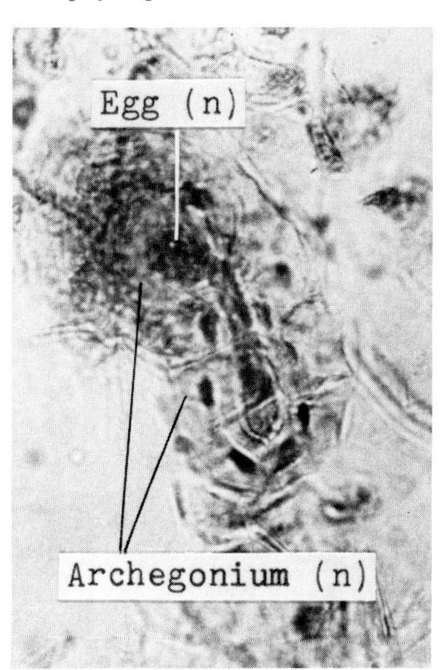

Fig. 33d Fern archegonium w.m. x430.

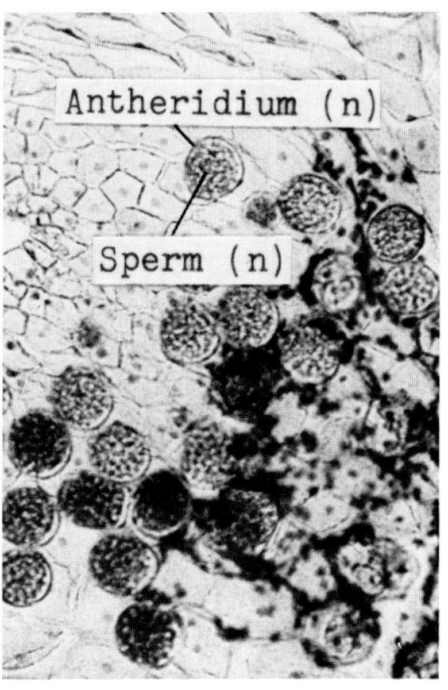

Fig. 33e Fern antheridia w.m. x100.

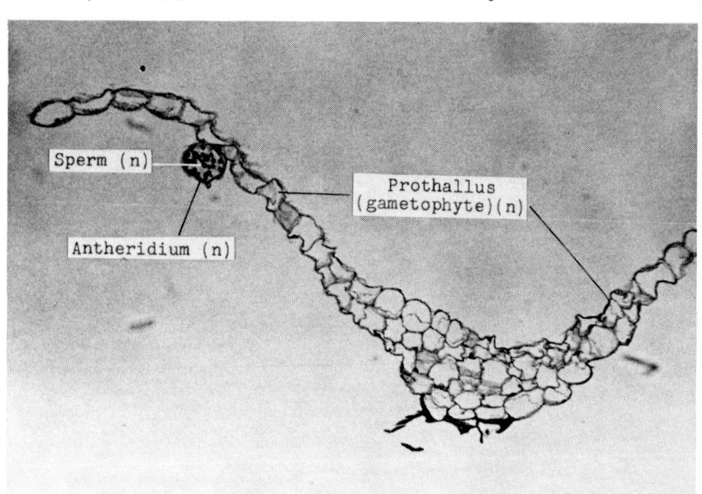

Fig. 34a Fern prothallus and antheridium x.s. x100.

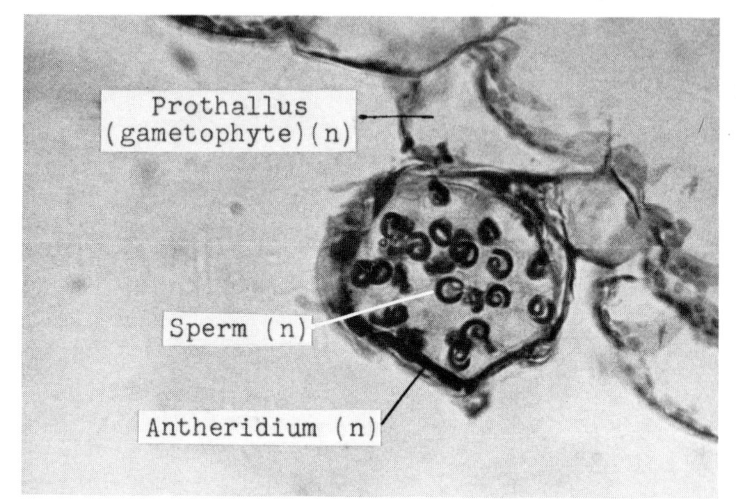

Fig. 34b Fern antheridium x.s. x430.

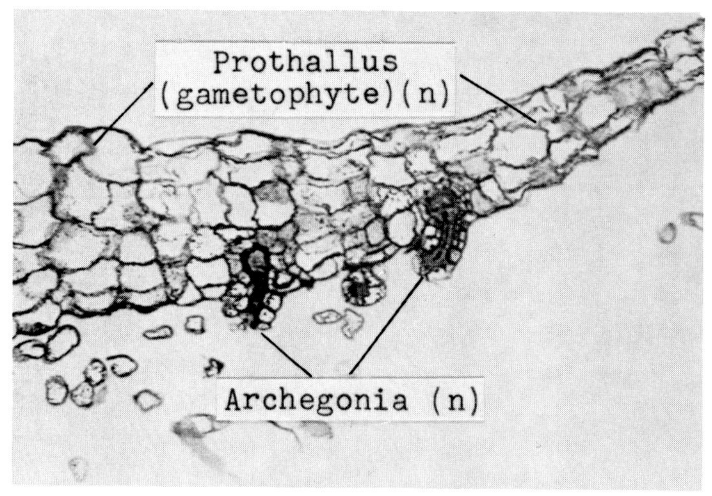

Fig. 34c Fern prothallus and archegonium x.s. x100.

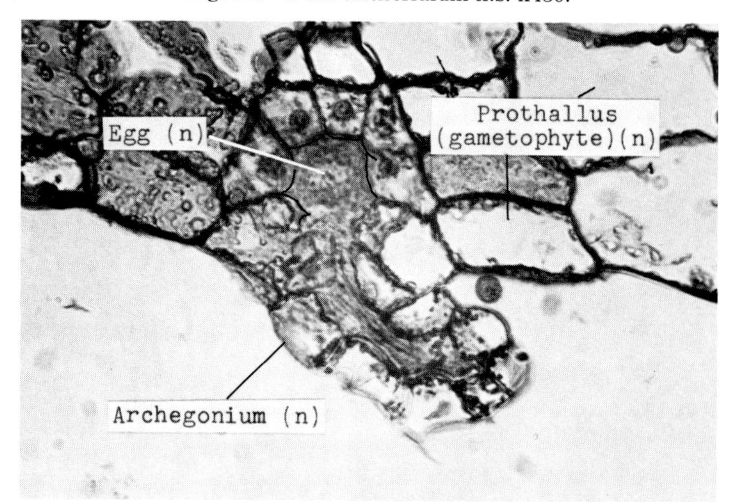

Fig. 34d Fern archegonium x.s. x430.

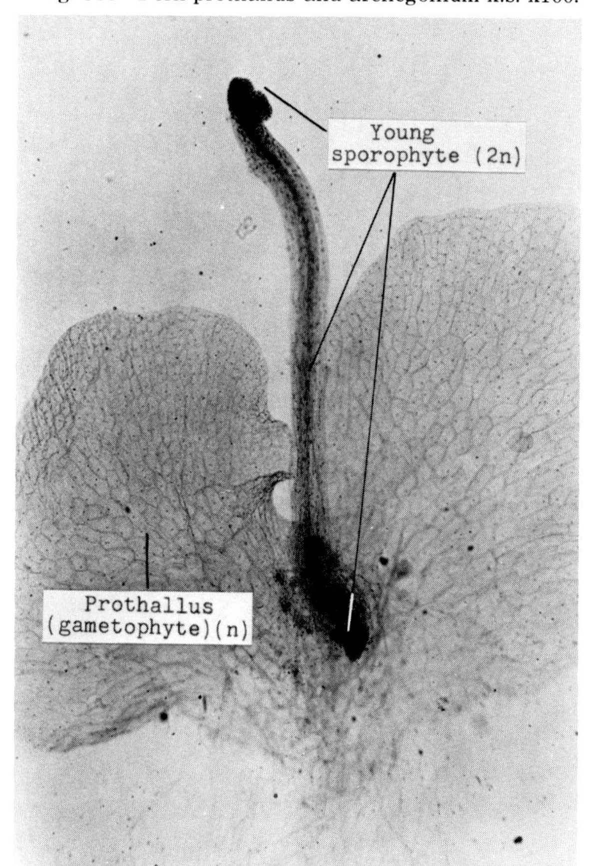

Fig. 34e Young sporophyte growing out of the prothallus w.m. x40.

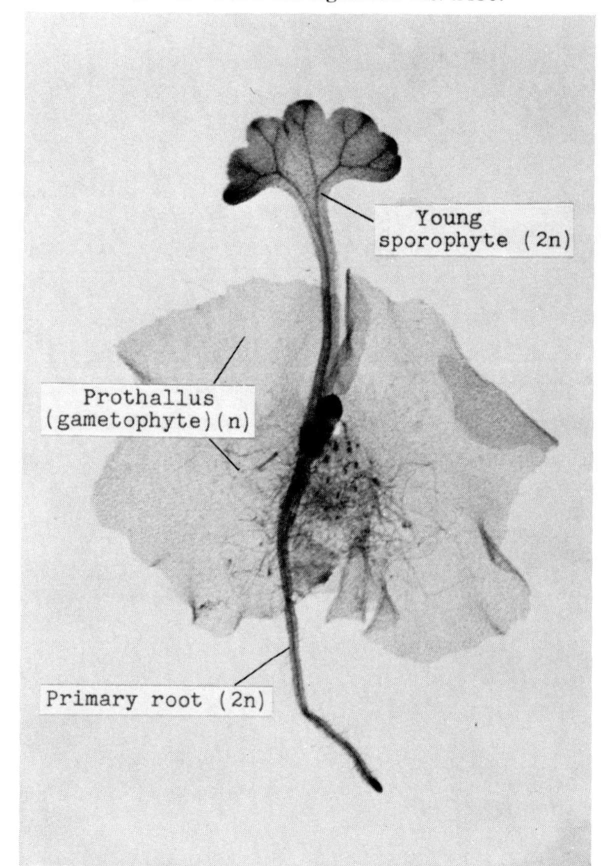

Fig. 34f Young sporophyte growing out of the prothallus w.m. x40. The sporophyte will become a new frond.

Fig. 35a Pine cones x1. A,B,C, and D are stages of increasing maturity of the female cone. The female cone is also known as the carpellate, pistillate, mature ovulate, megastrobilus or seed cone. (Redfish Lake, Idaho)

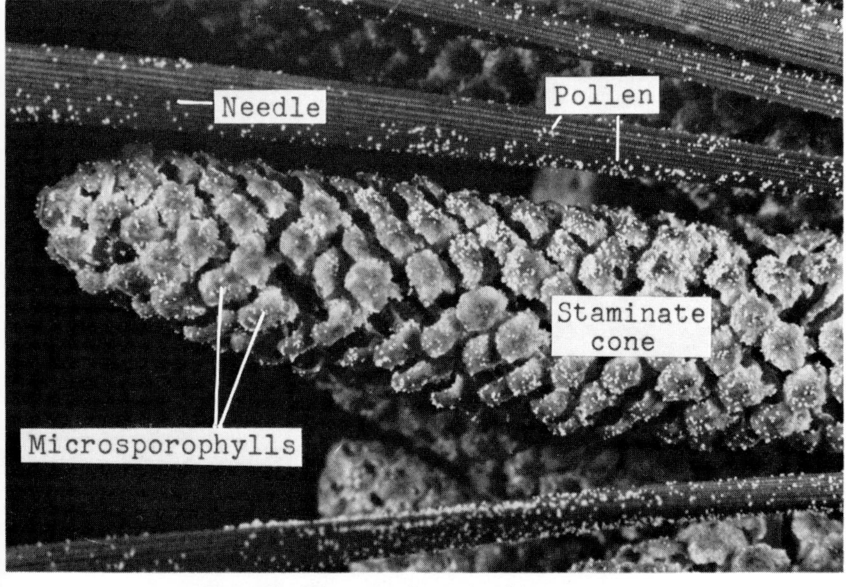

Fig. 35b Pine staminate (male) cone x10.

Fig. 35d Pine staminate cone x.s. x20.

Fig. 35c Pine staminate cones x2. (Ratcliff, Texas)

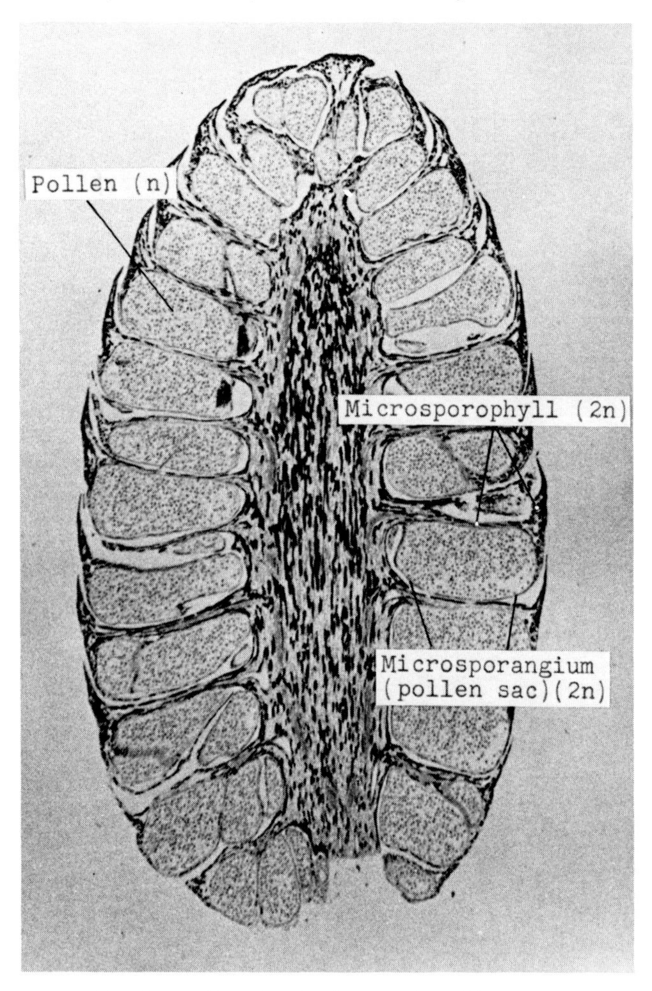

Fig. 36a Pine staminate cone l.s. x20.

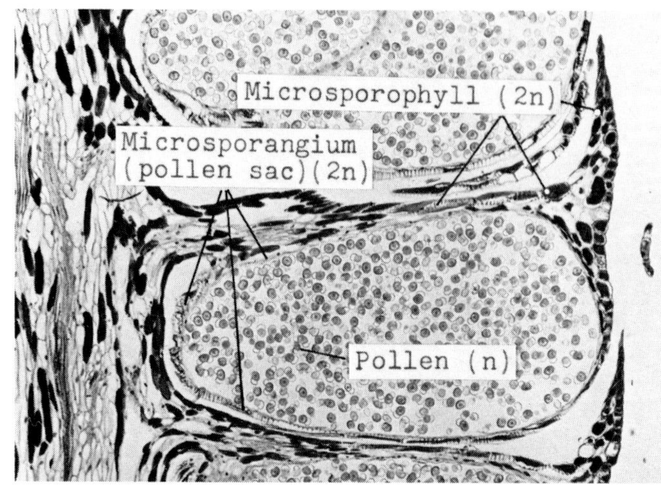

Fig. 36b Microsporophyll of staminate cone l.s. x100.

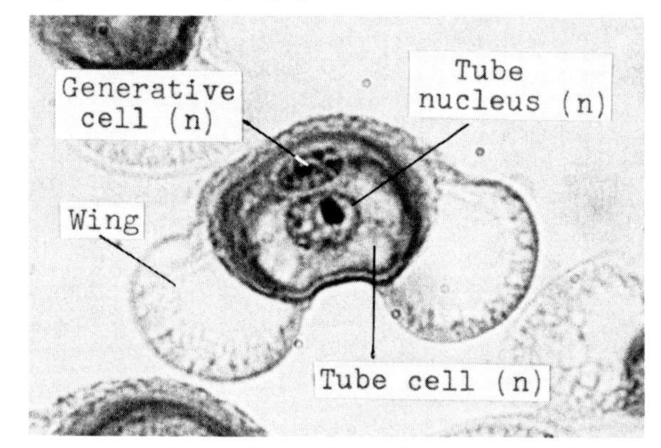

Fig. 36c Pine pollen grain (male gametophyte) w.m. x430.

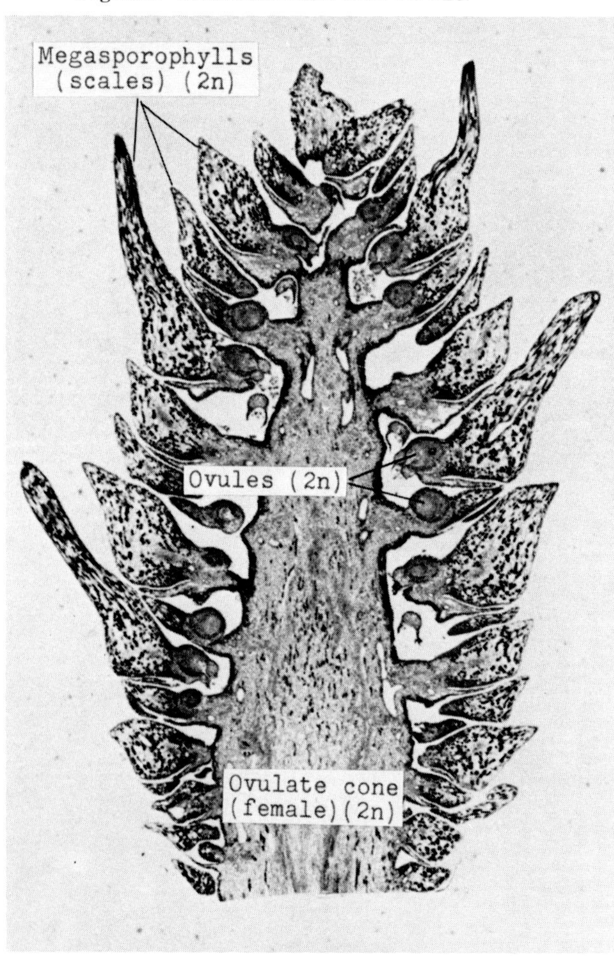

Fig. 36d Pine ovulate cone l.s. x20.

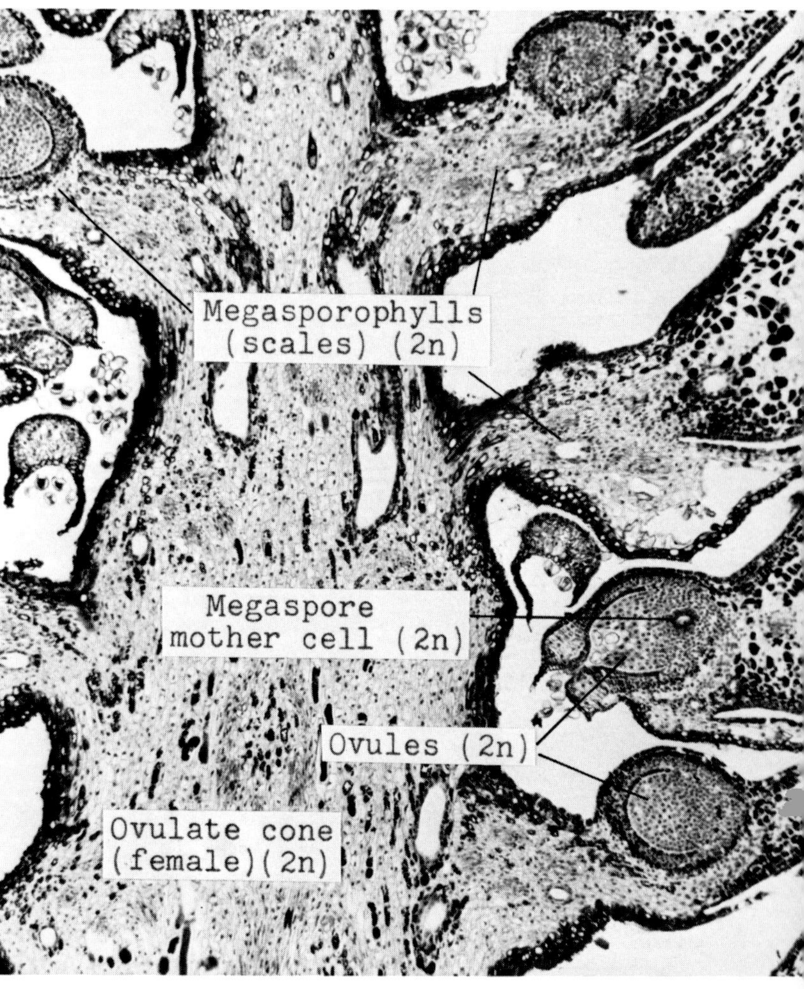

Fig. 36e Pine ovulate cone l.s. x40.

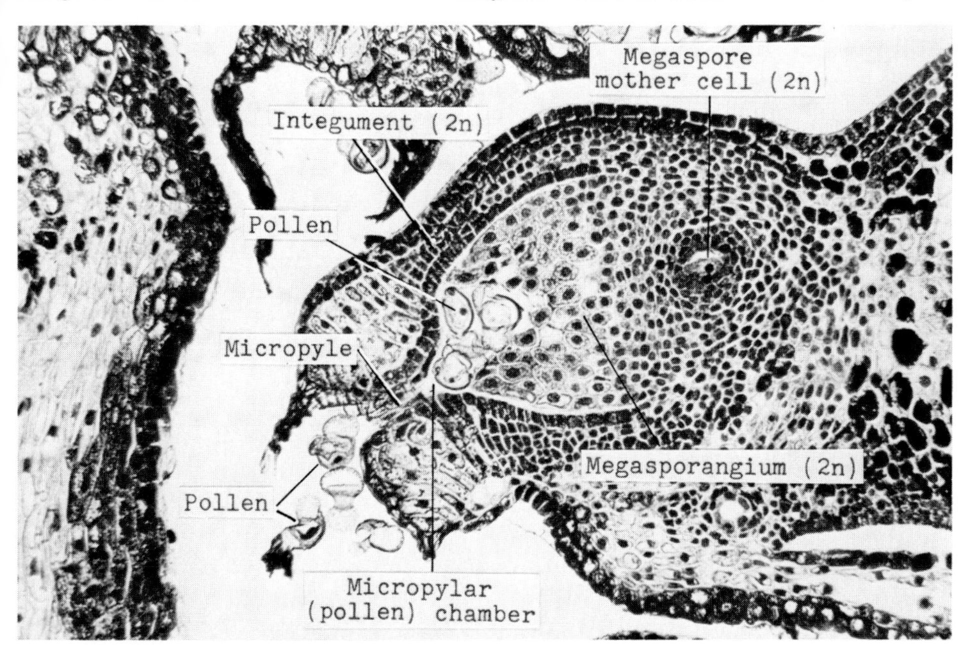

Fig. 37a Megaspore mother cell in ovulate cone l.s. x100.

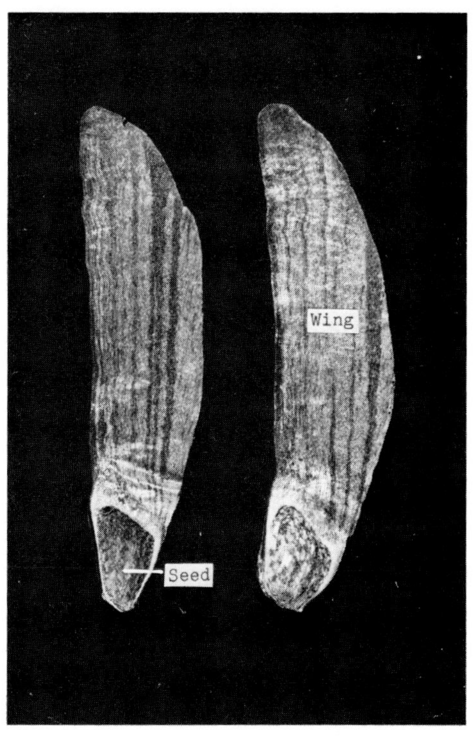

Fig. 37d Pine seeds x5.

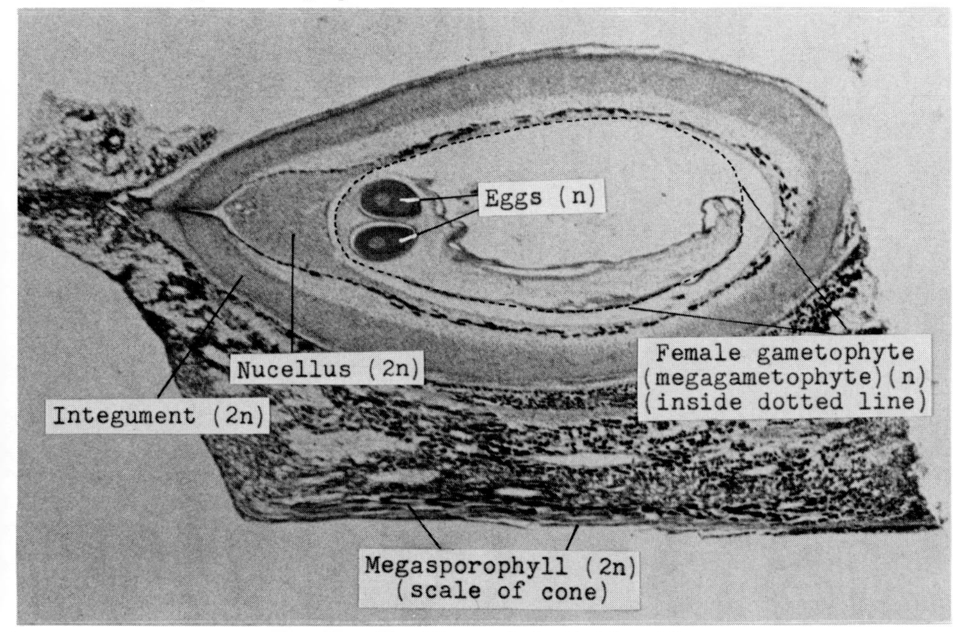

Fig. 37b Pine ovule (entire) l.s. x10.

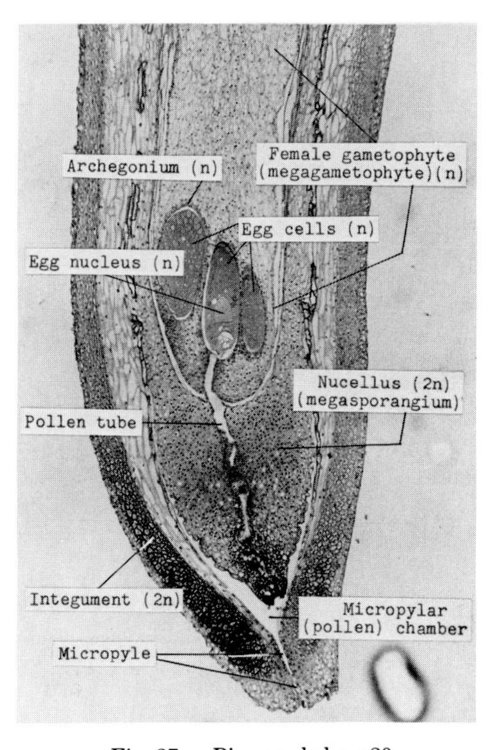

Fig. 37e Pine ovule l.s. x20.

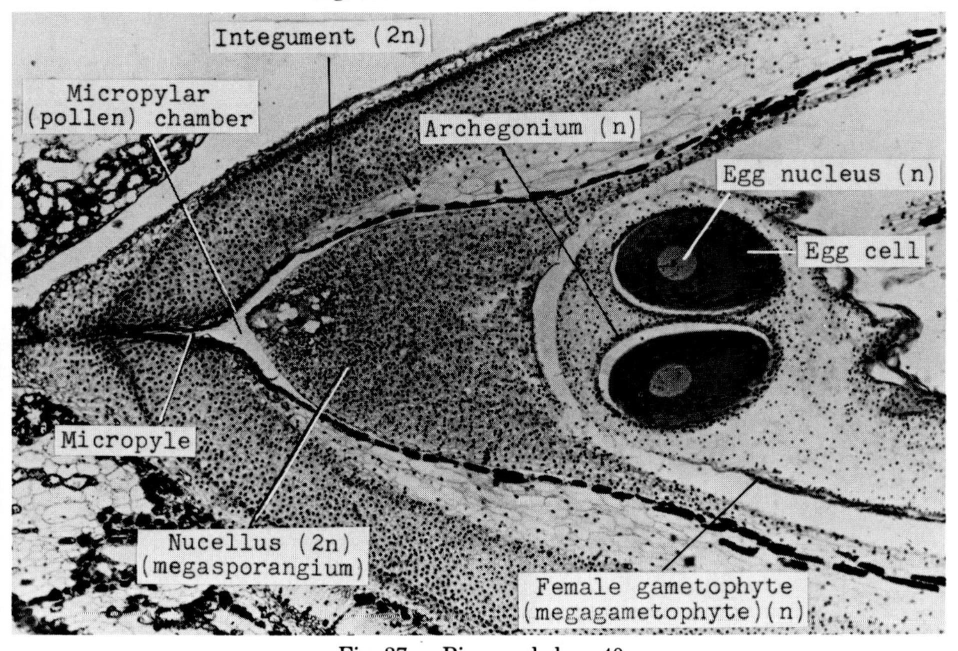

Fig. 37c Pine ovule l.s. x40.

Notice that the structures in 37a become or give rise to the structures in 37b & c. The megaspore mother cell in 37a gives rise to a megaspore by meiosis which divides many times to form the haploid female gametophyte in 37b & c. All other diploid structures came from the original diploid ovule cells in 37a.

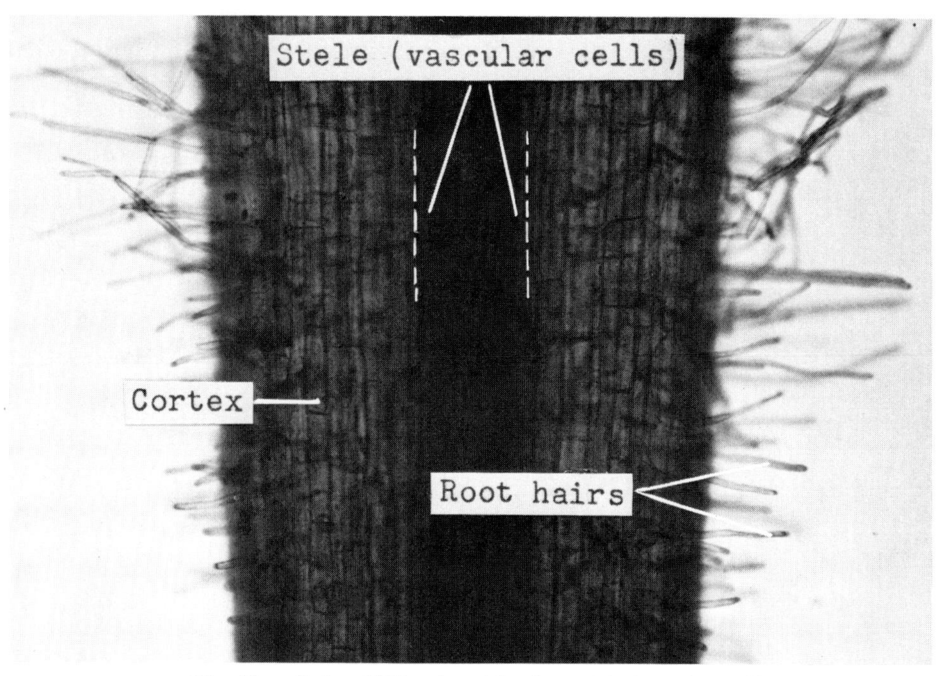

Fig. 38a Onion (Allium) root l.s. in root hair region x40.

Fig. 38b Root hairs on radish seedlings x2.

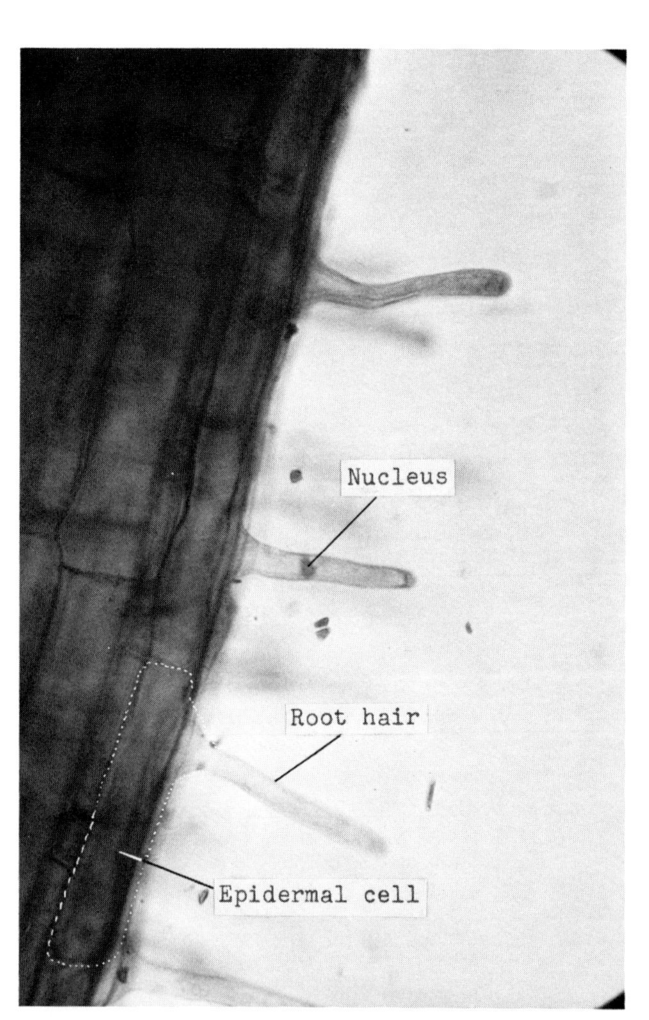

Fig. 38c Onion root hairs l.s. x430. Notice that the root hairs are merely extensions of epidermal cells.

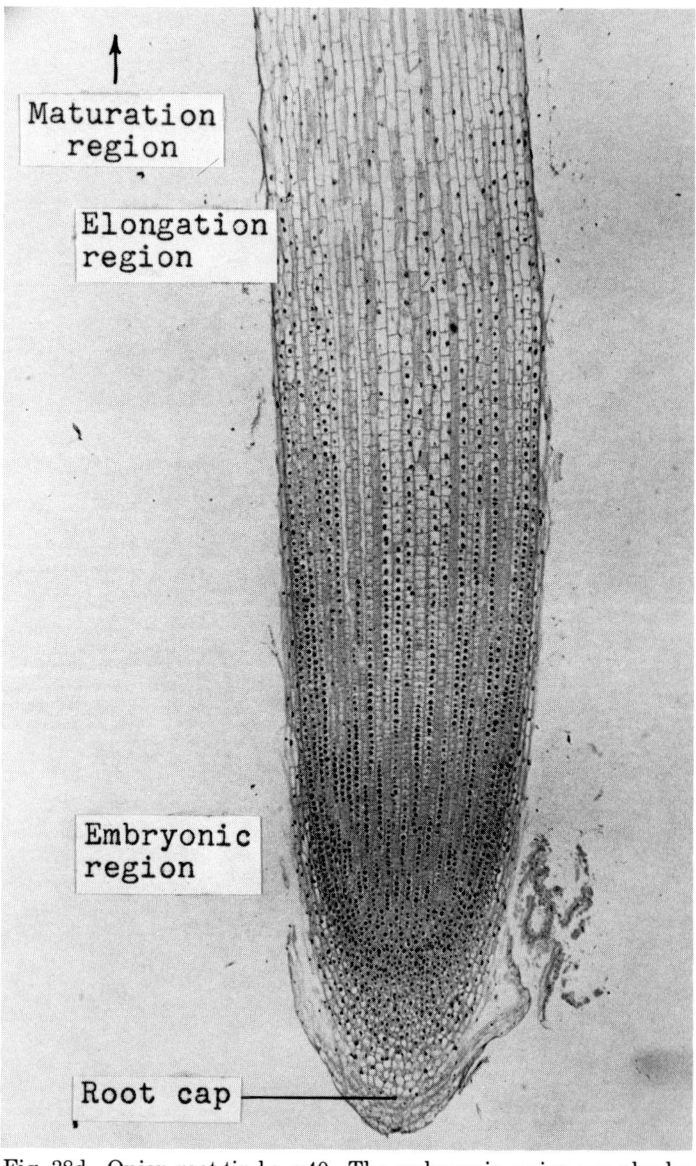

Fig. 38d Onion root tip l.s. x40. The embryonic region can also be called the meristematic region. See Fig. 38a for the maturation region (where differentiation of cells occurs).

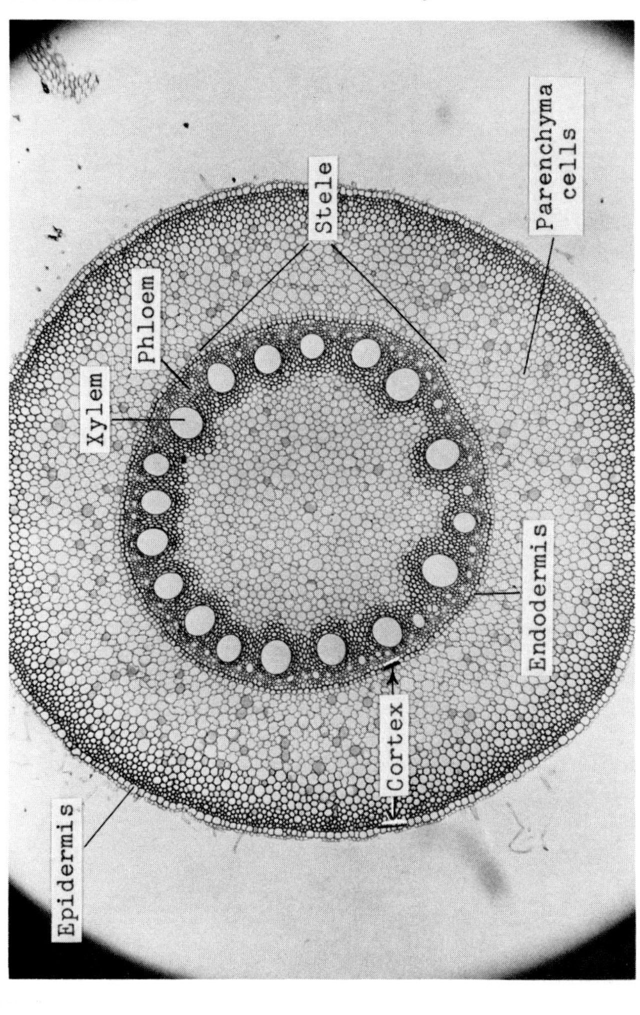

Fig. 39b Monocot root x.s. x40.

Fig. 39d Dicot root stele x.s. x430.

Fig. 39a Monocot root x.s. x40.

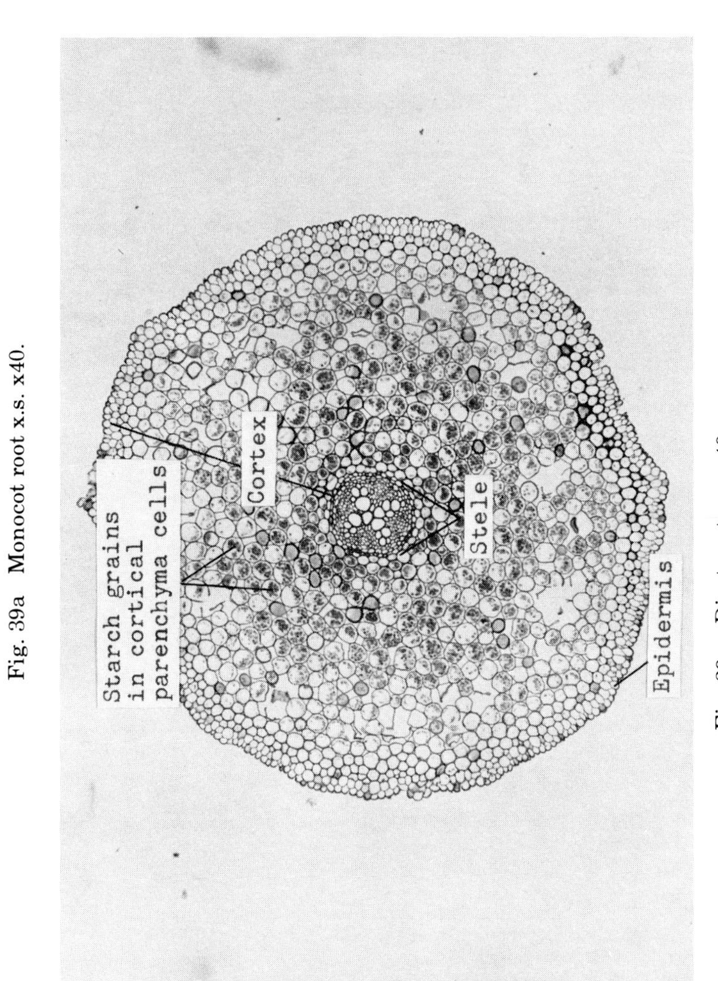

Fig. 39c Dicot root x.s. x40.

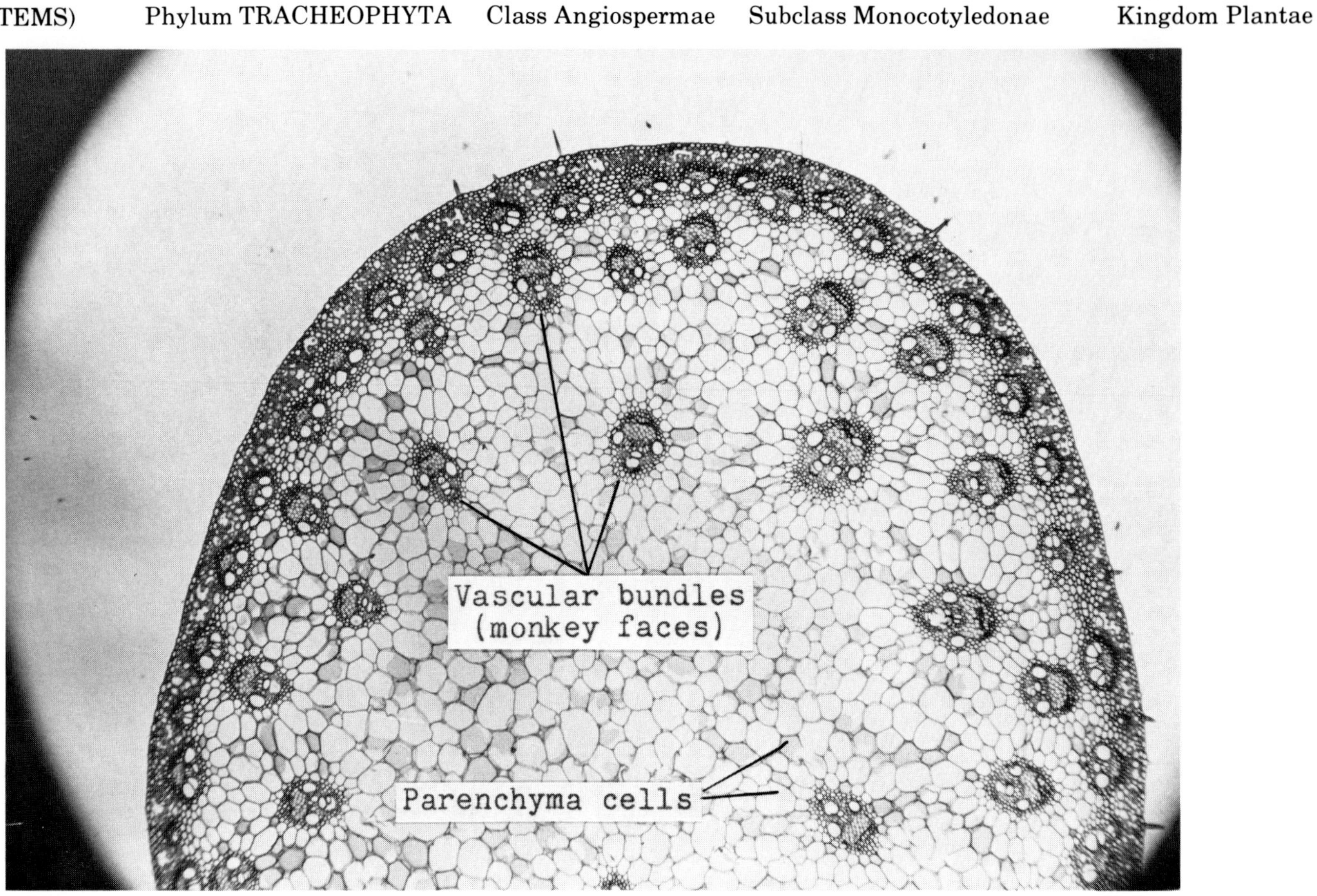

Fig. 40a Monocot stem x.s. x40.

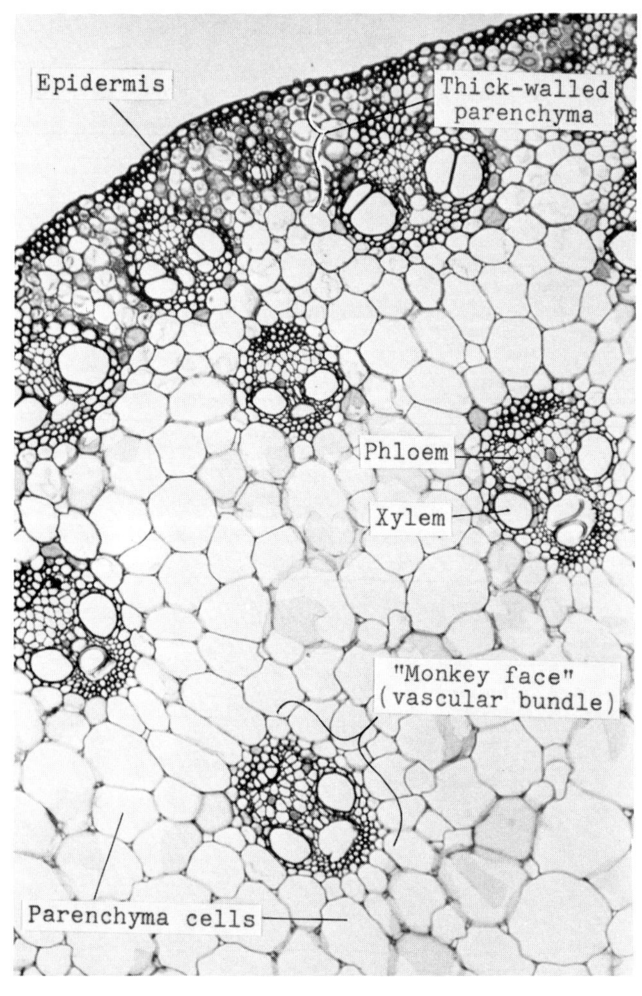

Fig. 40b Monocot stem x.s. x100.

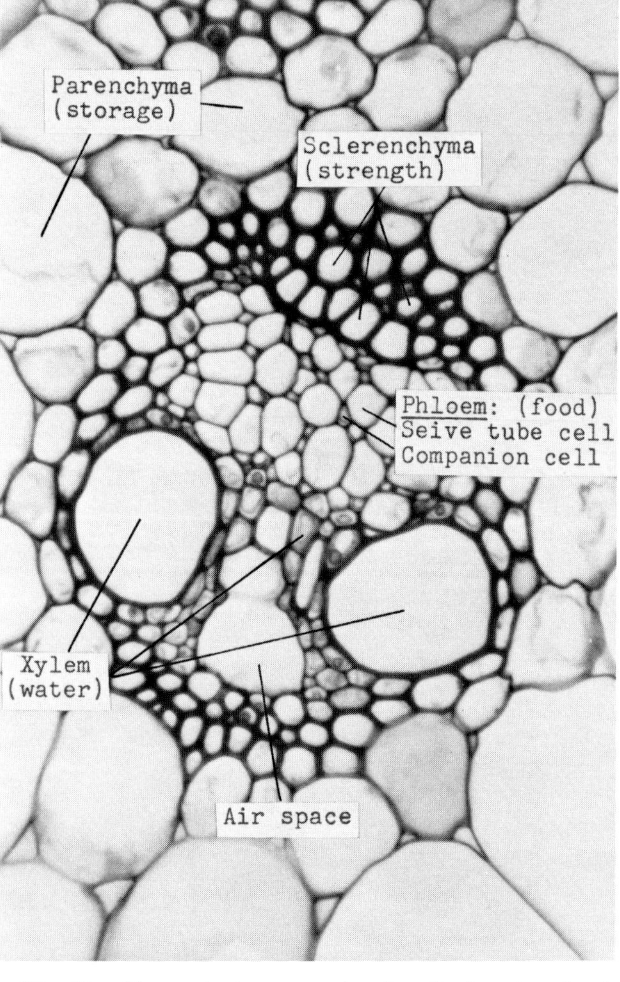

Fig. 40c Monocot stem vascular bundle (monkey face) x.s. x430.

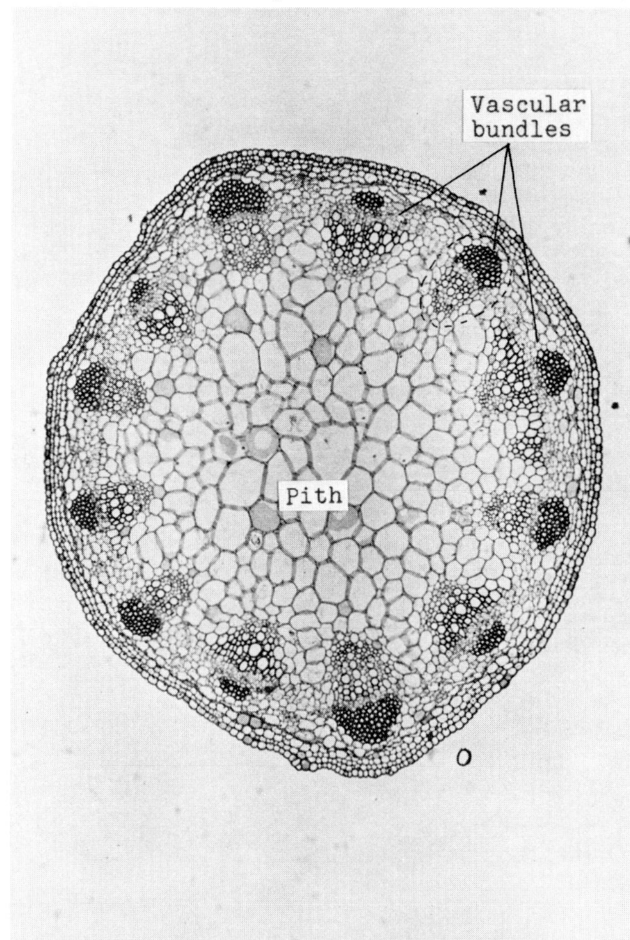

Fig. 41a Herbaceous dicot stem x.s. x40.

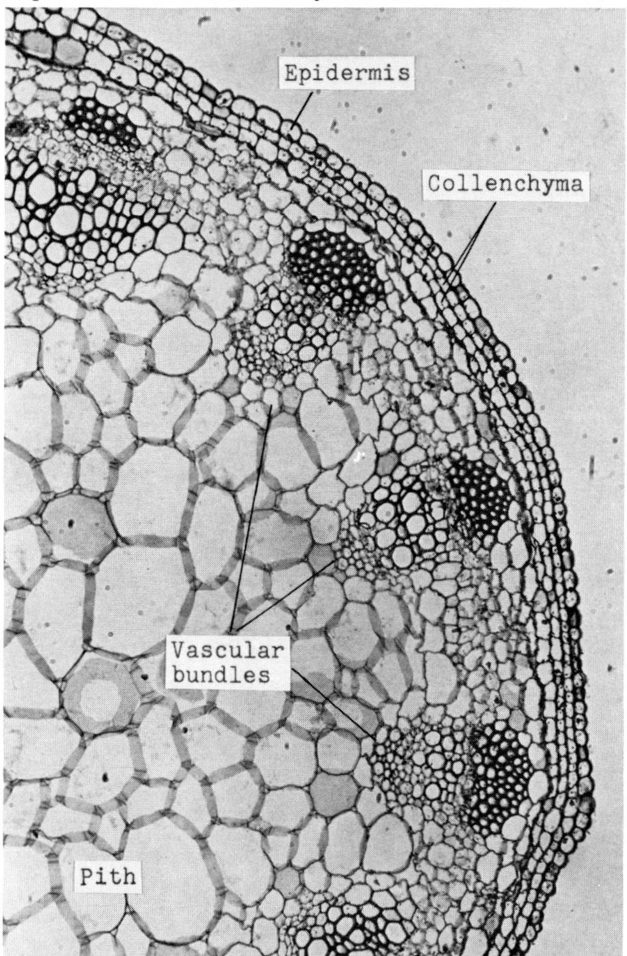

Fig. 41b Herbaceous dicot stem x.s. x100.

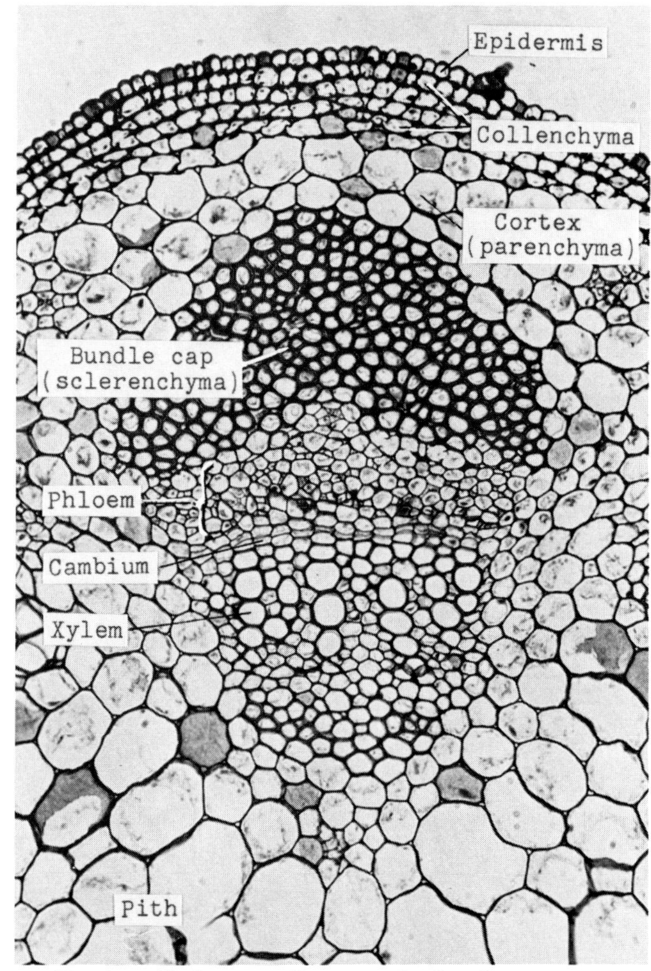

Fig. 41c Herbaceous dicot vascular bundle x.s. x430.

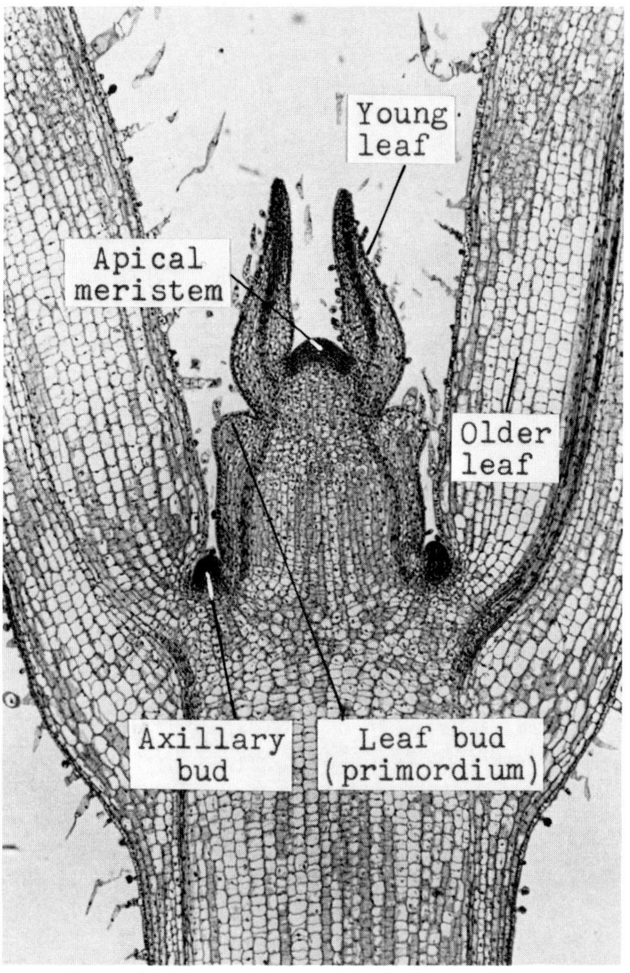

Fig. 41d Stem tip with apical meristem l.s. x40.

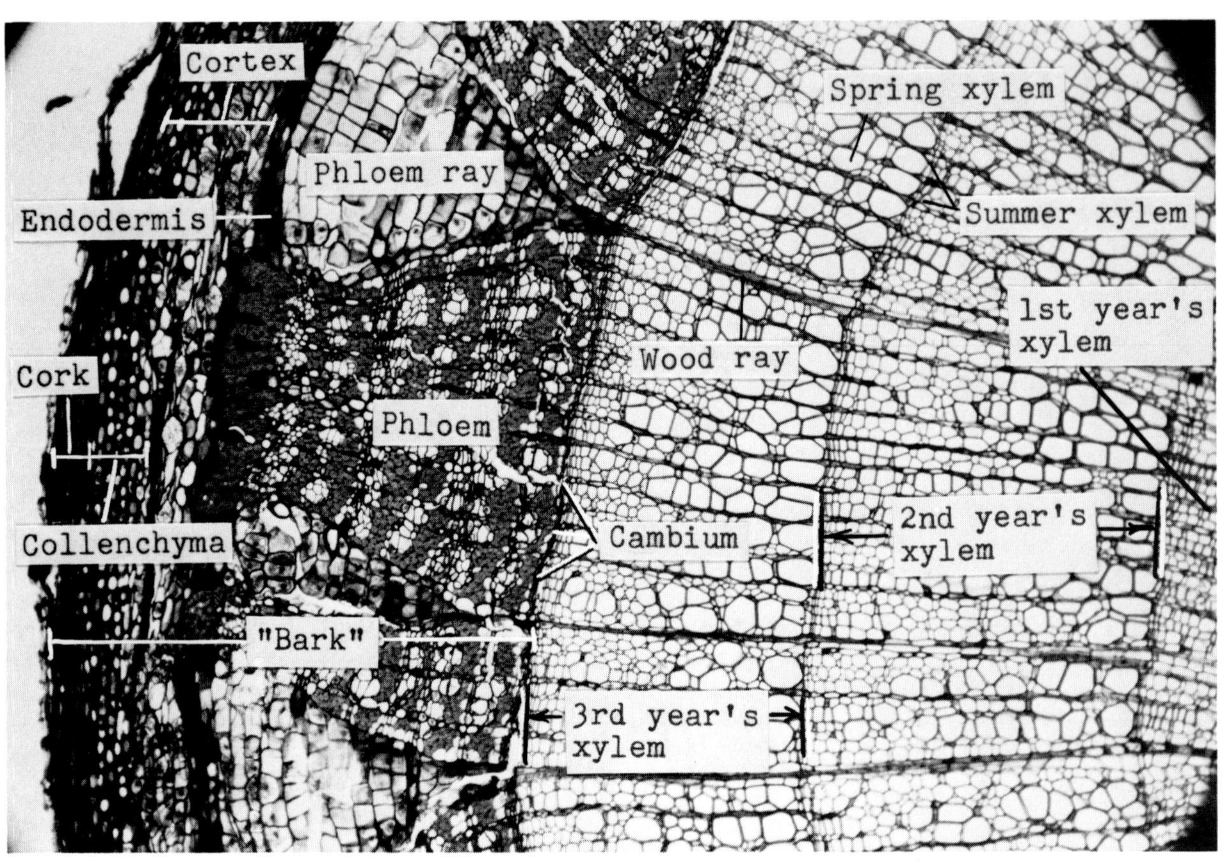

Fig. 42a Woody dicot stem (3 year old Tilia) x.s. x40.

Fig. 42b Woody dicot stem x.s. x1.

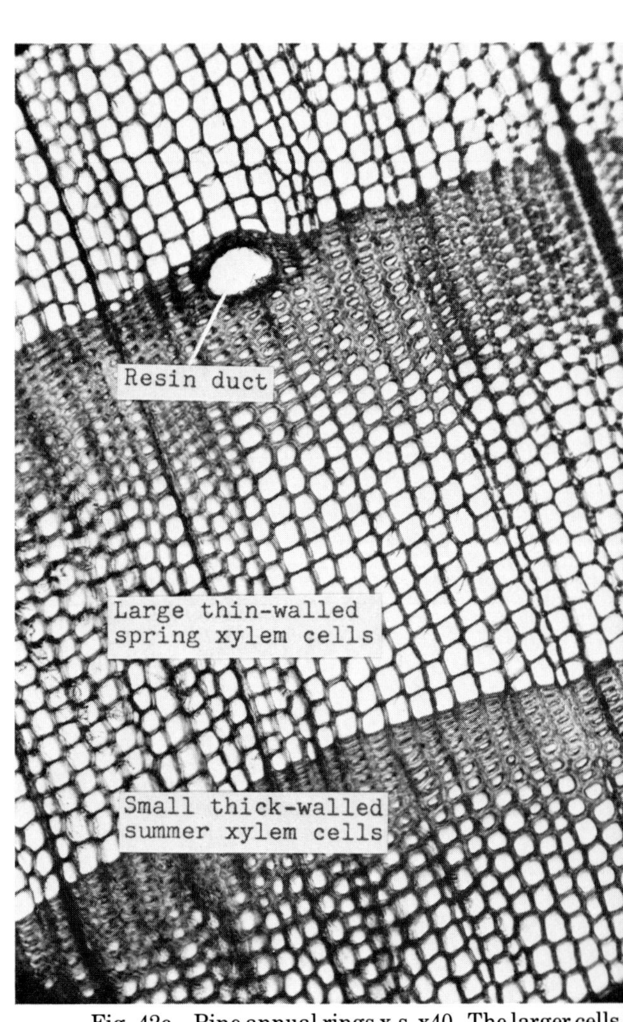

Fig. 42c Pine annual rings x.s. x40. The larger cells are made in the spring when water is plentiful. The smaller cells are made in late summer and fall when growing conditions are less favorable.

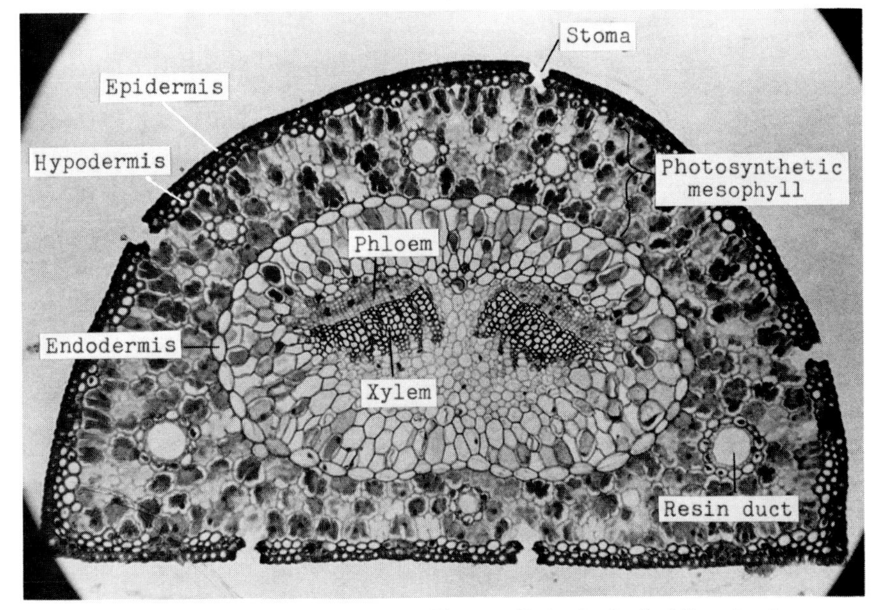

Fig. 43a Pine needle x.s. x40. The needle is the leaf of the pine tree.

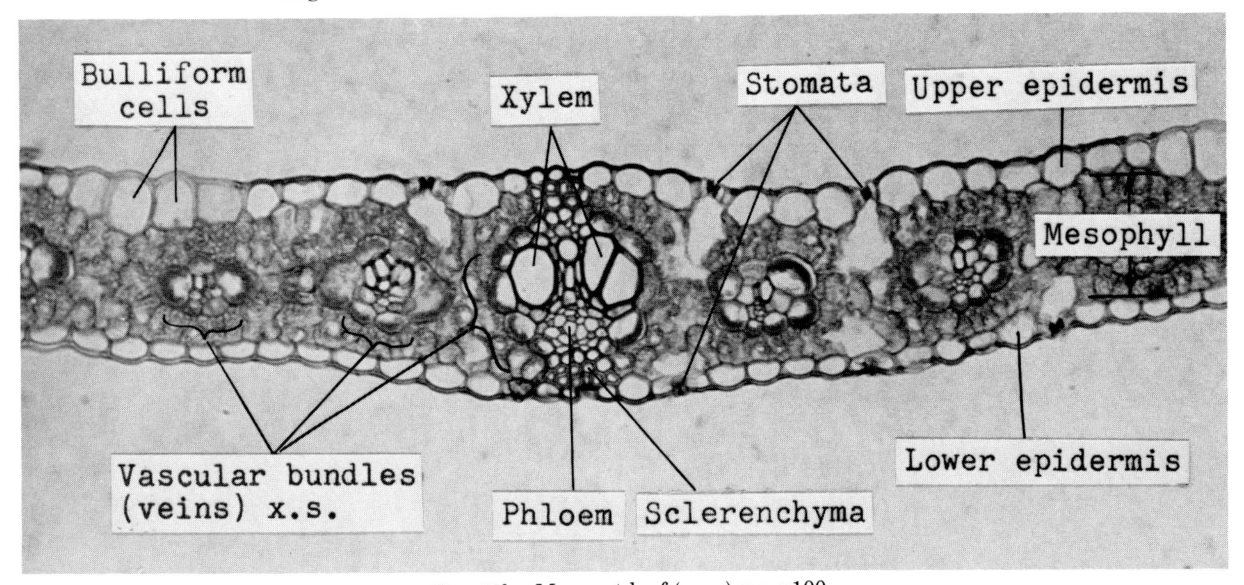

Fig. 43b Monocot leaf (corn) x.s. x100.

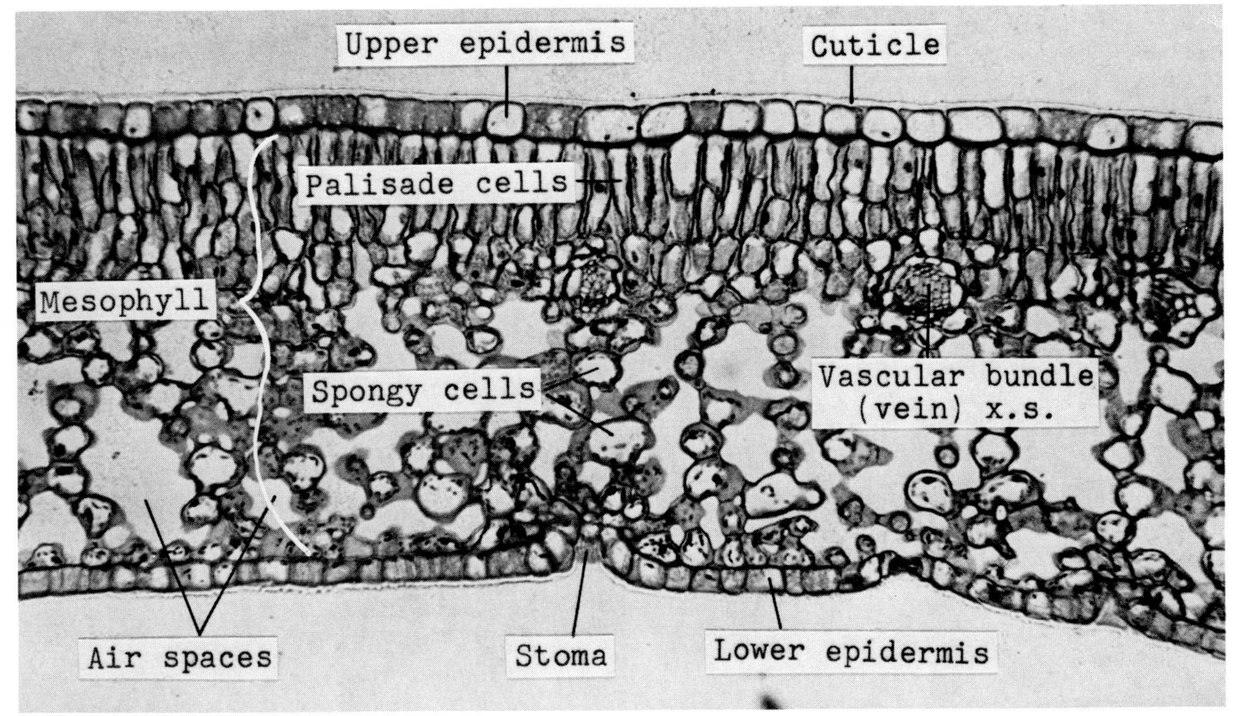

Fig. 43c Dicot leaf x.s. x100.

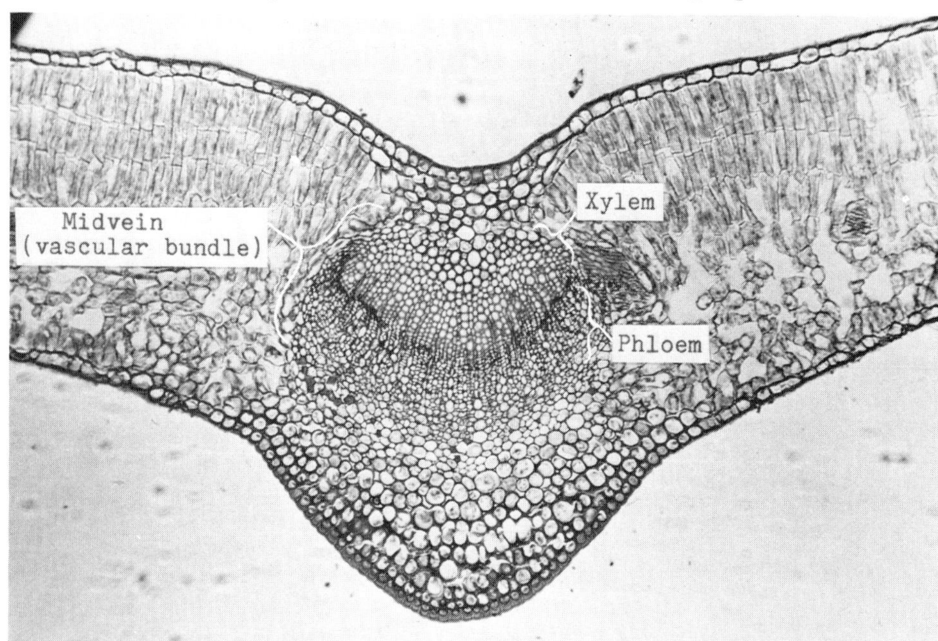

Fig. 44a Dicot leaf mid-vein x.s. x100.

Fig. 44d Leaf "skeleton" x1. The softer tissues of the leaf have decayed away leaving only the tougher vascular tissues.

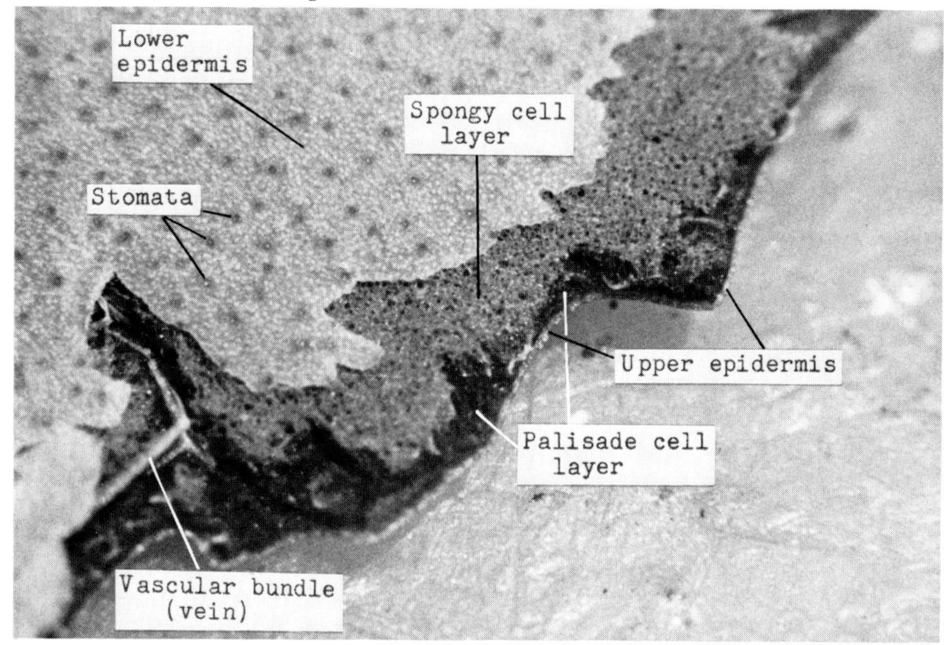

Fig. 44b Torn edge of a Ligustrum leaf x10. This leaf is upside down.

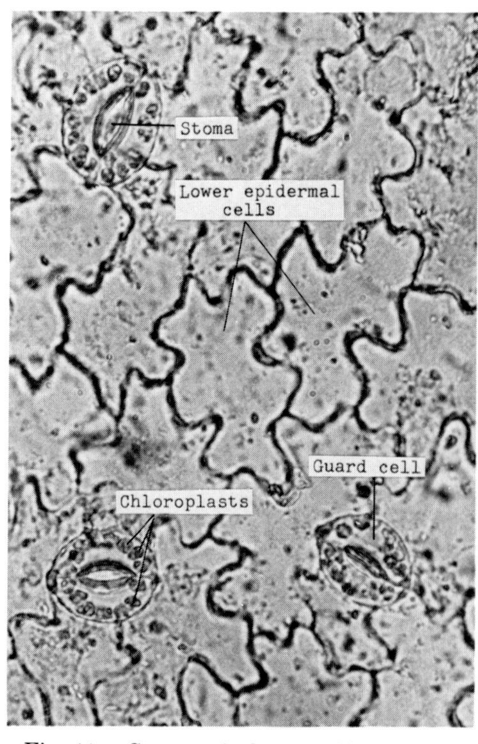

Fig. 44e Stomata in lower epidermis of a geranium leaf w.m. x430.

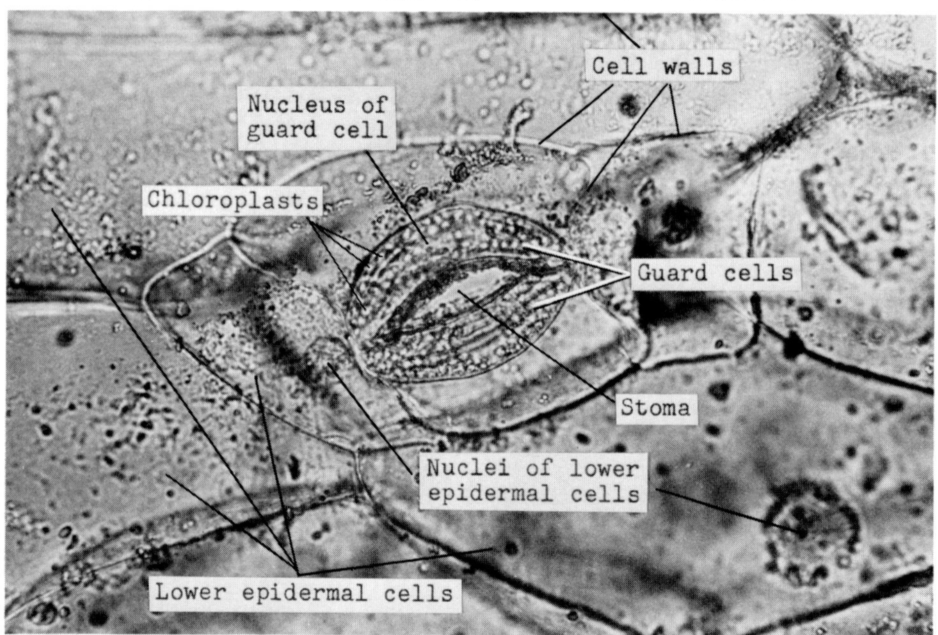

Fig. 44c Stoma in lower epidermis of Tradescantia leaf w.m. x430.

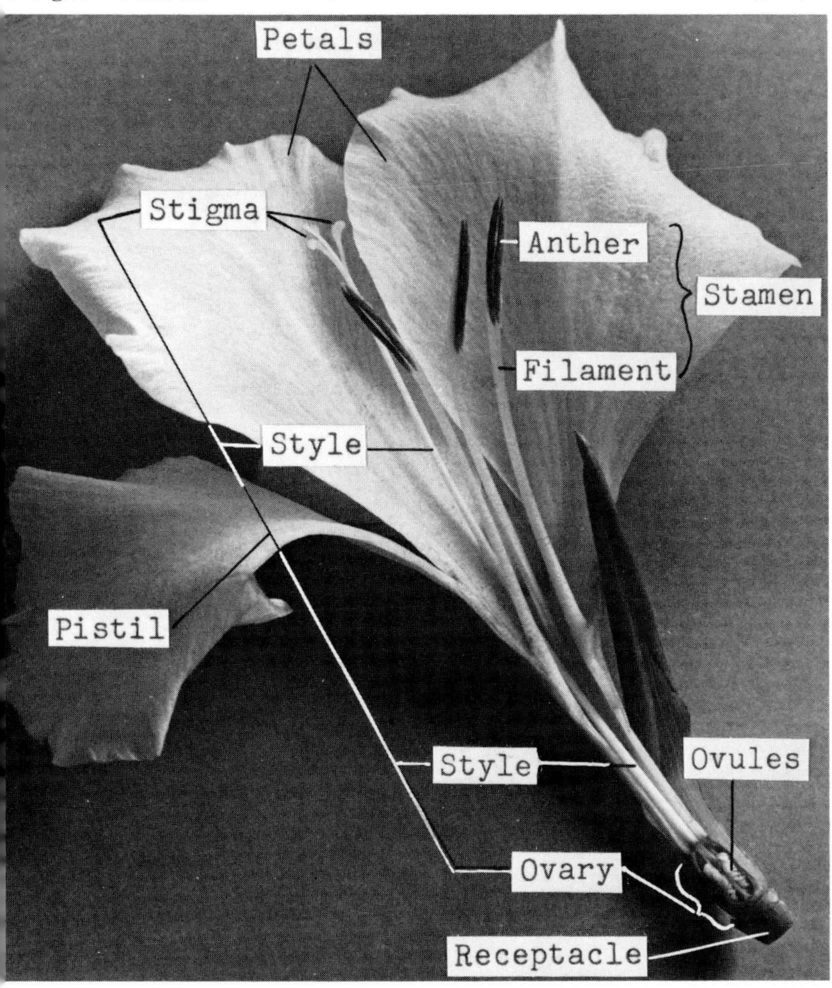

Fig. 45a Gladiola flower dissection l.s. x1.

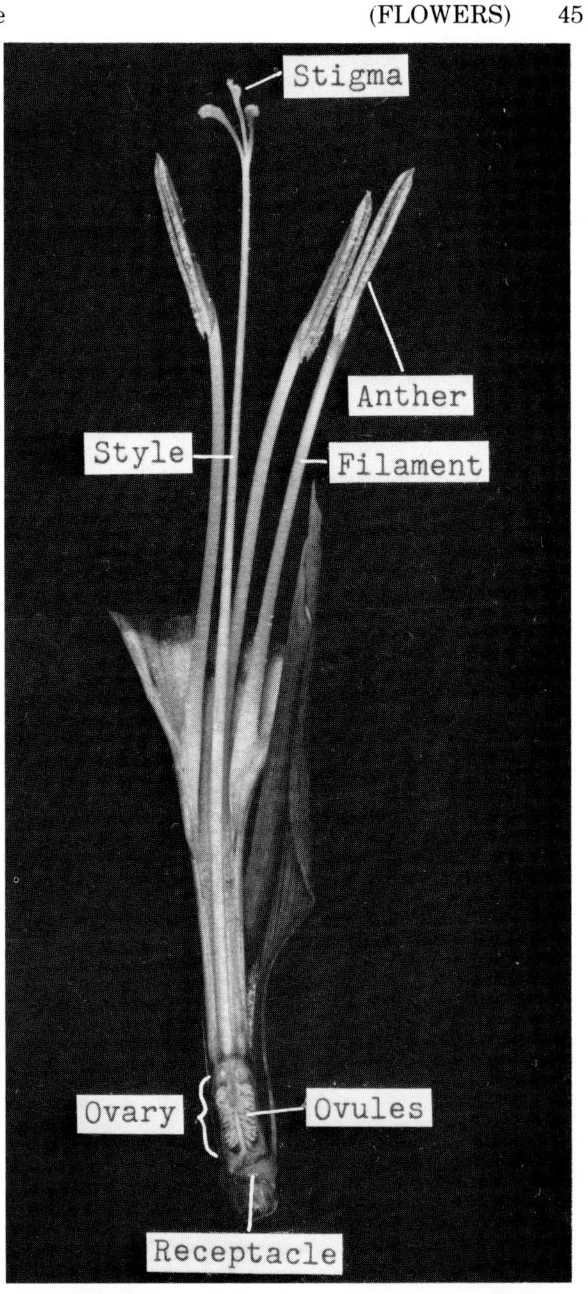

Fig. 45b Gladiola flower dissection with petals removed l.s. x1.

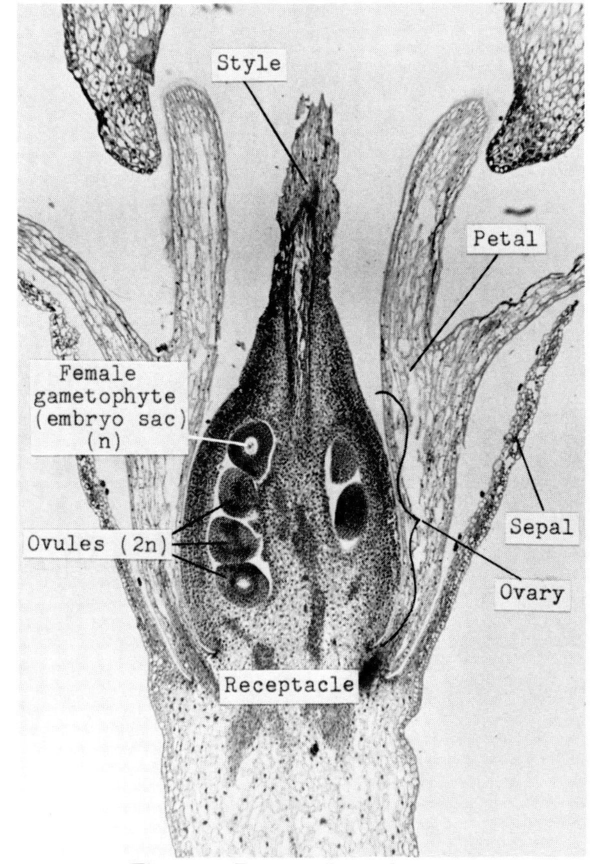

Fig. 45c Tomato flower l.s. x40.

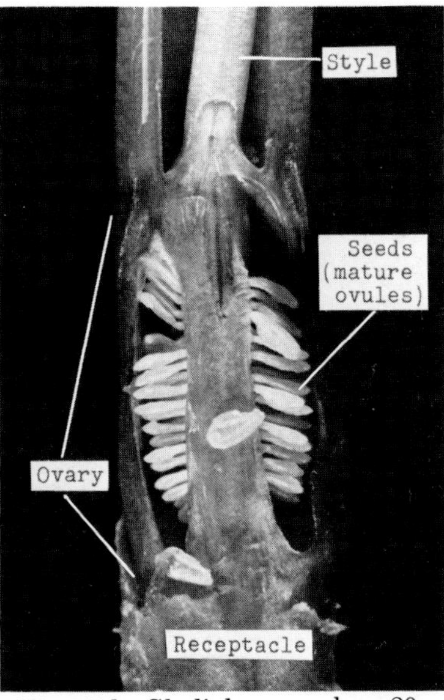

Fig. 45d Gladiola ovary l.s. x20.

Fig. 45e Gladiola anthers, stigma and style x10.

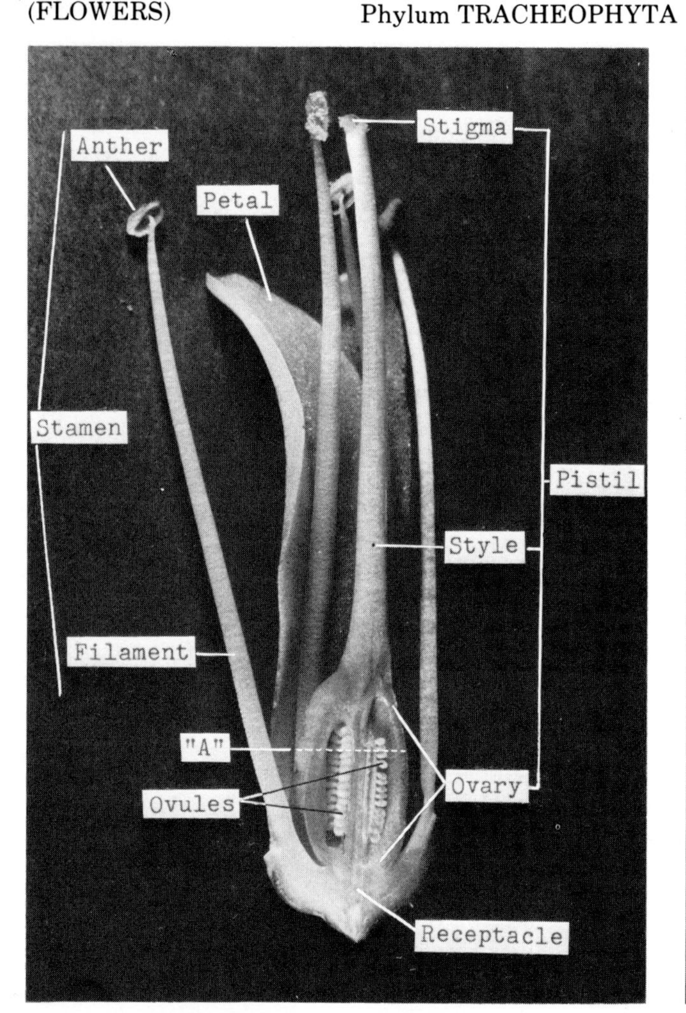

Fig. 46a Yucca flower l.s. x3. A section cut at point "A" would look like Fig. 47a.

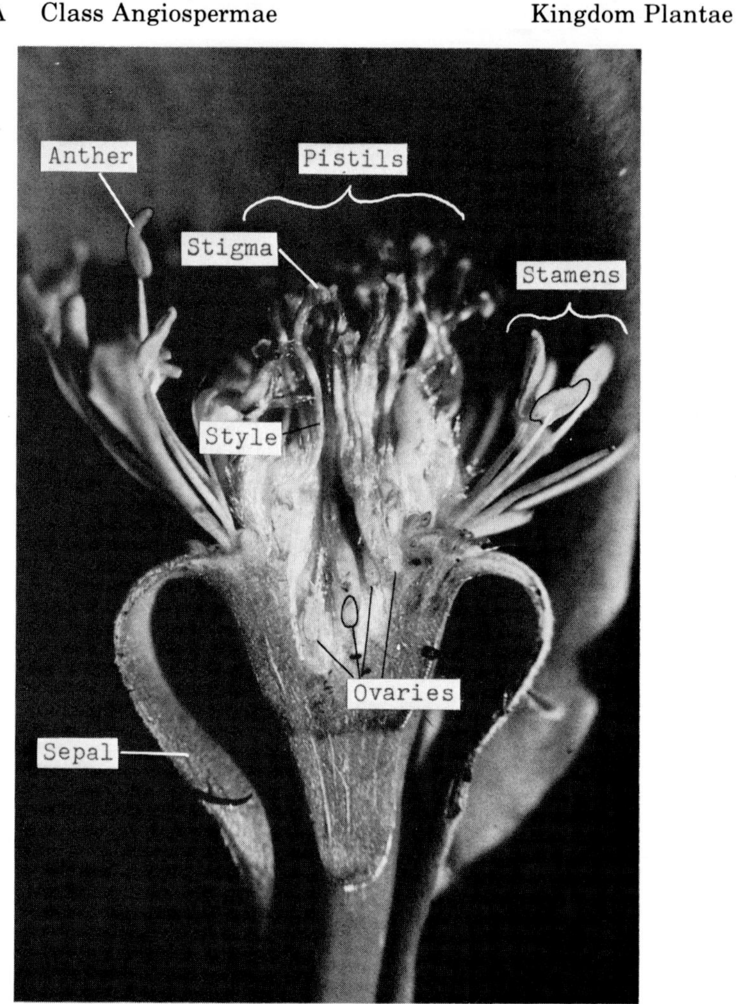

Fig. 46b Rose flower l.s. x3.

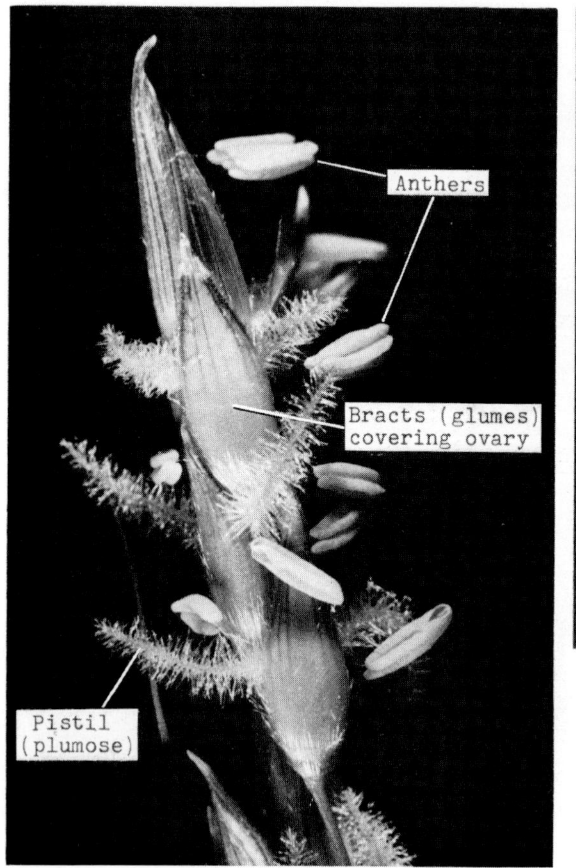

Fig. 46c Johnson grass flowers x20. A typical grass (monocot) flower.

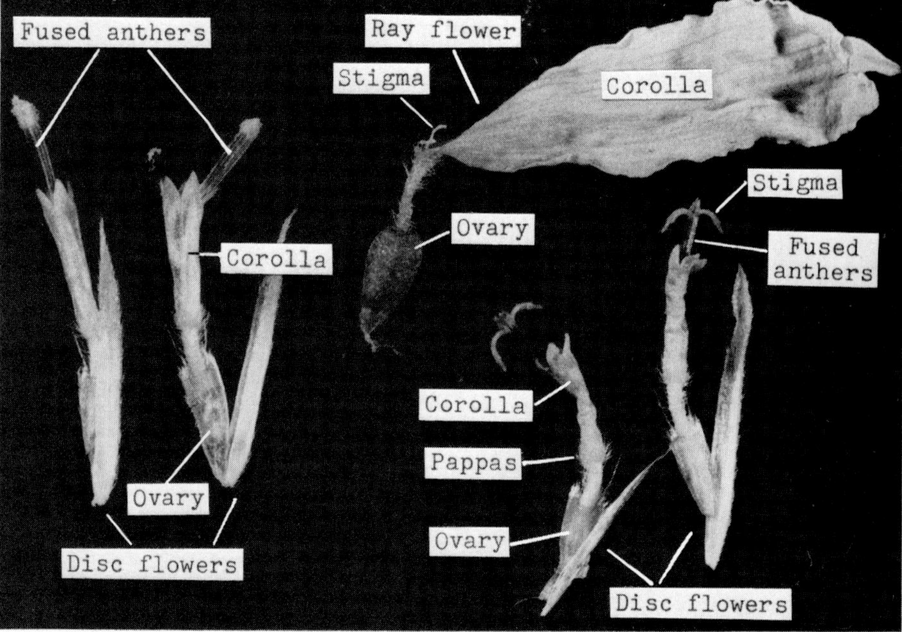

Fig. 46d Disc and ray flowers from a composite flower x10.

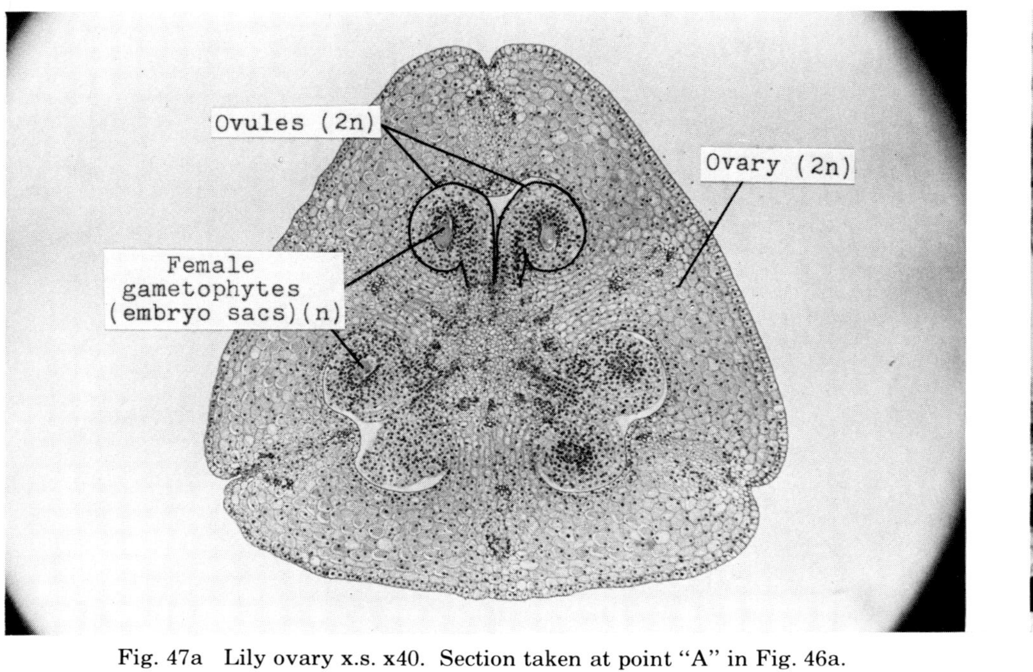

Fig. 47a Lily ovary x.s. x40. Section taken at point "A" in Fig. 46a.

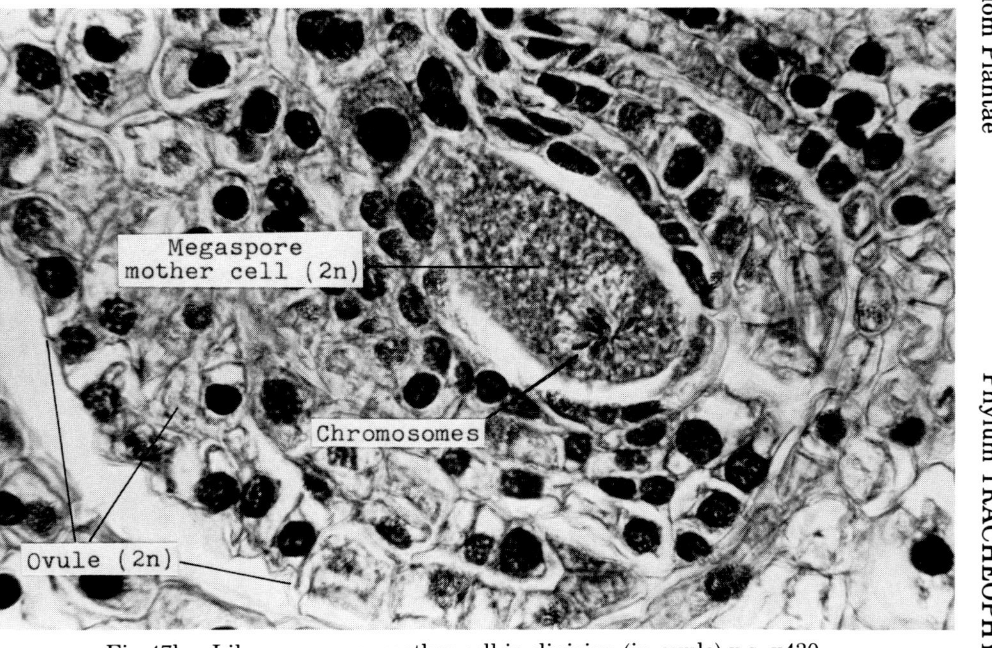

Fig 47b Lily megaspore mother cell in division (in ovule) x.s. x430.

Fig. 47c Lily ovule containing female gametophyte (embryo sac) x.s. x100.

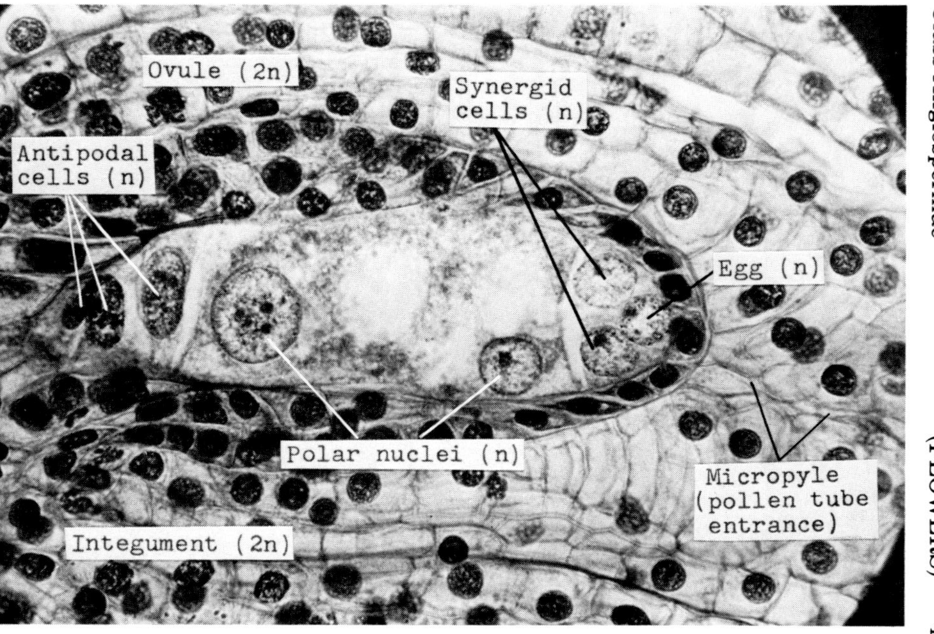

Fig. 47d Lily female gametophyte (embryo sac) 8-nuclei stage l.s. x430.

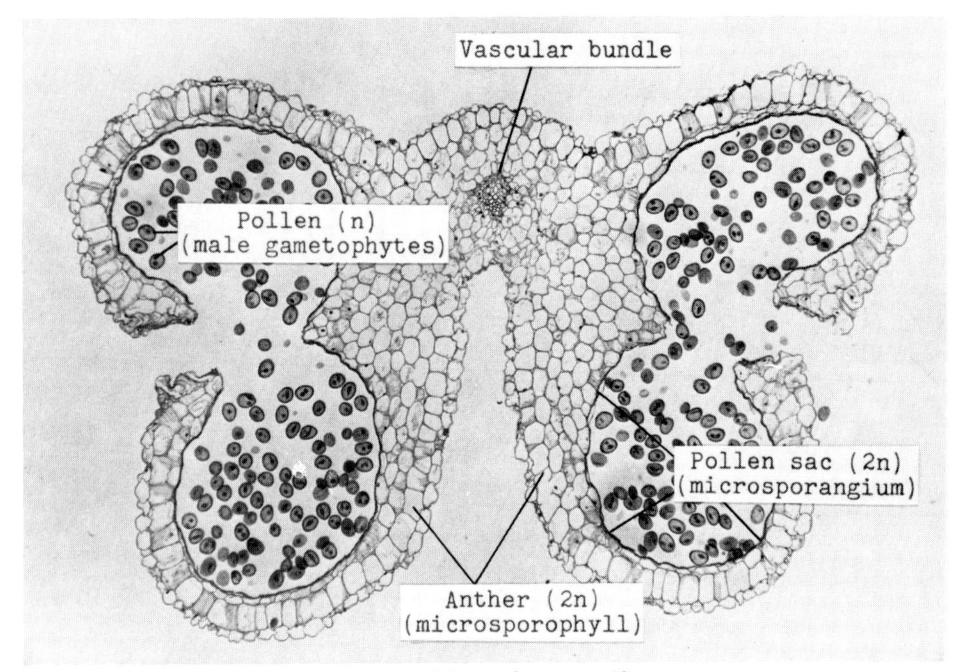

Fig. 48a Lily anther x.s. x40.

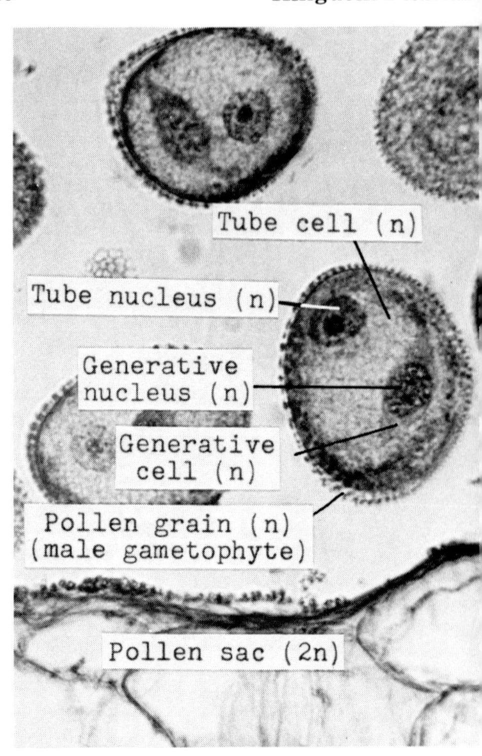

Fig. 48b Lily pollen grains (male gametophytes) x.s. x430. The generative nucleus divides to form the two sperm nuclei.

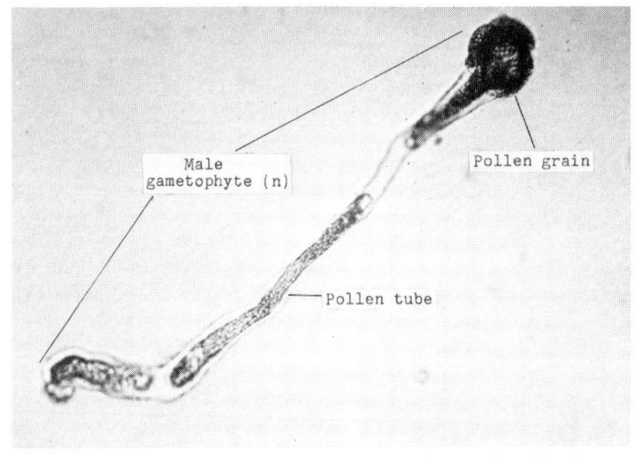

Fig. 48c Pollen grain with pollen tube (live) w.m. x430.

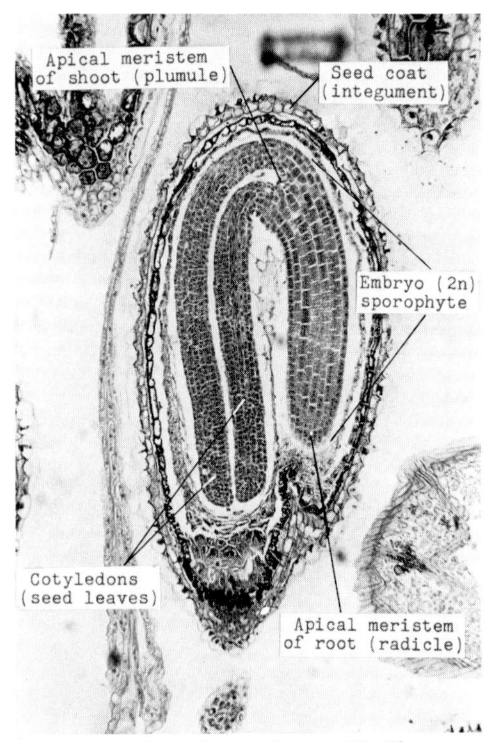

Fig. 48d Capsella seed l.s. x100. The two cotyledons show this is a dicot.

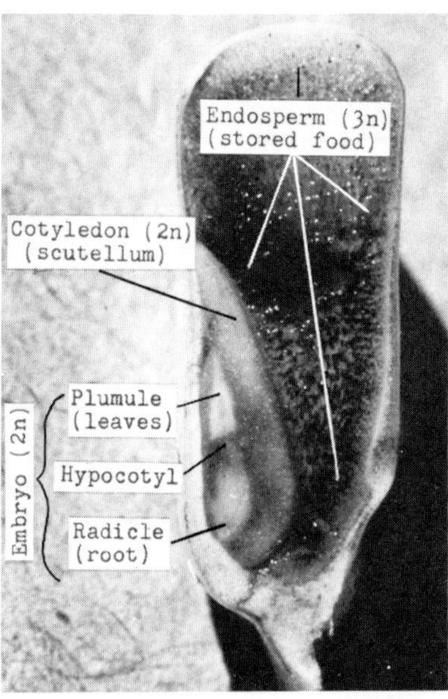

Fig. 48e Corn (monocot seed) stained with iodine to show location of starch l.s. x20. Starch is stored in the endosperm in monocots.

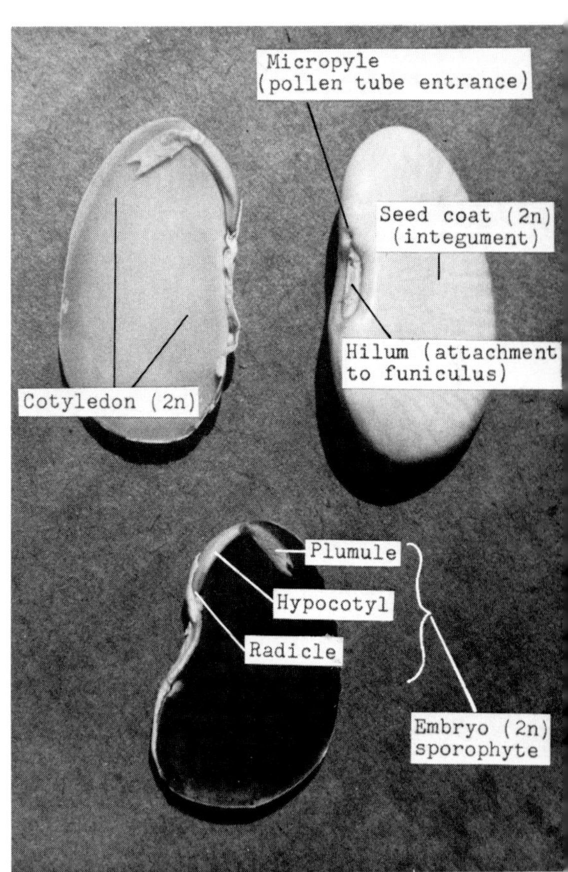

Fig. 48f Lima bean (dicot seed) l.s. & w.m. x1. The lower bean is stained with iodine for starch. Starch is stored in the cotyledons in dicots.

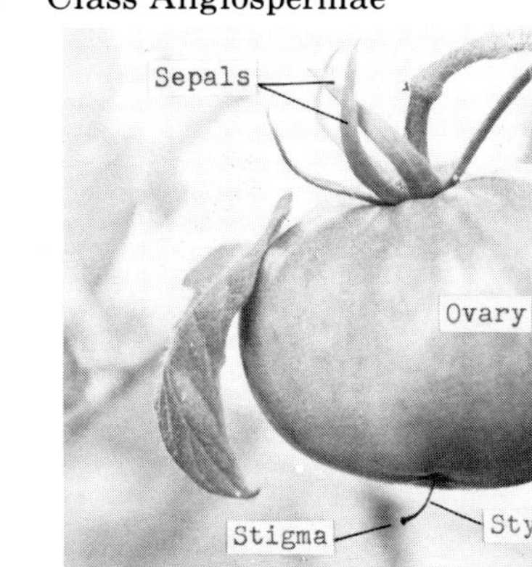

Fig. 49a Tomato (fruit) l.s. x1.

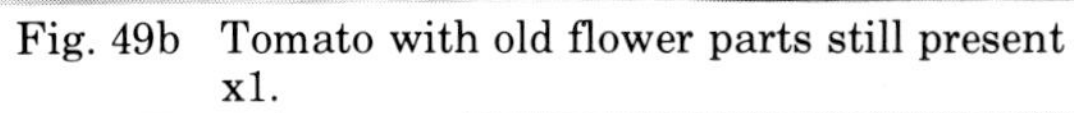

Fig. 49b Tomato with old flower parts still present x1.

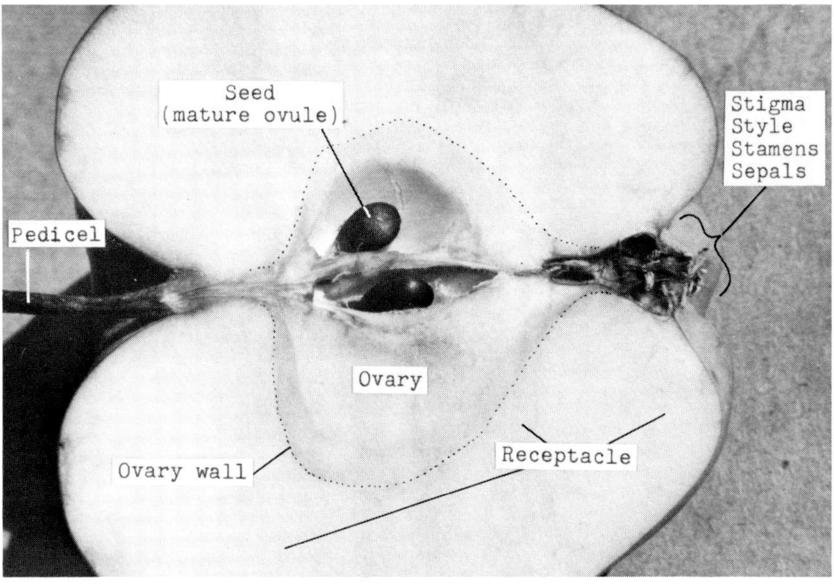

Fig. 49c Apple l.s. x1.

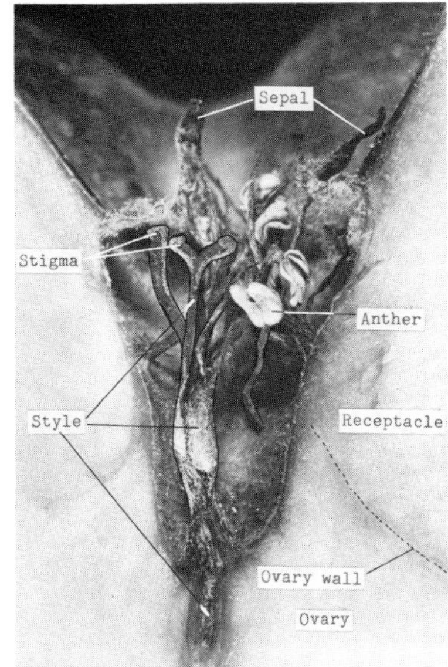

Fig. 49d Remains of old flower parts in the apple l.s. x5.

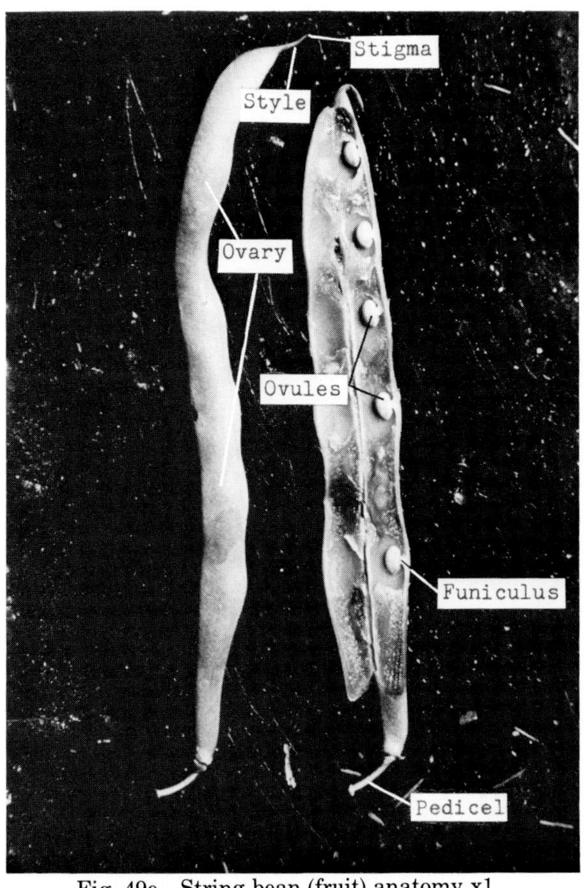

Fig. 49e String bean (fruit) anatomy x1.

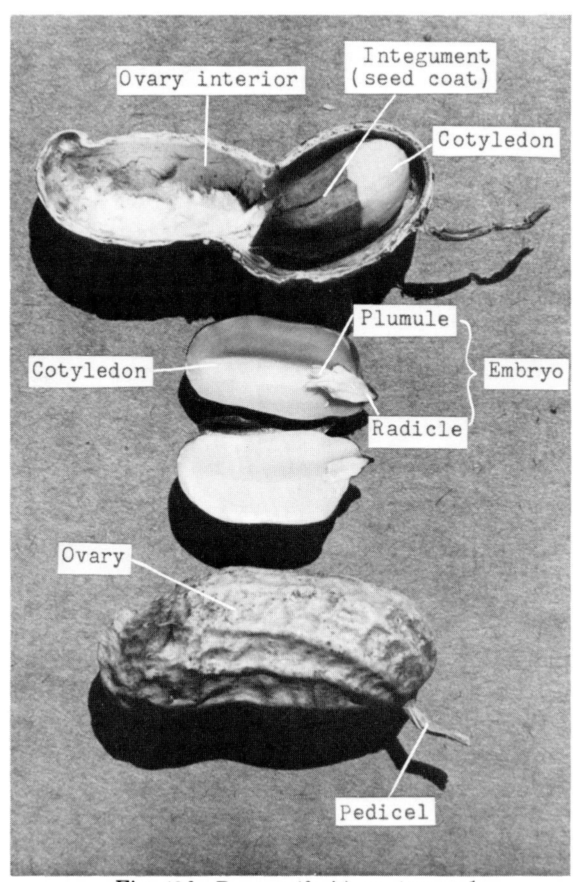

Fig. 49f Peanut (fruit) anatomy x1.

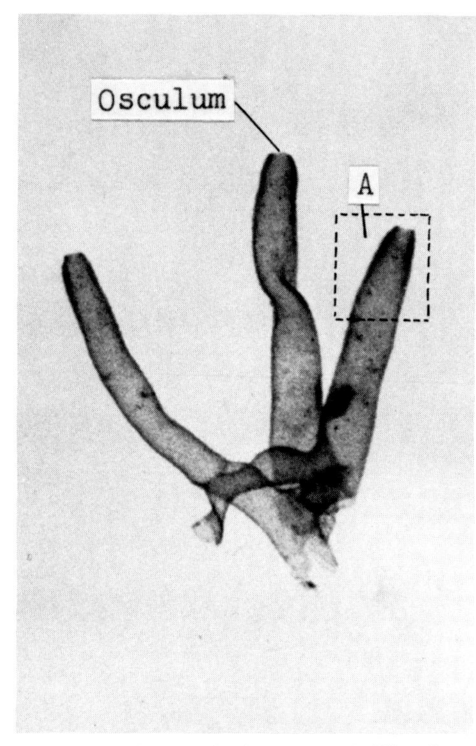

Fig. 50a Leucosolenia w.m. x10. The simplest type of sponge (asconoid). Box "A" is enlarged in Fig. 50b.

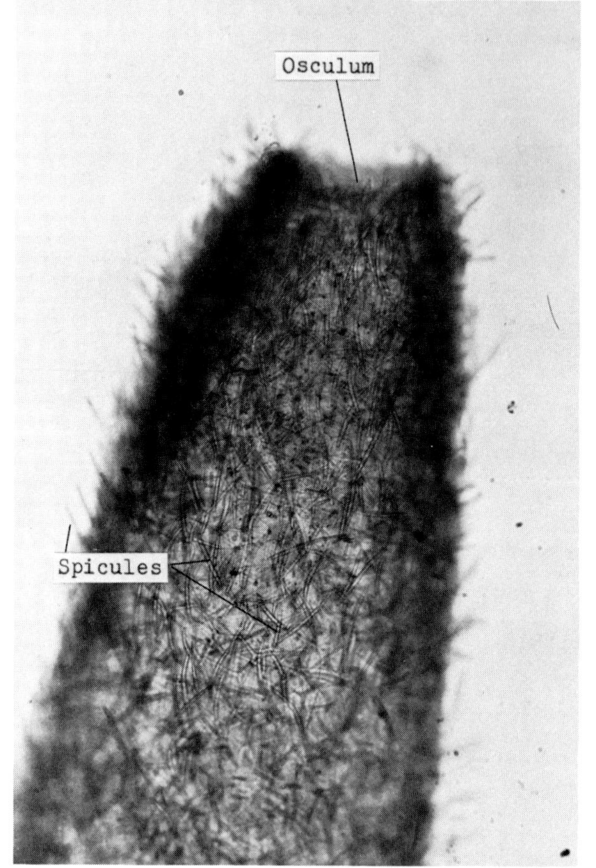

Fig. 50b Leucosolenia w.m. x40.

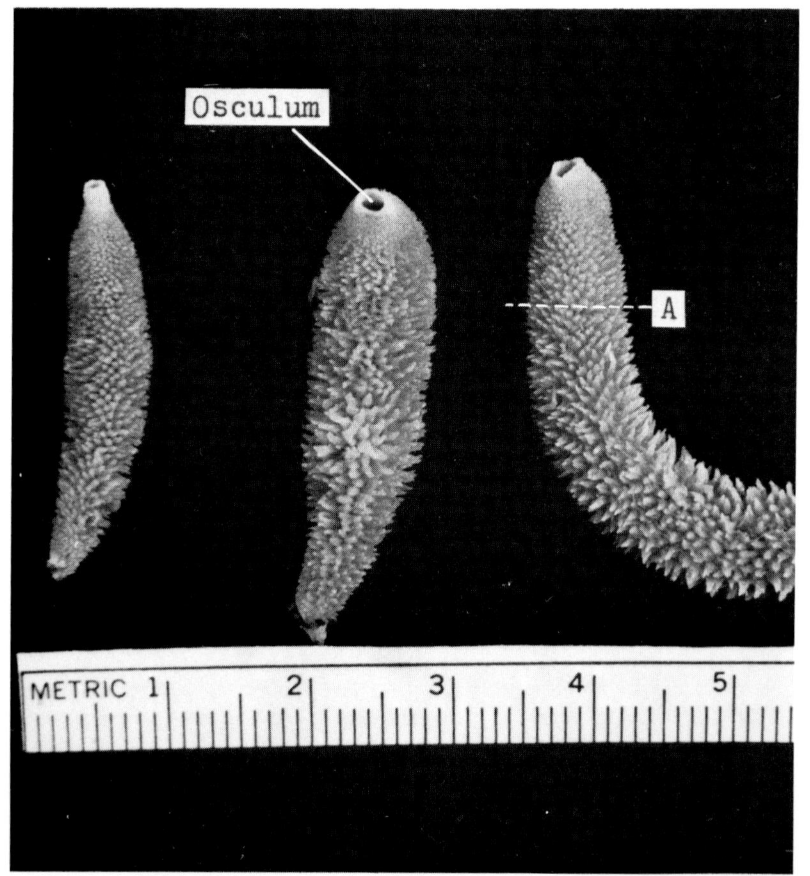

Fig. 50c Scypha (formerly known as Grantia) w.m. x1. This is a sponge of intermediate structural complexity (syconoid). A section cut at point "A" would look like Fig. 51b.

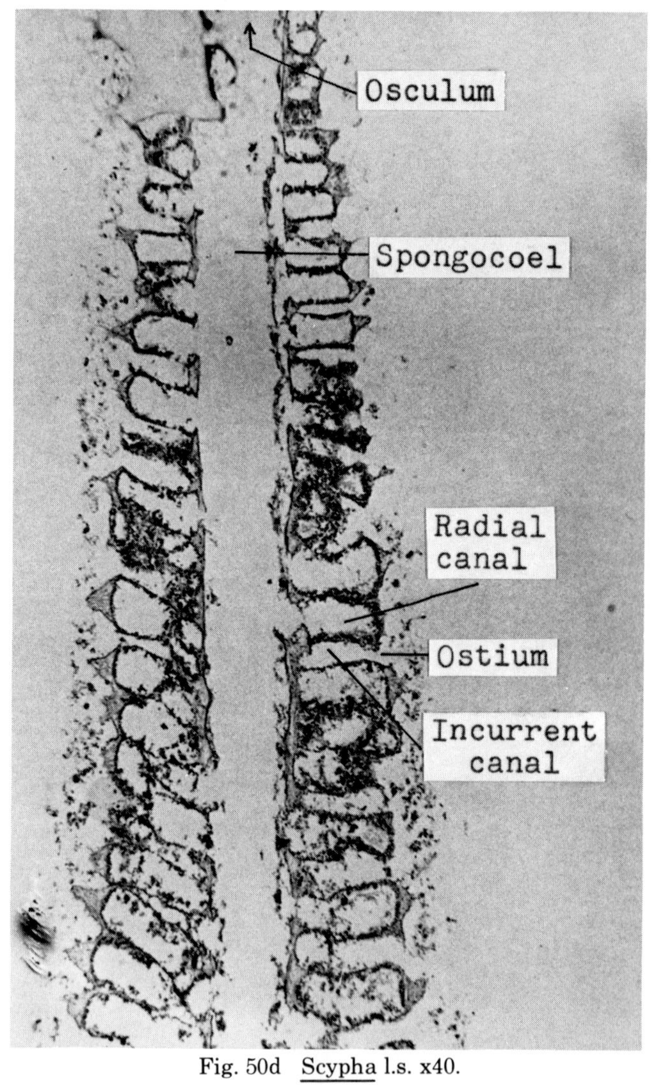

Fig. 50d Scypha l.s. x40.

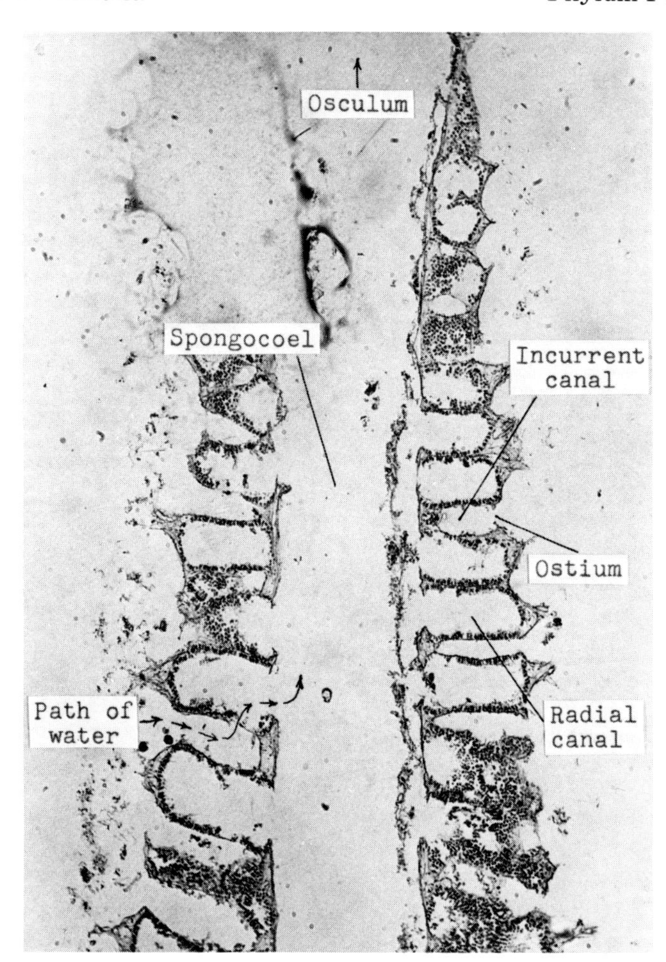

Fig. 51a Scypha l.s. x100.

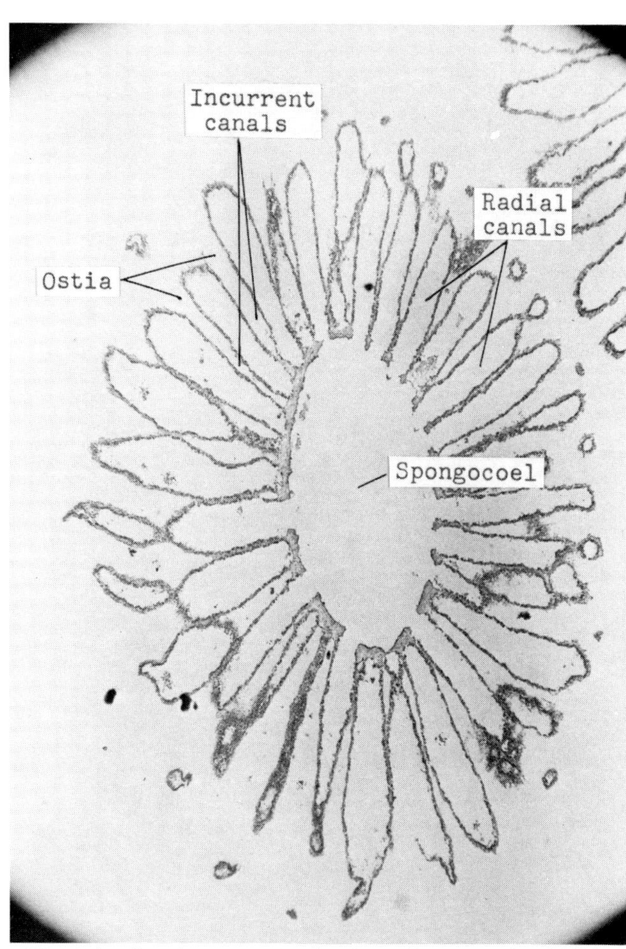

Fig. 51b Scypha x.s. x40. Section taken at point "A" in Fig. 50c.

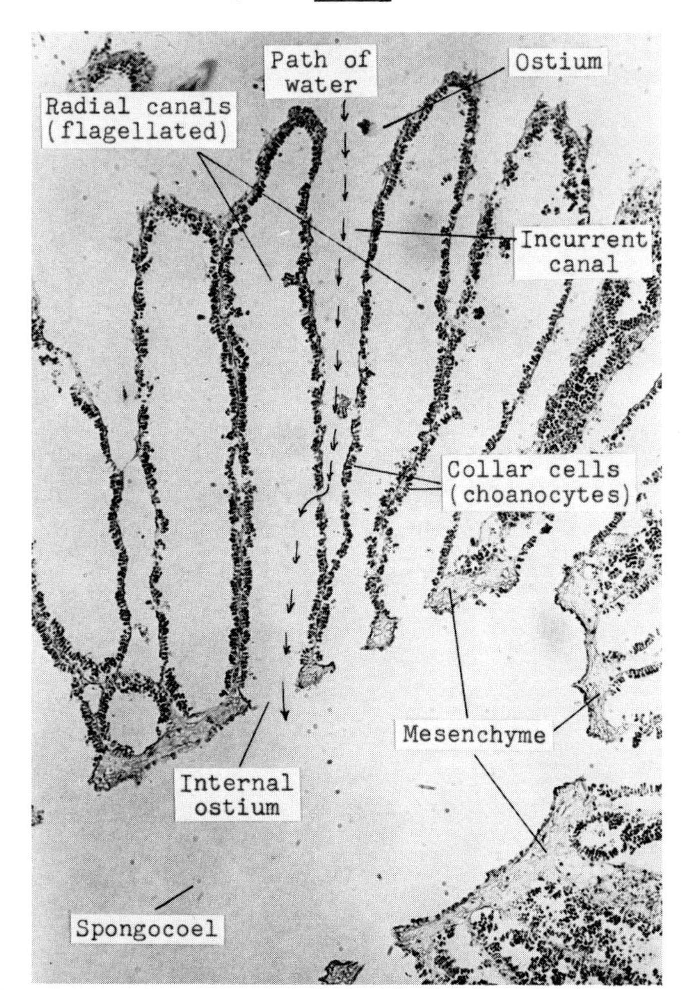

Fig. 51c Scypha x.s. x100.

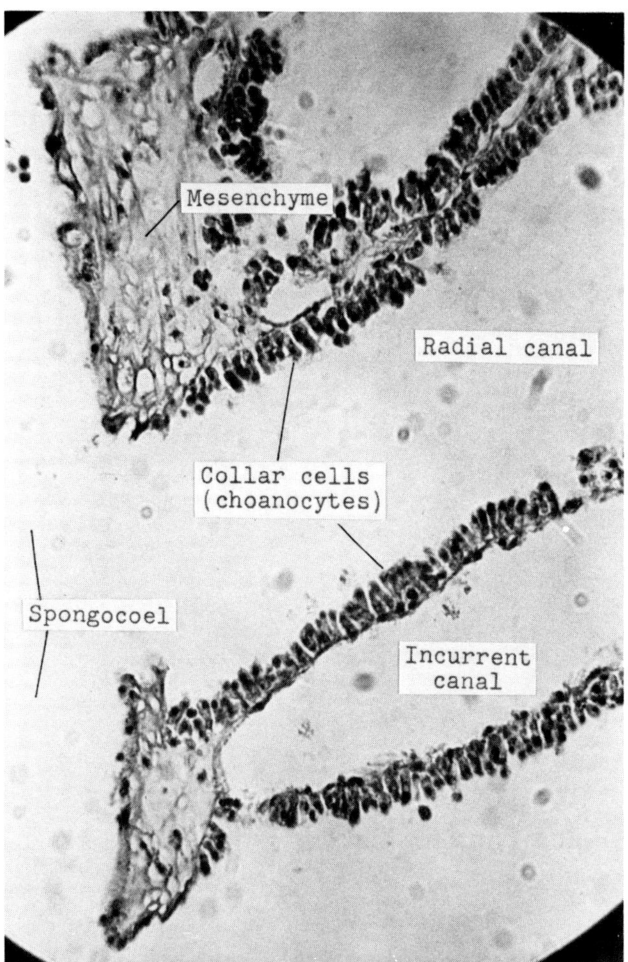

Fig. 51d Scypha x.s. x430.

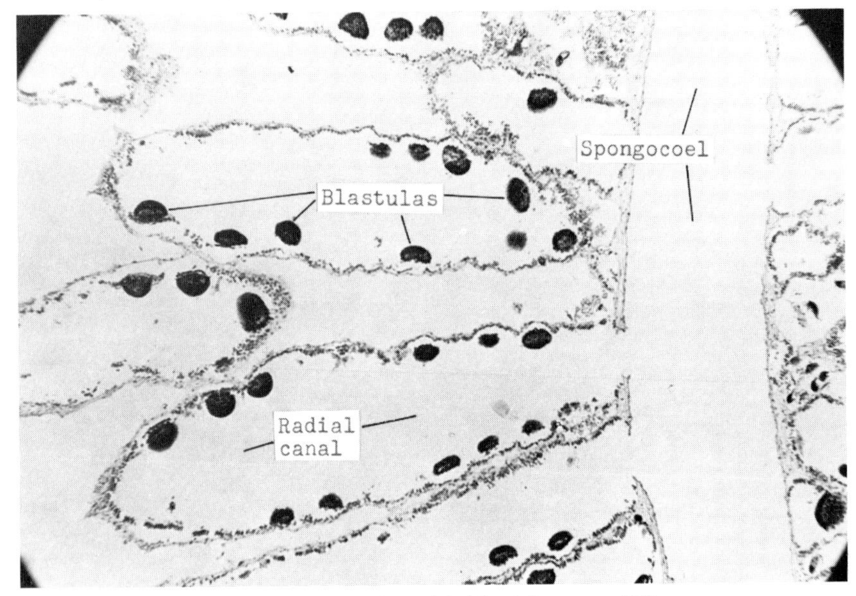

Fig. 52a Scypha with blastulas x.s. x100.

Fig. 52d A "squash" preparation of a portion of Scypha to show spicules and cells w.m. x100.

Fig. 52b Spicules w.m. x100.

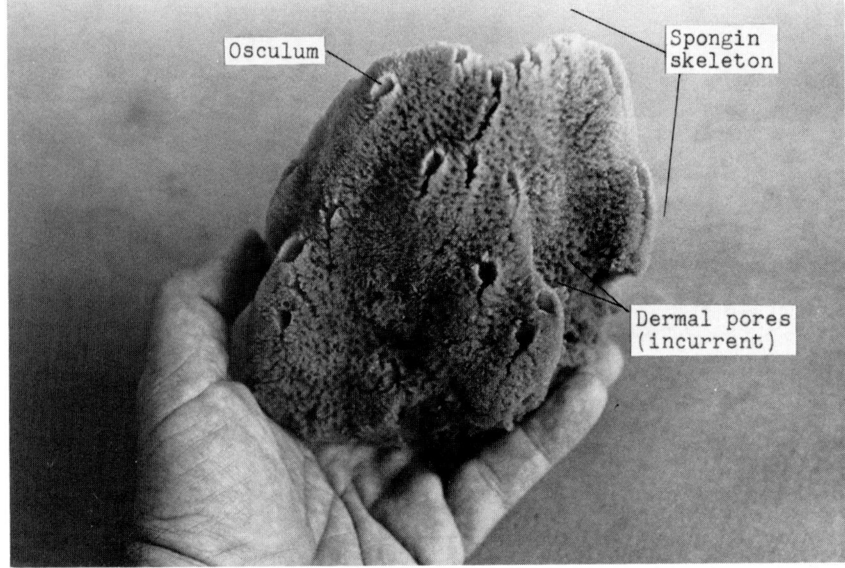

Fig. 52c Bath sponge x½. The cells have all been removed leaving only the spongin skeleton in this photo. This type of sponge exhibits the highest complexity of structure (leuconoid).

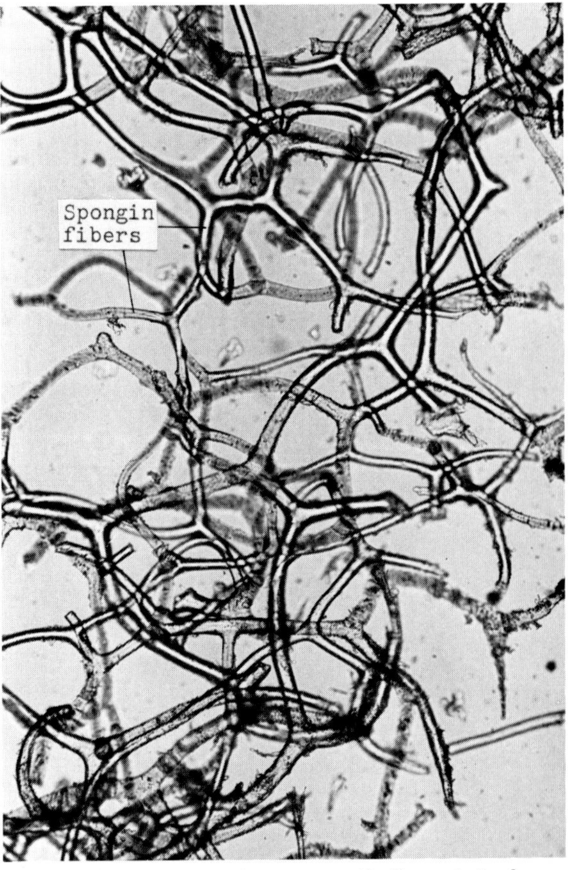

Fig. 52e Spongin fibers w.m. x40. Spongin is chemically a type of protein (scleroprotein).

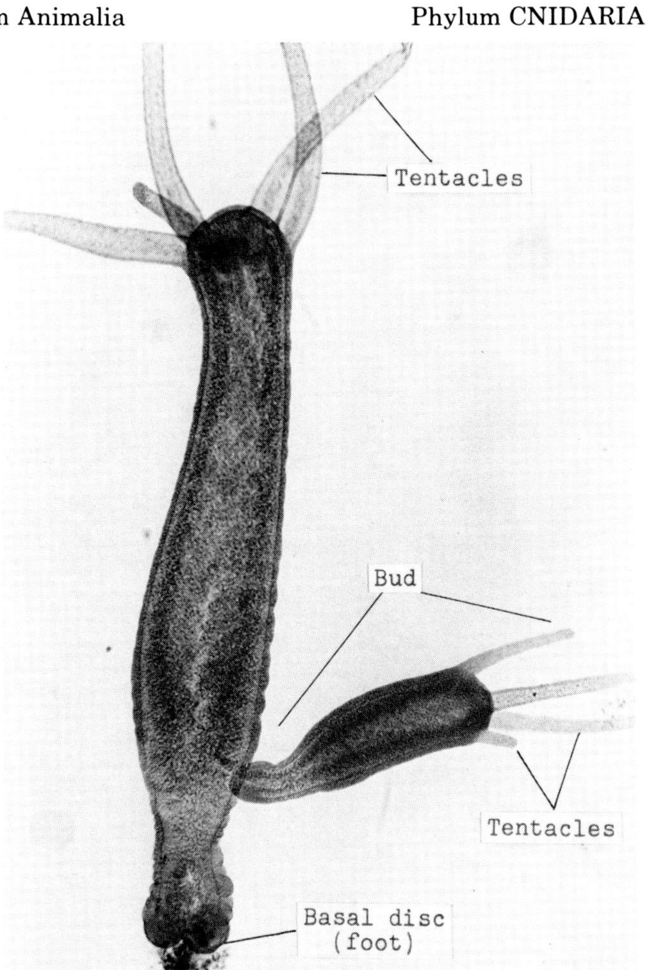

Fig. 53a Hydra with bud w.m. x40.

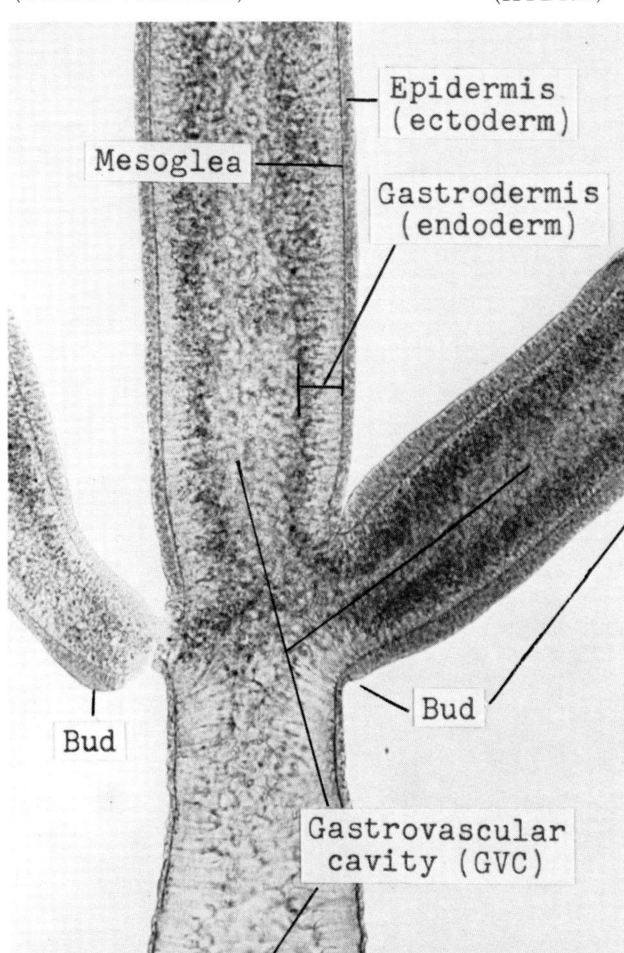

Fig. 53b Hydra body wall with bud w.m. x100. Notice the continuous GVC between the adult and the bud.

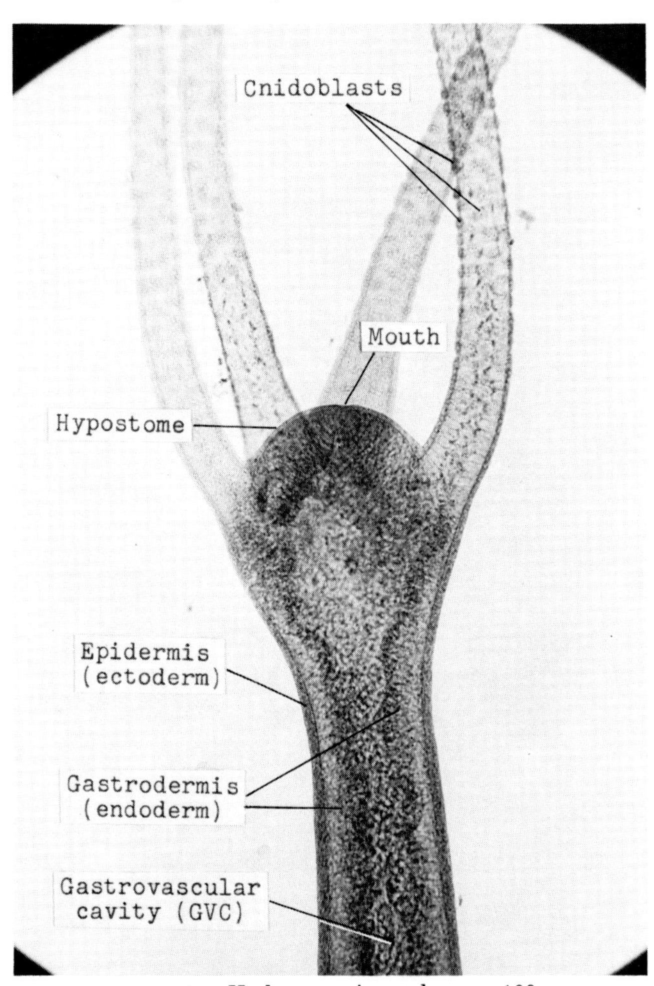

Fig. 53c Hydra anterior end w.m. x100.

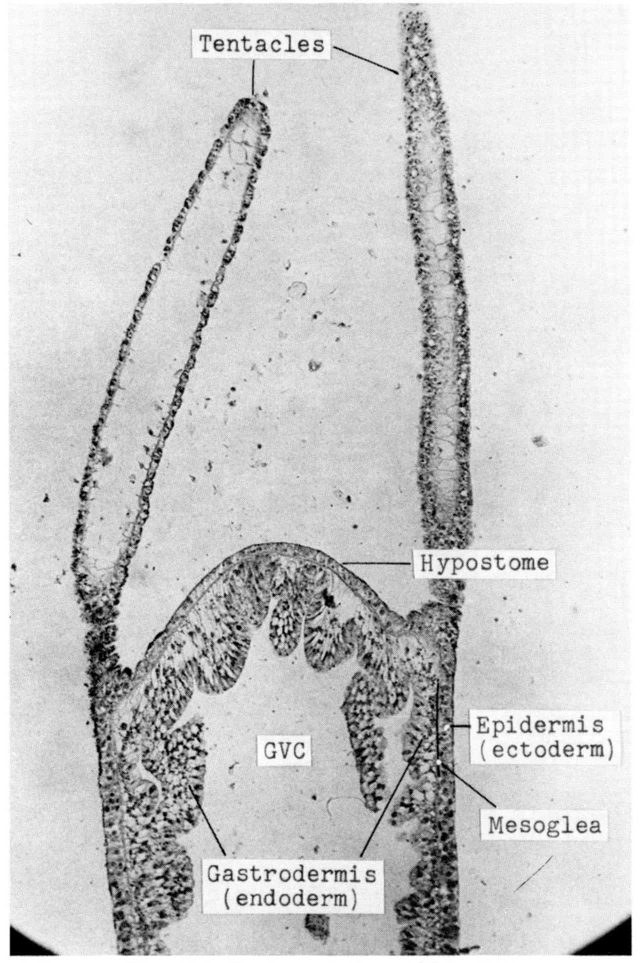

Fig. 53d Hydra anterior end l.s. x100.

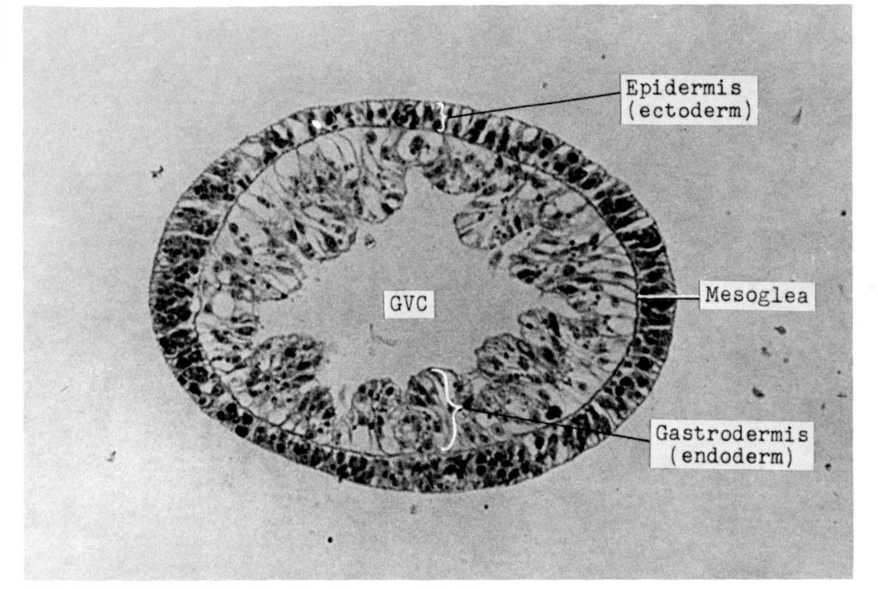

Fig. 54a Hydra x.s. x430.

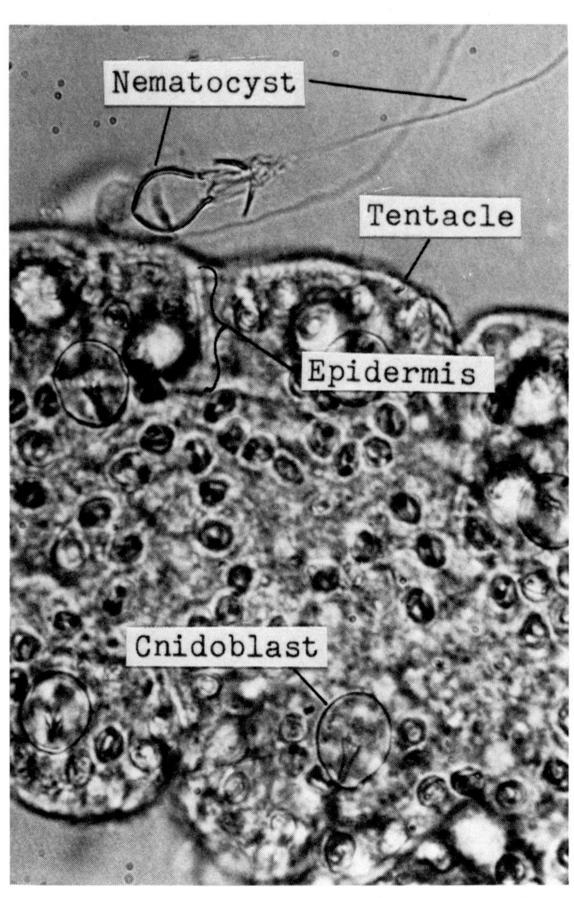

Fig. 54b Hydra nematocyst (live) w.m. x100.

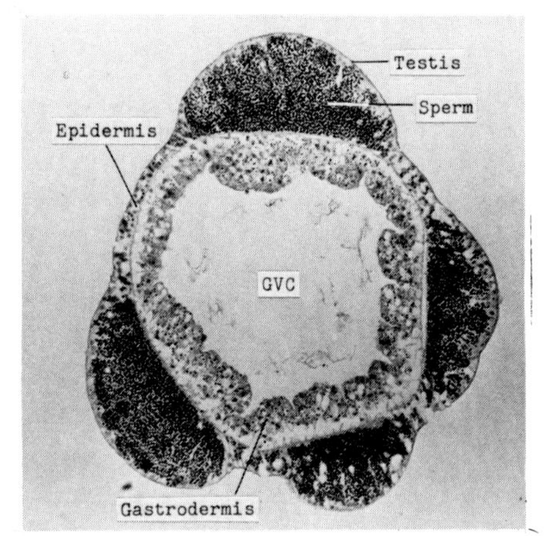

Fig. 54c Hydra w/testes x.s. x430.

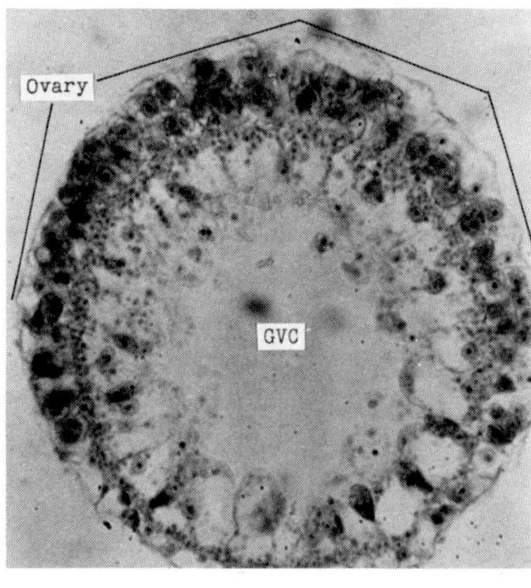

Fig. 54d Hydra w/ovary x.s. x430.

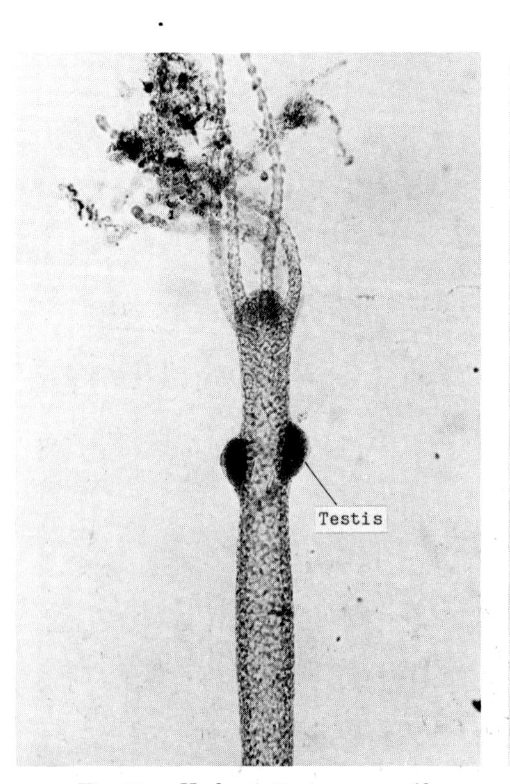

Fig. 54e Hydra w/testes w.m. x40.

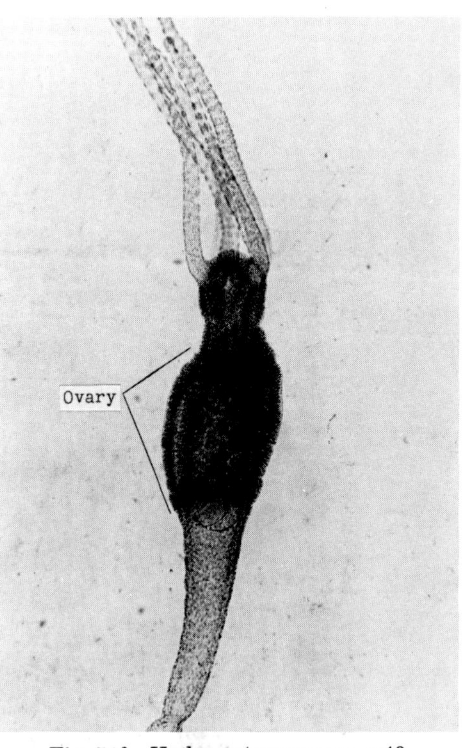

Fig. 54f Hydra w/ovary w.m. x40.

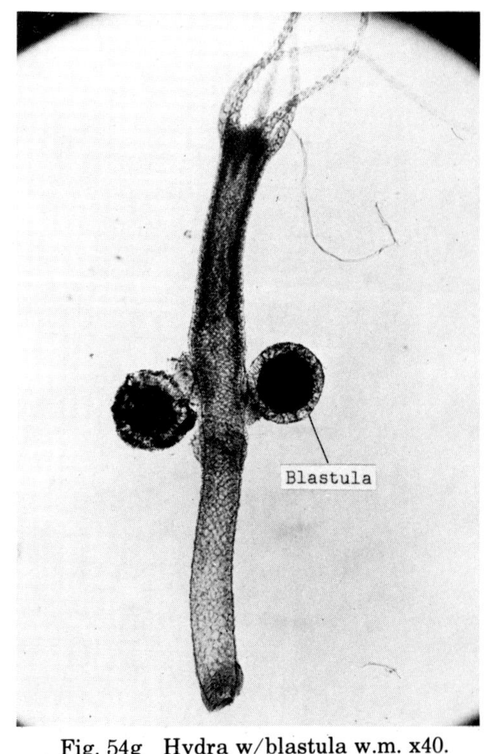

Fig. 54g Hydra w/blastula w.m. x40.

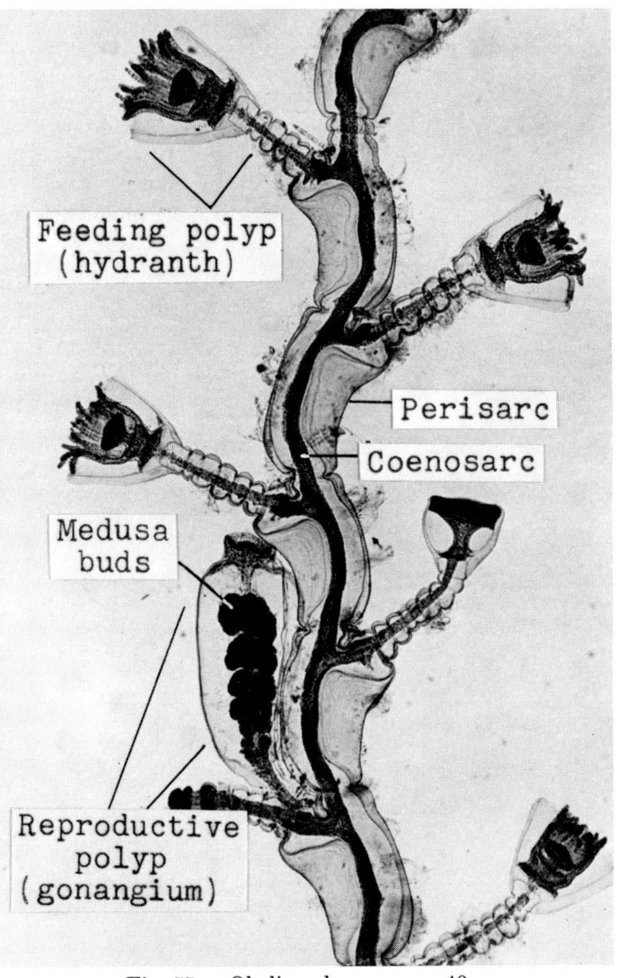

Fig. 55a Obelia colony w.m. x40.

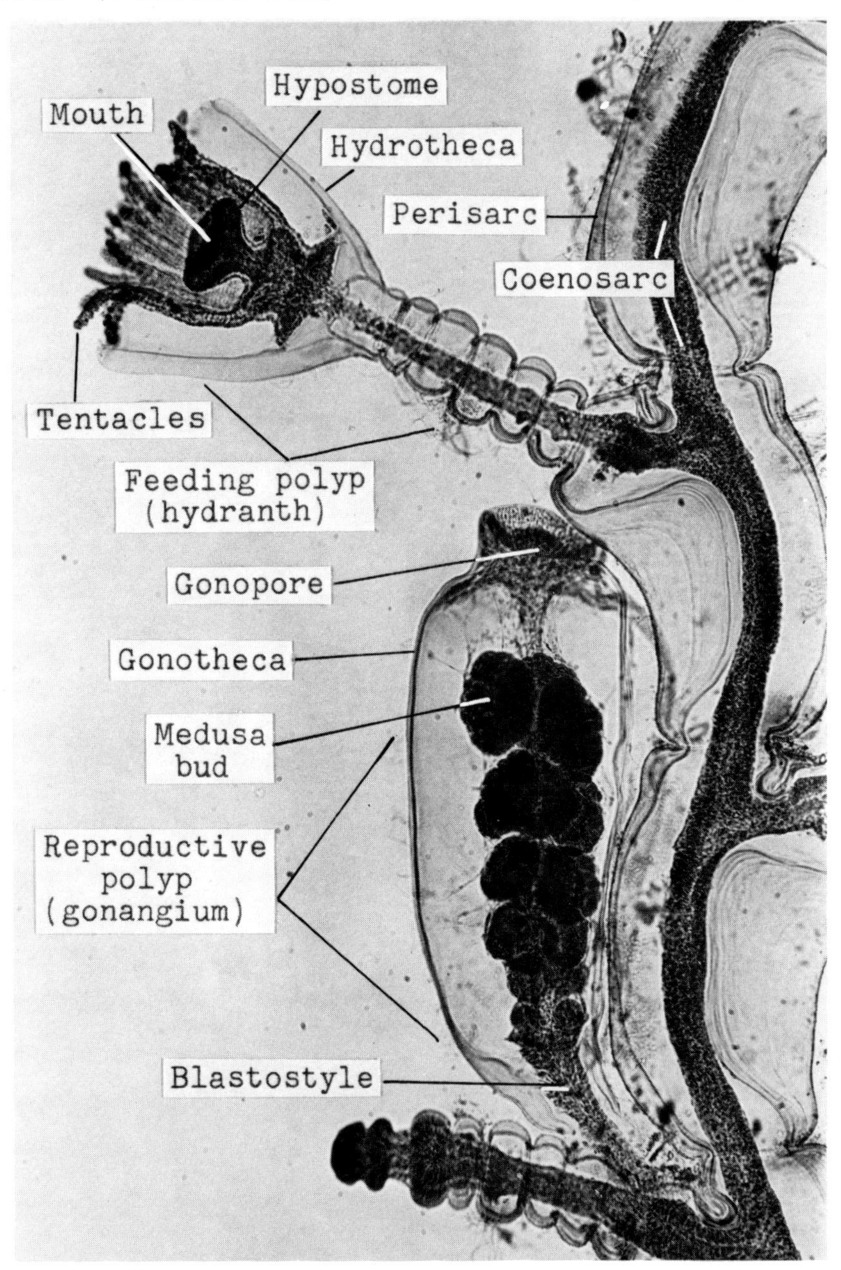

Fig. 55b Obelia colony w.m. x100.

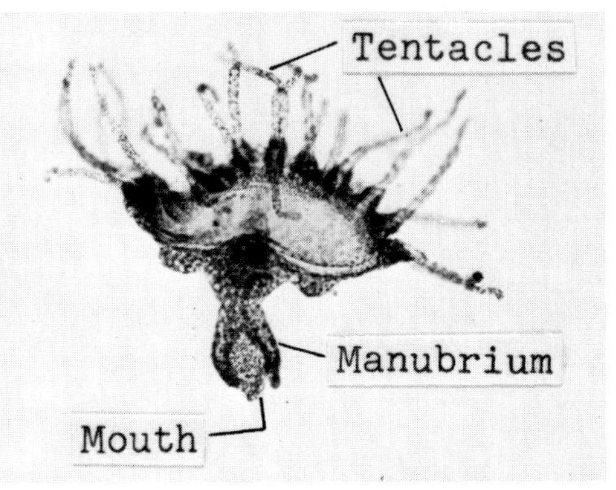

Fig. 55c Obelia medusa w.m. x100. Feeding position. The medusa slowly sinks in this position with tentacles outspread to capture food.

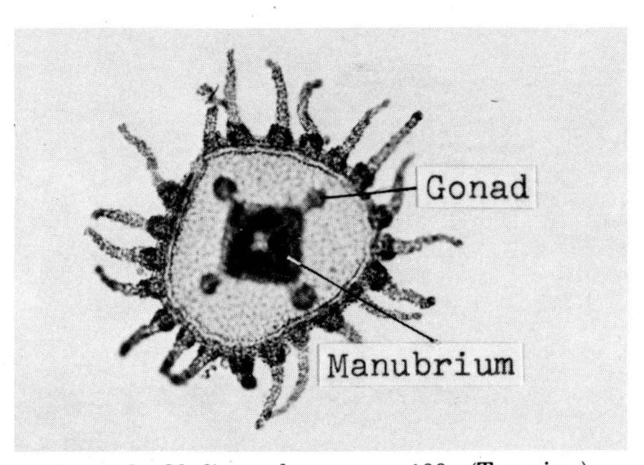

Fig. 55d Obelia medusa w.m. x100. (Top view)

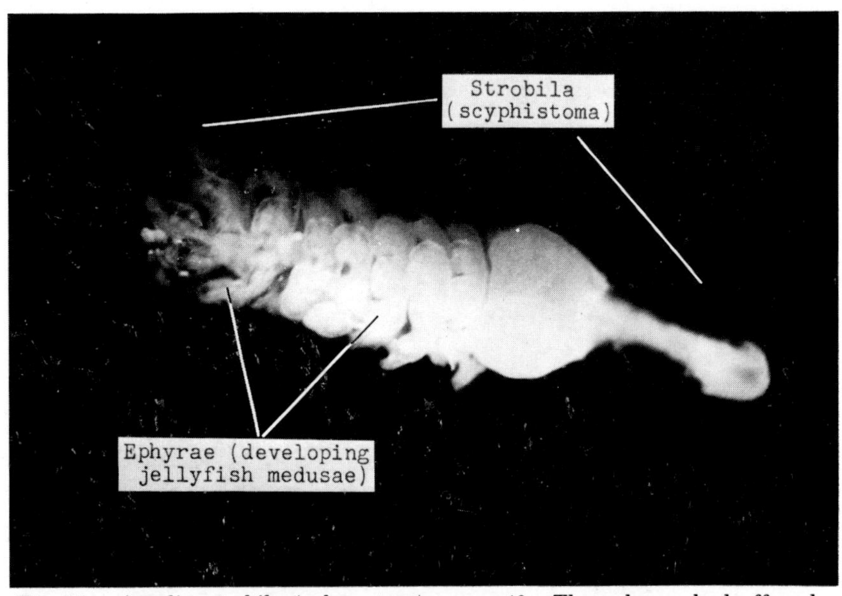

Fig. 55e Aurelia strobila (polyp stage) w.m. x40. The ephyrae bud off and become adult Aurelia medusae (jellyfish).

Fig. 56a Physalia, the Portuguese man-o-war x¼. This organism is a colony of modified medusae and polyps exhibiting a high division of labor.

Fig. 56b Coral polyps (live) x10. Gorgonian coral (Pt. Aransas, Texas). Commonly known as soft coral.

Fig. 56c Sea Anemone and Clown fish (live) x1. A symbiotic relationship where the fish attracts food organisms for the anemone but is not affected by the anemone's nematocysts and receives protection among the tentacles.

Fig. 56d Sea Anemone (live) x¼. (Point Lobos, Calif.)

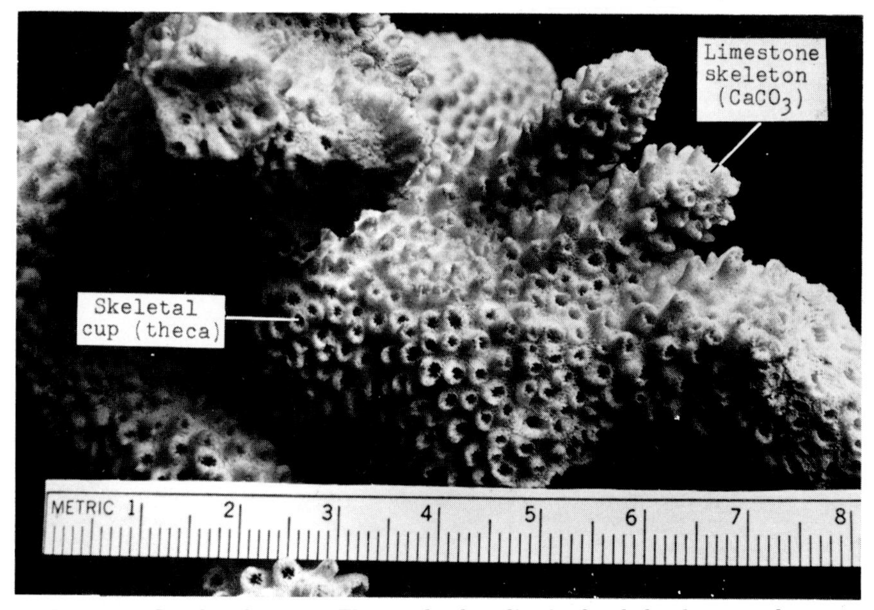

Fig. 56e Coral skeleton x1. The coral polyps live in the skeletal cups and secrete $CaCO_3$ which further builds up the limestone skeleton.

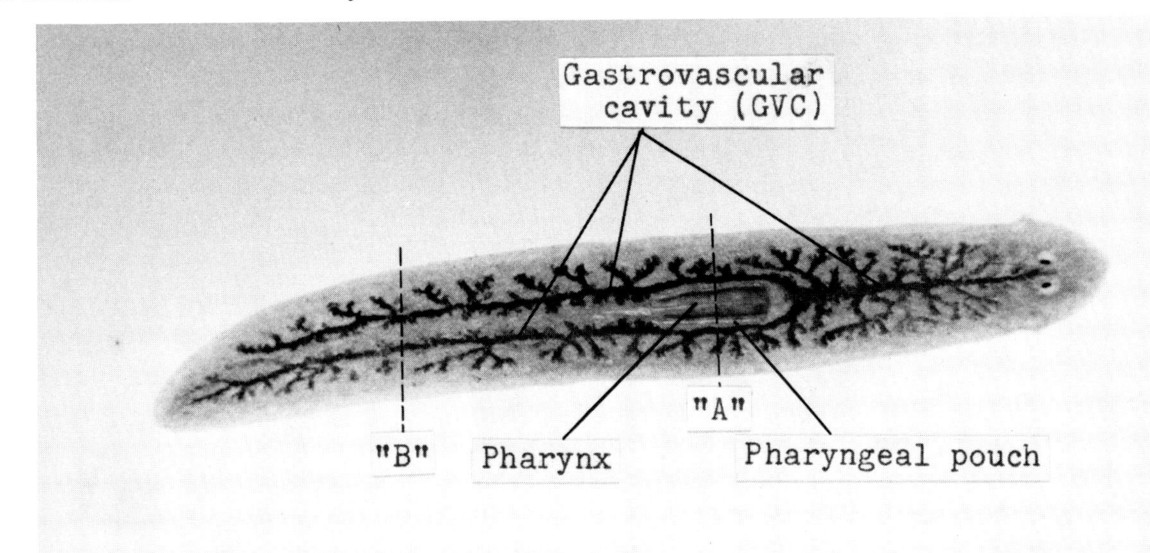

Fig. 57a Planaria w.m. x40. (Cedar Falls, Iowa)

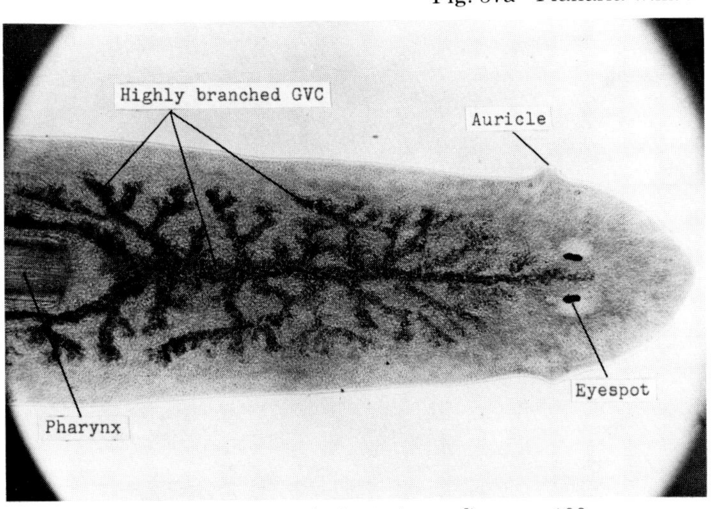

Fig. 57b Planaria (anterior end) w.m. x100.

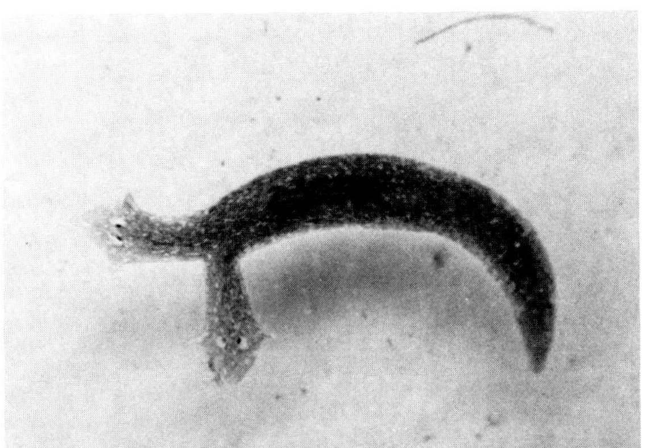

Fig. 57c Two-headed planarian (live) produced by cutting the head lengthwise and allowing regeneration to occur x20. (San Antonio College)

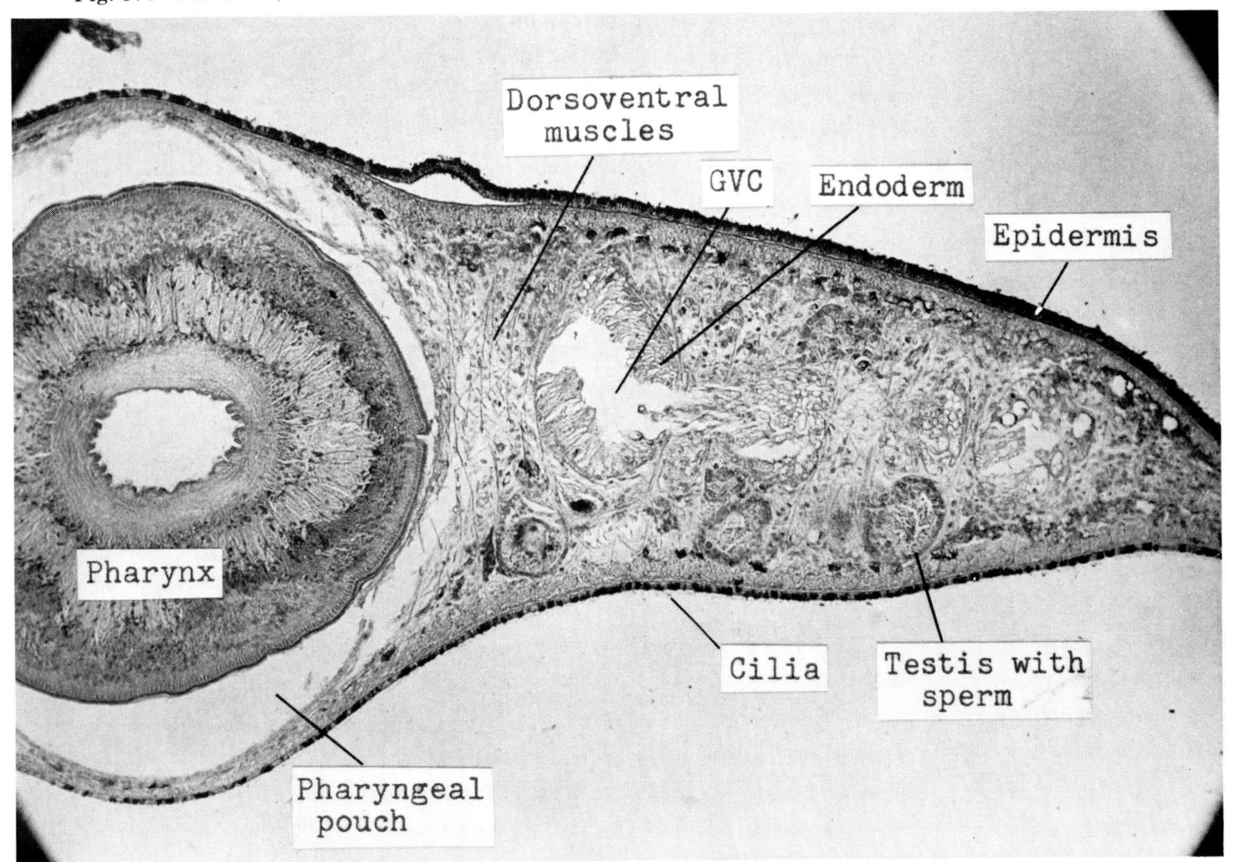

Fig. 57d Planaria x.s. through pharynx region x100. Section taken at point "A" in Fig. 57a.

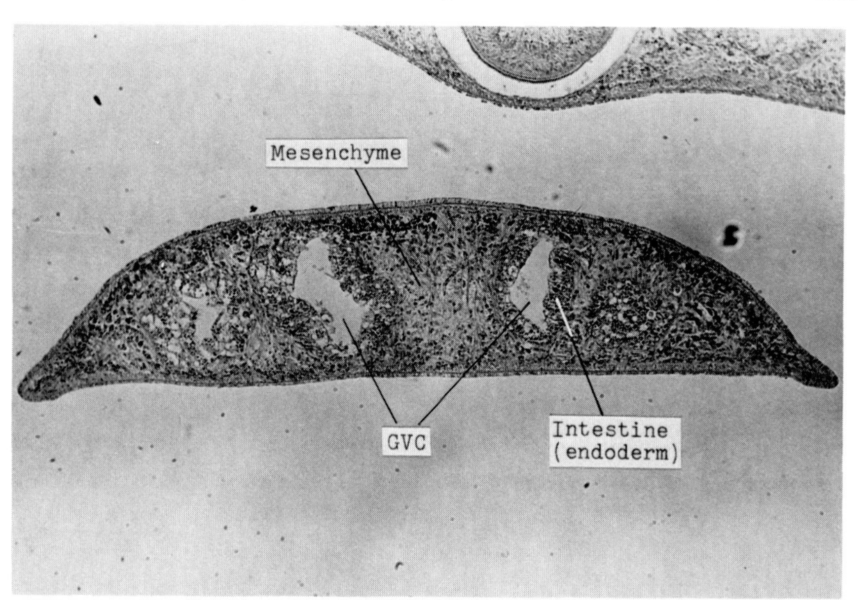

Fig. 58a Planaria x.s. through posterior region x100. Section taken at point "B" in Fig. 57a.

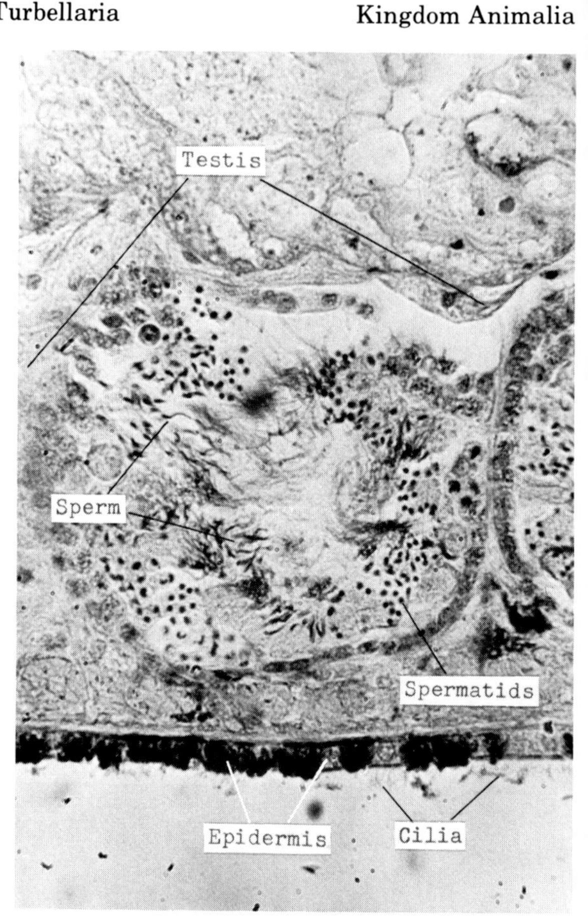

Fig. 58d Planaria testis x.s. x430.

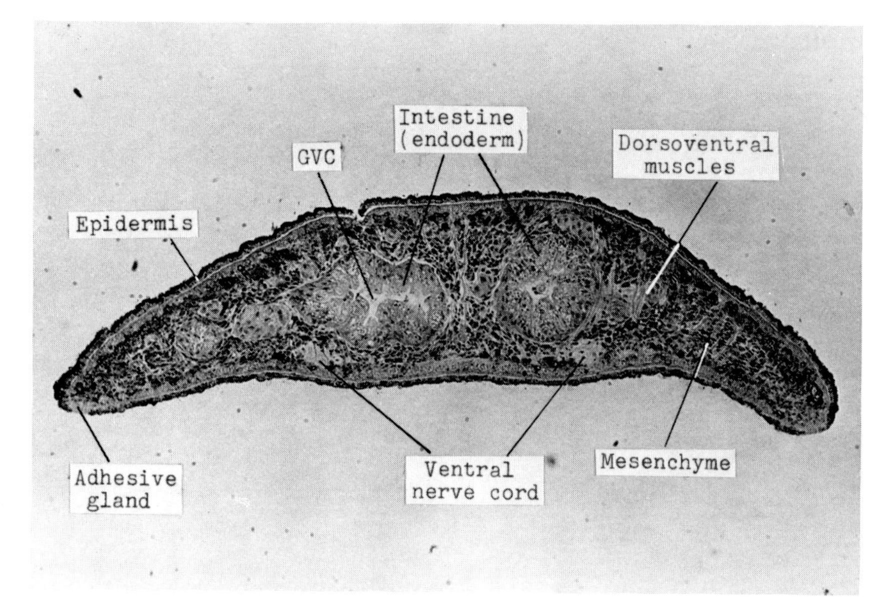

Fig. 58b Planaria x.s. through posterior region x100. Section taken near point "B" in Fig. 57a.

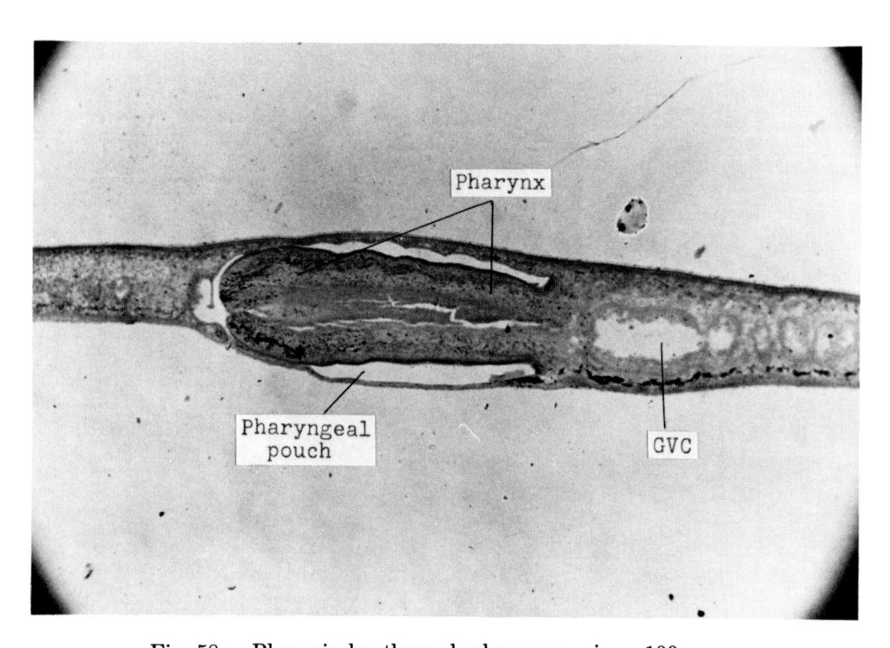

Fig. 58c Planaria l.s. through pharynx region x100.

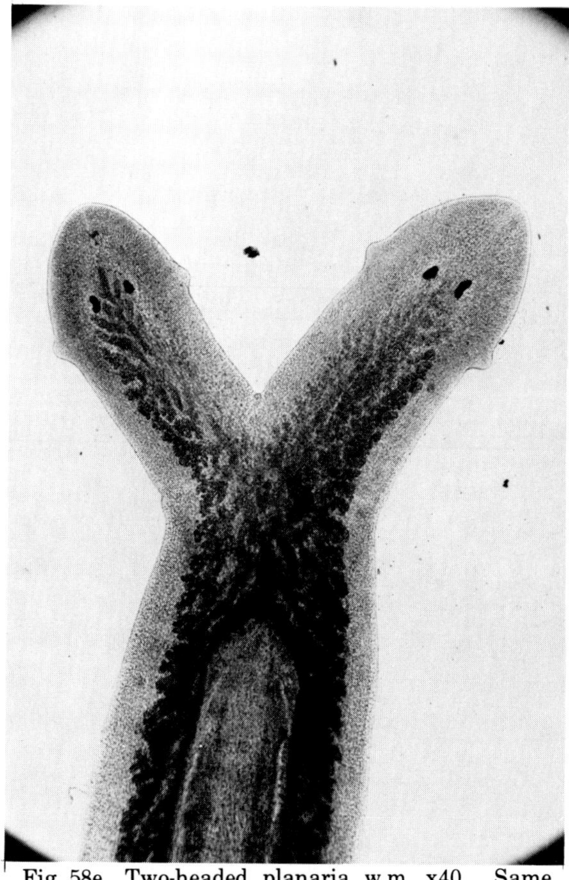

Fig. 58e Two-headed planaria w.m. x40. Same organism as in Fig. 57c when mounted on a slide.

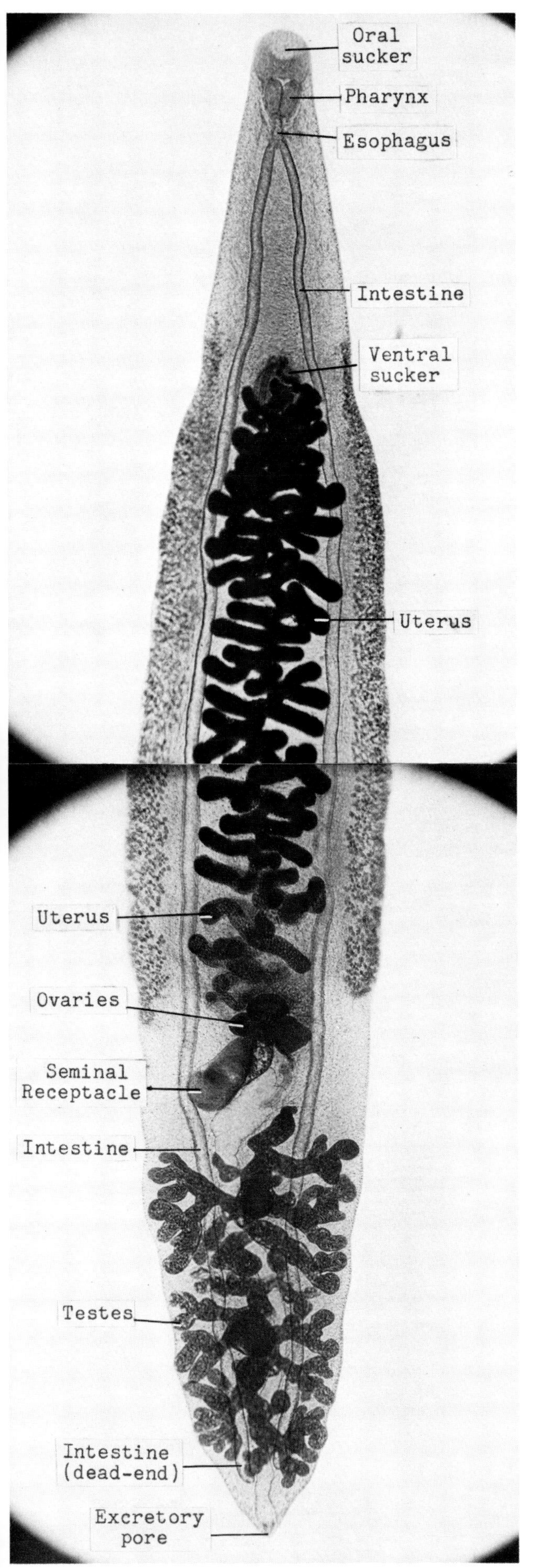

Fig. 59a Chinese liver fluke (Opisthorchis sinensis w.m. x40. (Formerly Clonorchis)

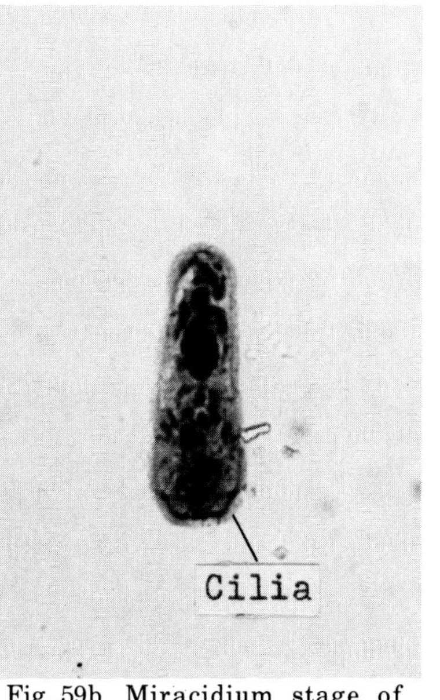

Fig. 59b Miracidium stage of Opisthorchis w.m. x430.

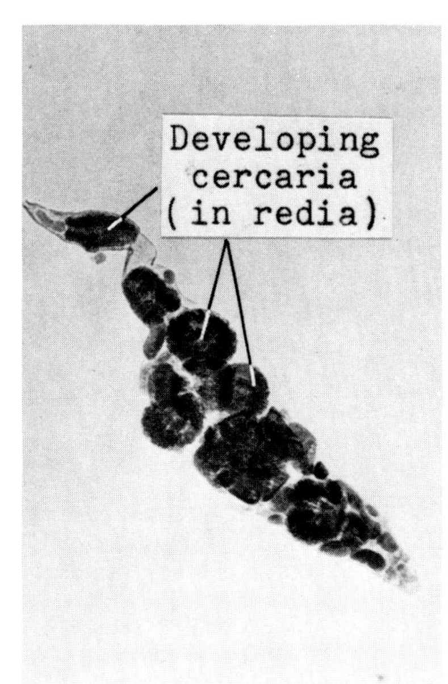

Fig. 59c Redia stage of Opisthorchis w.m. x430.

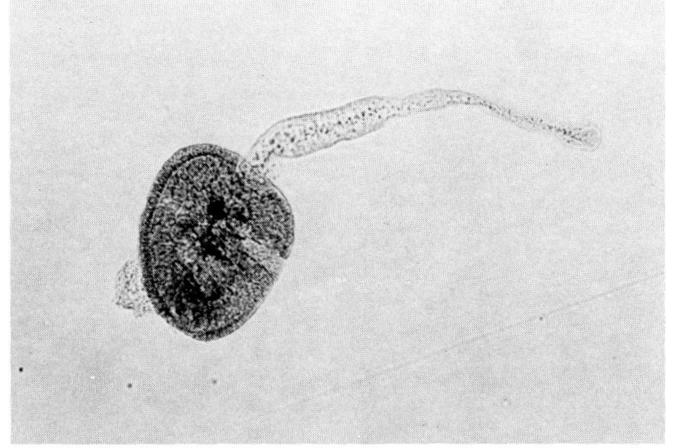

Fig. 59d Cercaria stage of Opisthorchis w.m. x430.

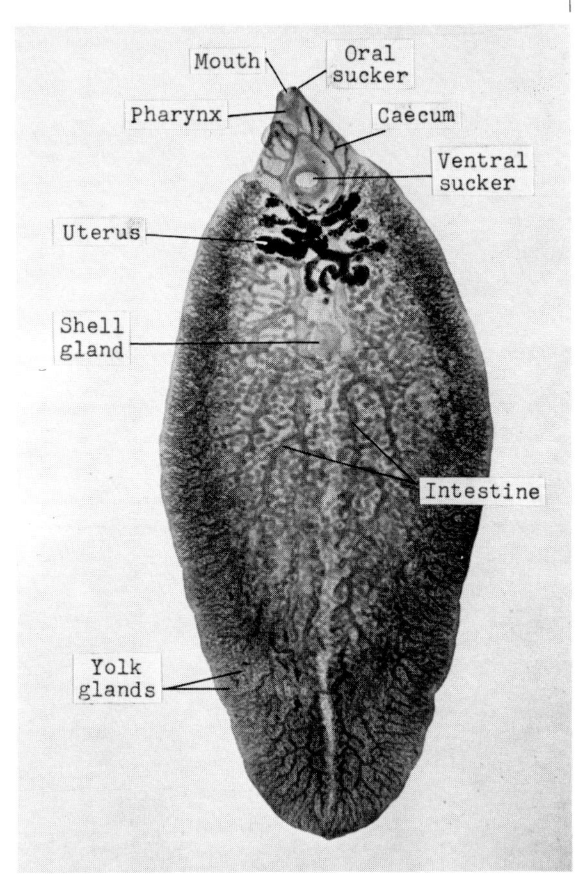

Fig. 59e Fasicola, sheep liver fluke w.m. x10.

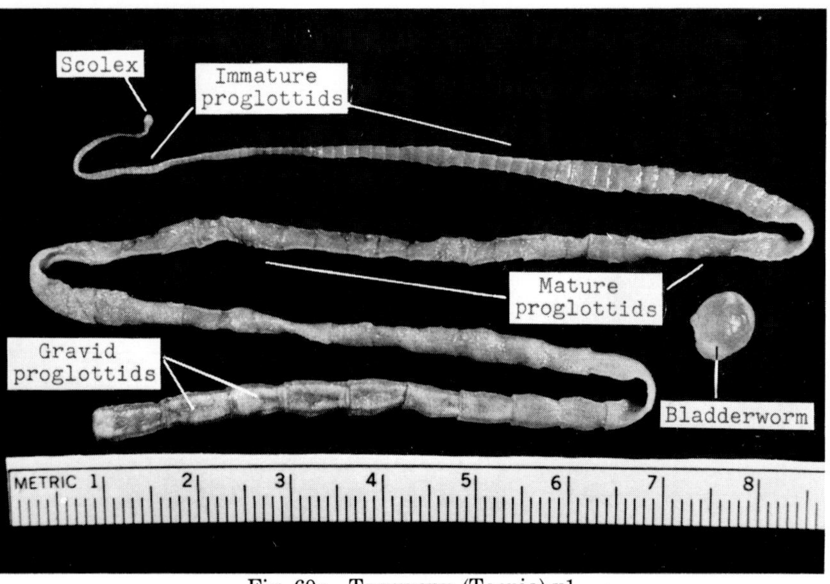

Fig. 60a Tapeworm (Taenia) x1.

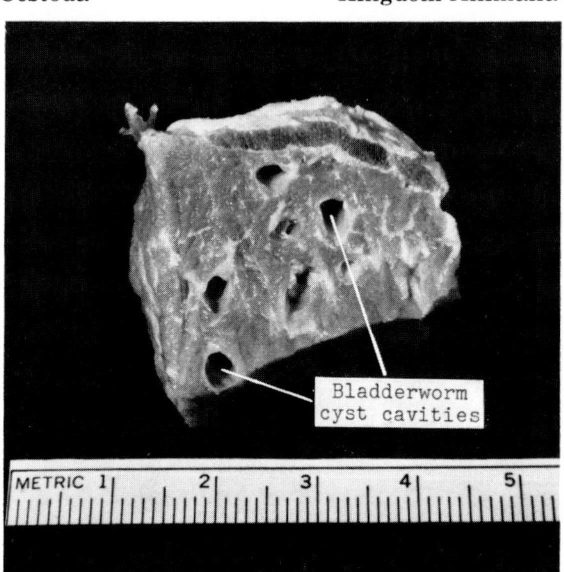

Fig. 60b Measly beef with cavities where bladderworm cysts were x1.

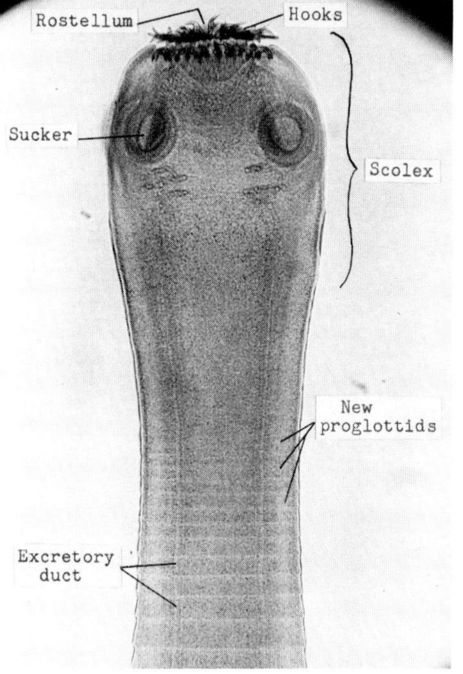

Fig. 60c Scolex of Taenia w.m. x40. The suckers & hooks are used for attachment only.

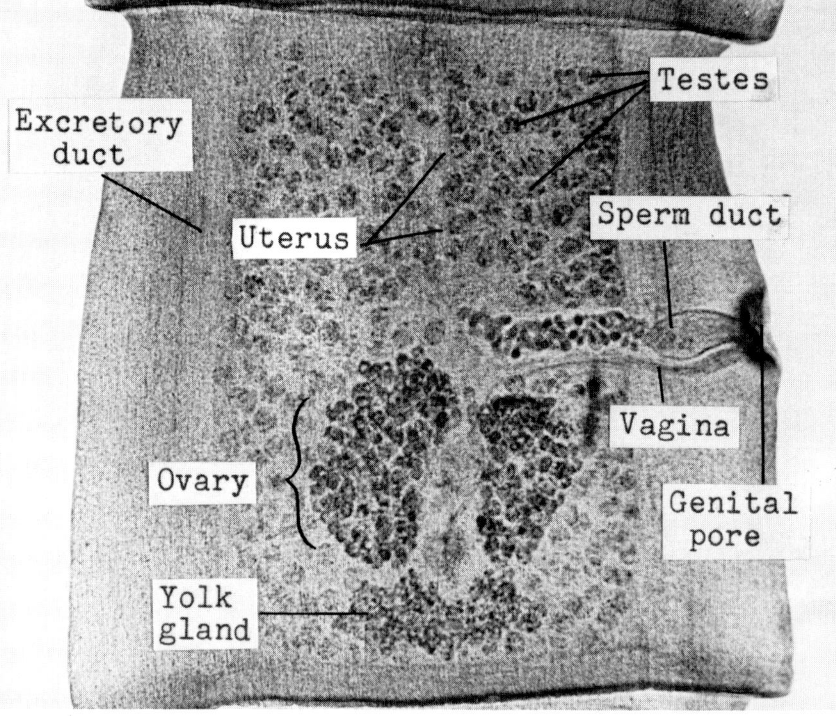

Fig. 60d Mature proglottid of Taenia w.m. x40. See Fig. 60f for detailed anatomy.

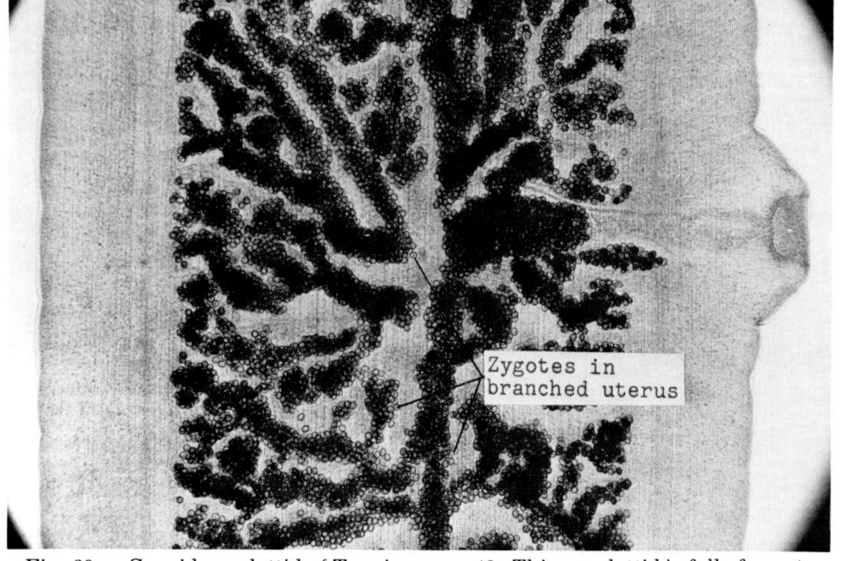

Fig. 60e Gravid proglottid of Taenia w.m. x40. This proglottid is full of zygotes and ready to pass out in the host's feces.

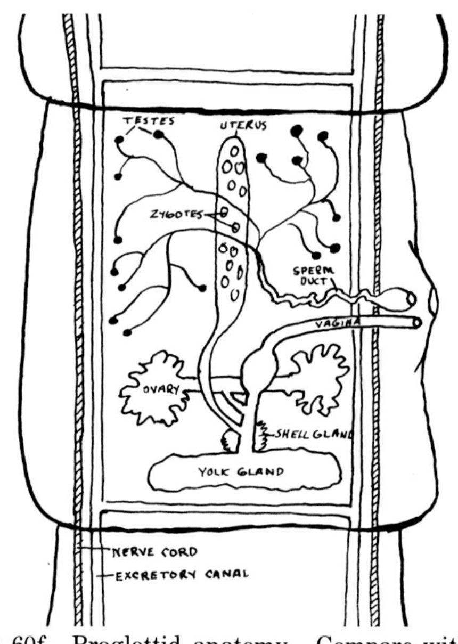

Fig. 60f Proglottid anatomy. Compare with Fig. 60d.

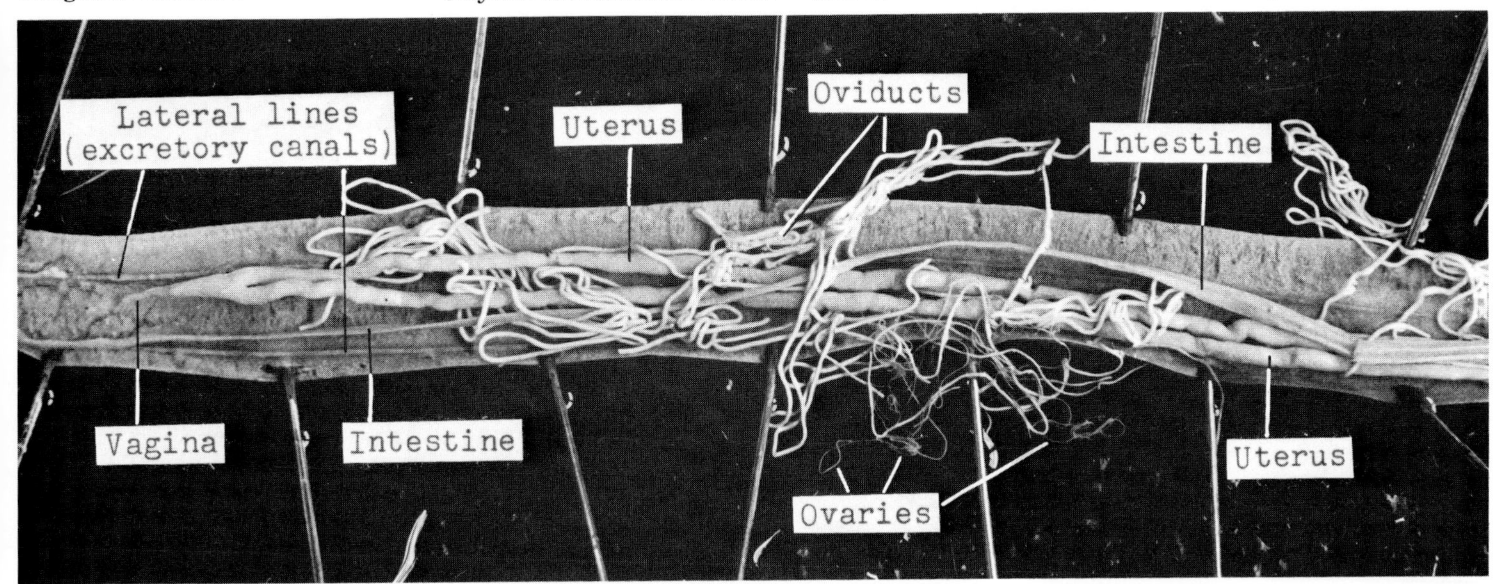

Fig. 61a Ascaris (female) x2. Dissection.

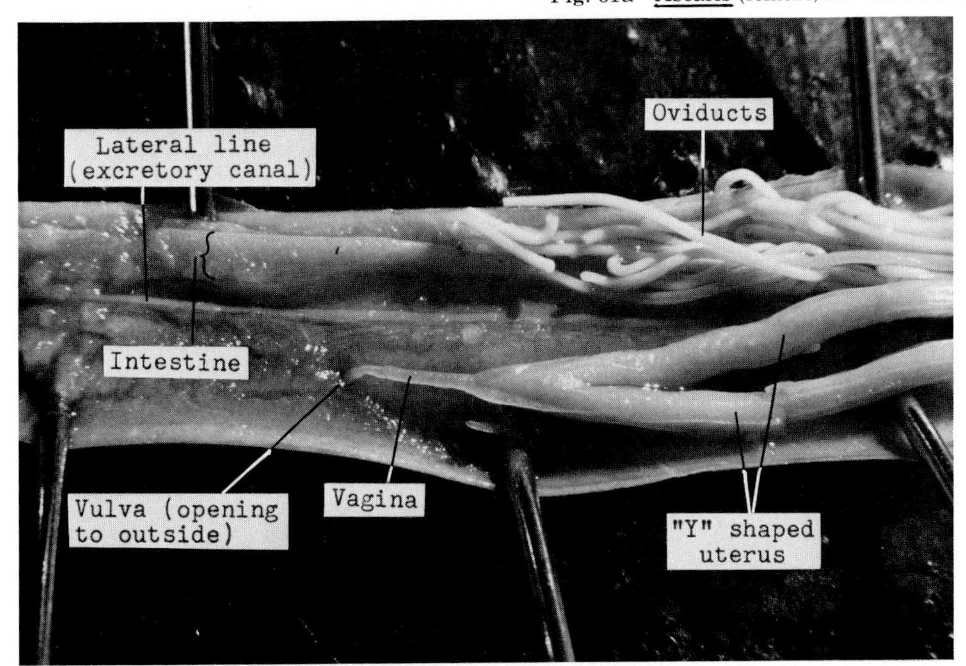

Fig. 61b Ascaris (female) x5. Dissection of anterior region.

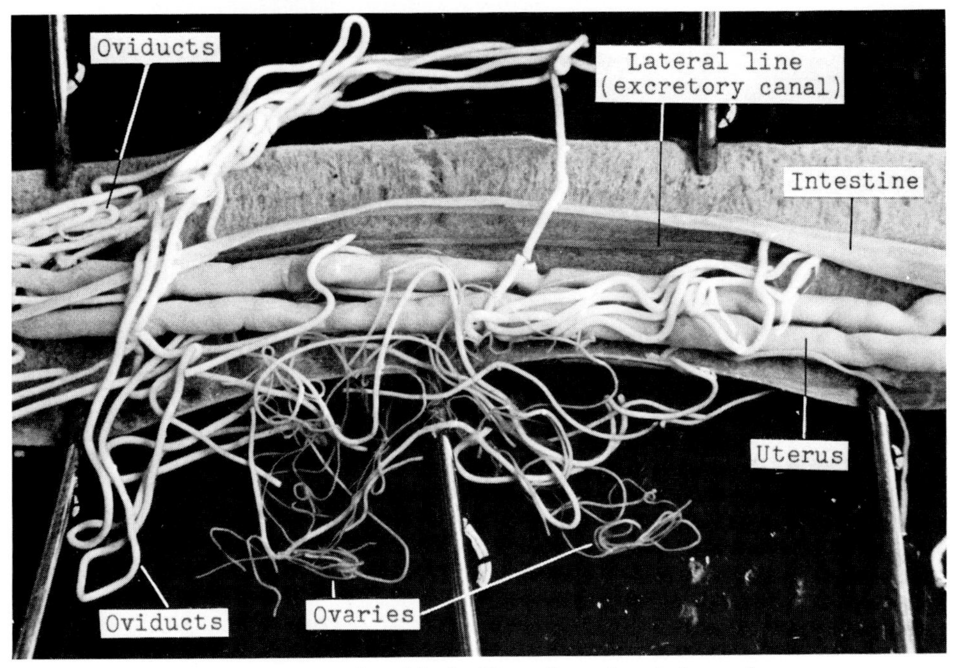

Fig. 61c Ascaris (female) x5. Dissection of posterior region.

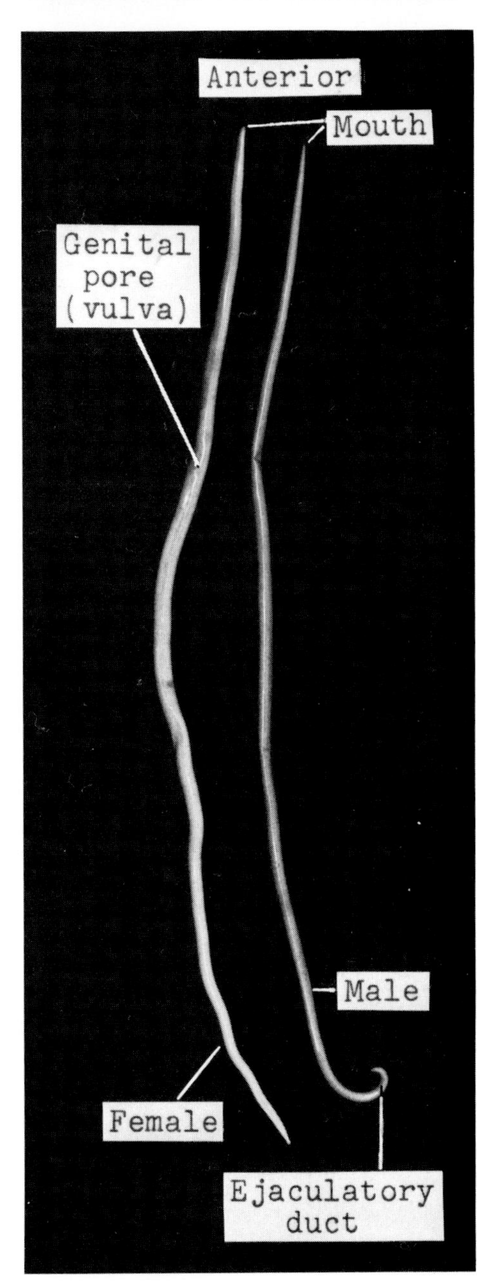

Fig. 61d Ascaris (male and female exterior) x½.

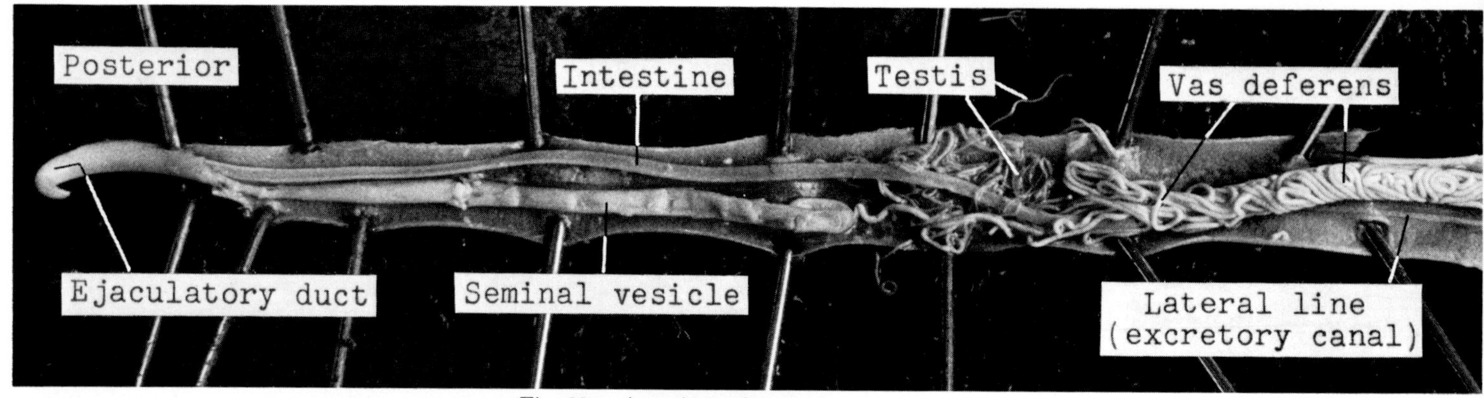

Fig. 62a Ascaris (male) x3. Dissection.

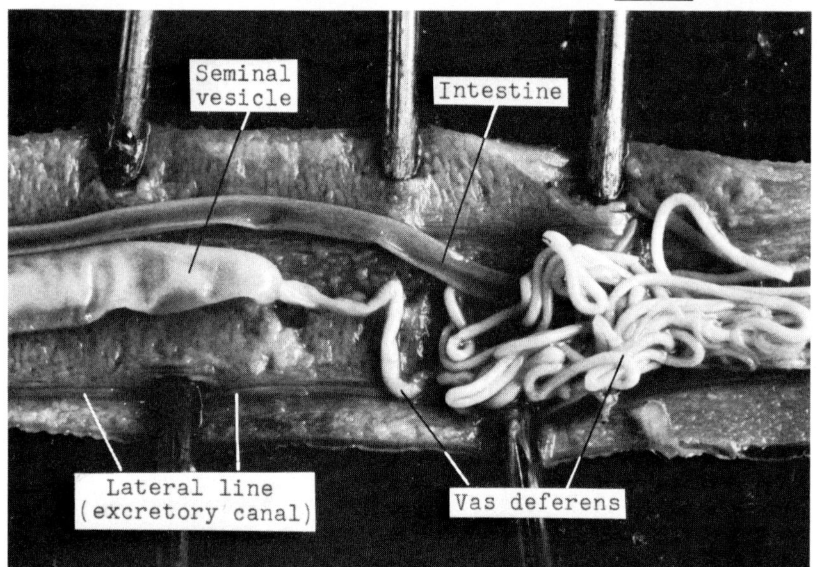

Fig. 62b Ascaris (male) x5. Dissection.

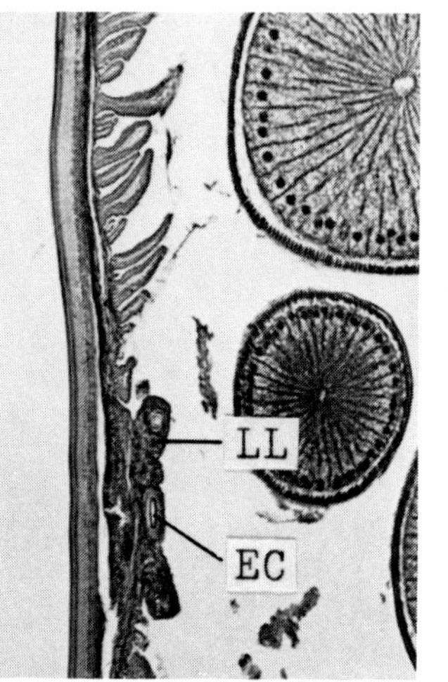

Fig. 62c Ascaris lateral line (LL) and excretory canal (EC) x.s. x100.

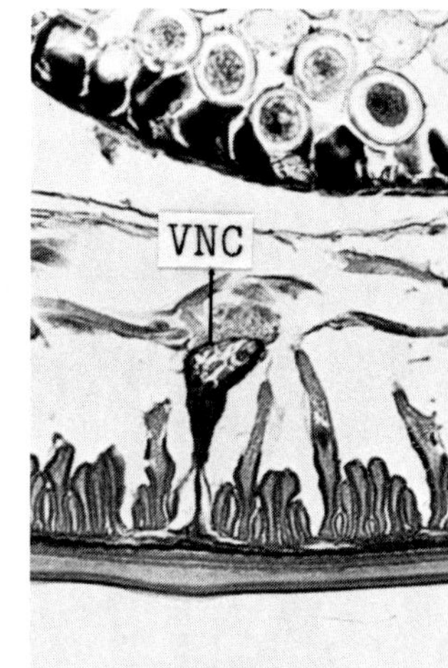

Fig. 62d Ascaris ventral nerve cord x.s. x100.

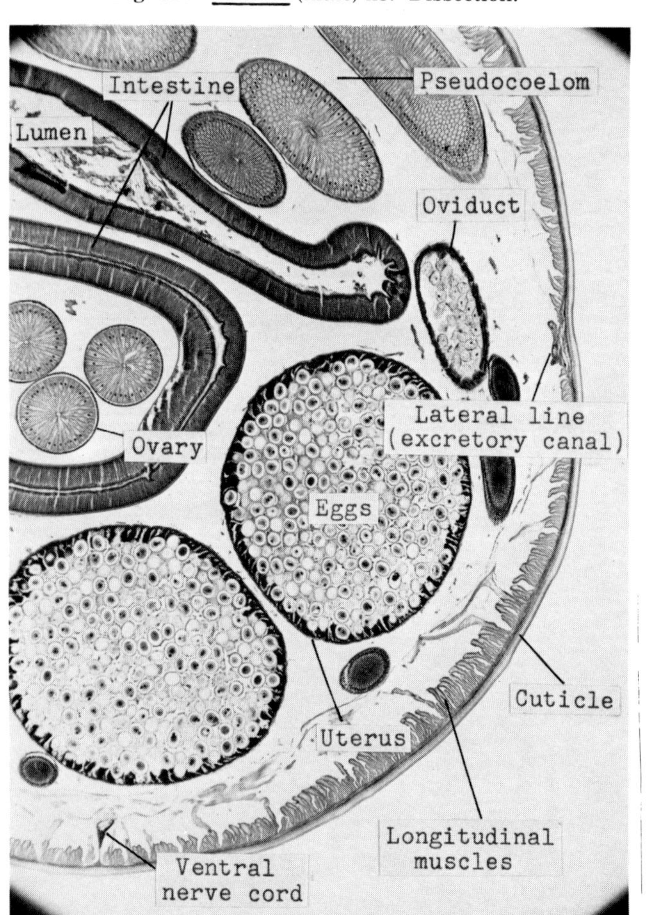

Fig. 62e Ascaris (female) x.s. x40.

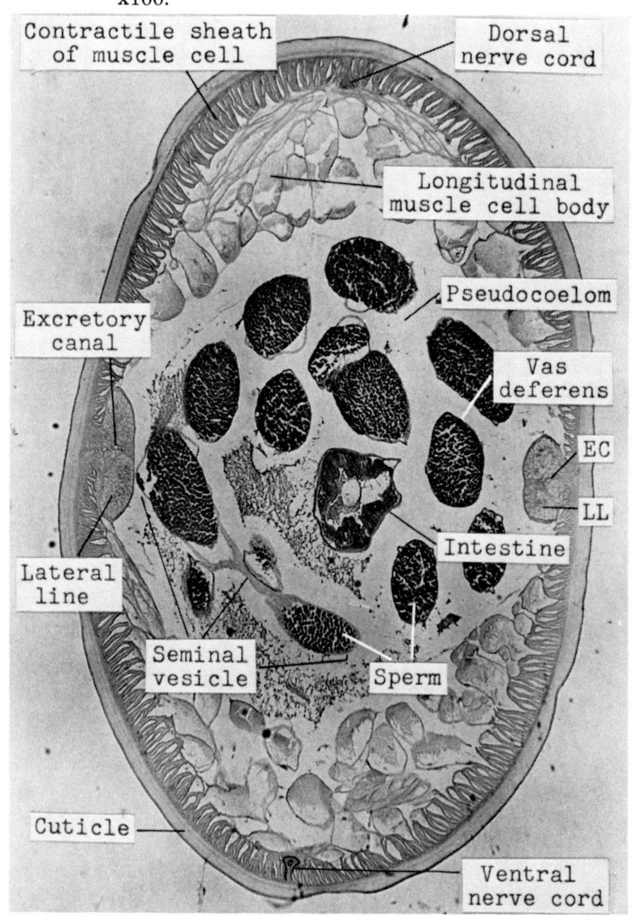

Fig. 62f Ascaris (male) x.s. x40.

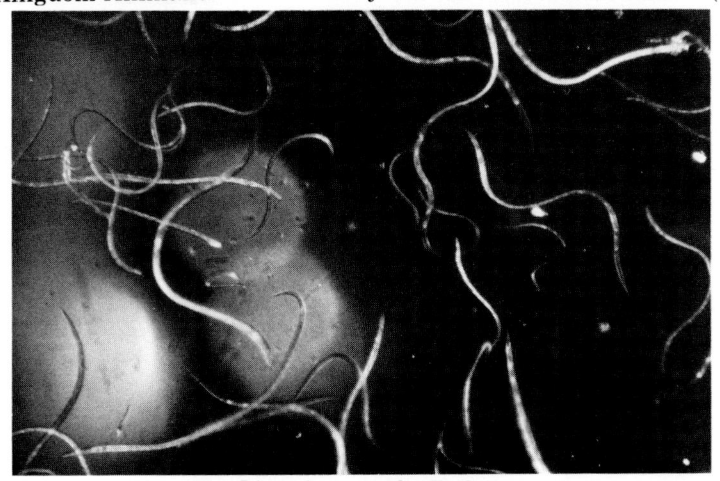

Fig. 63a Live vinegar eels (Turbatrix) x20.

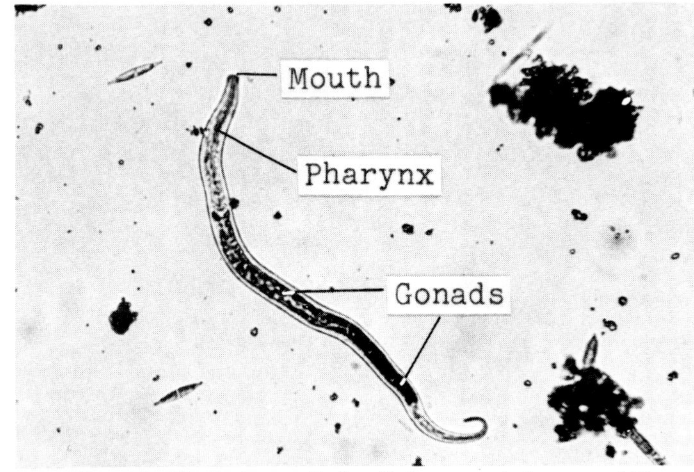

Fig. 63b Typical aquatic free-living nematode w.m. x100.

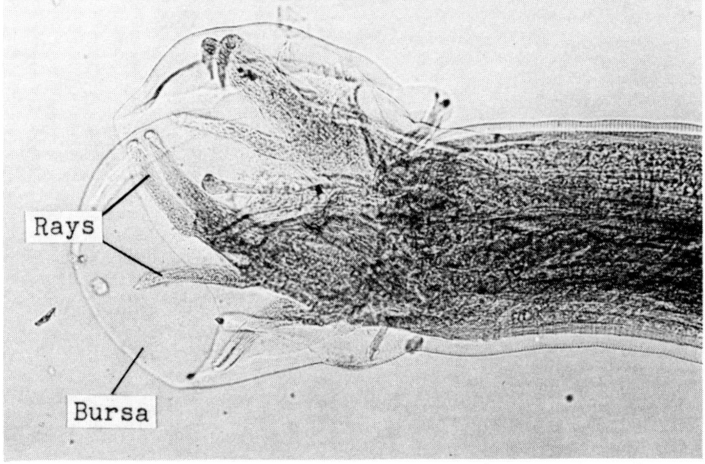

Fig. 63c Hookworm (Necator americanus) male, posterior end w.m. x40. The bursae and rays are used to hold the female during copulation.

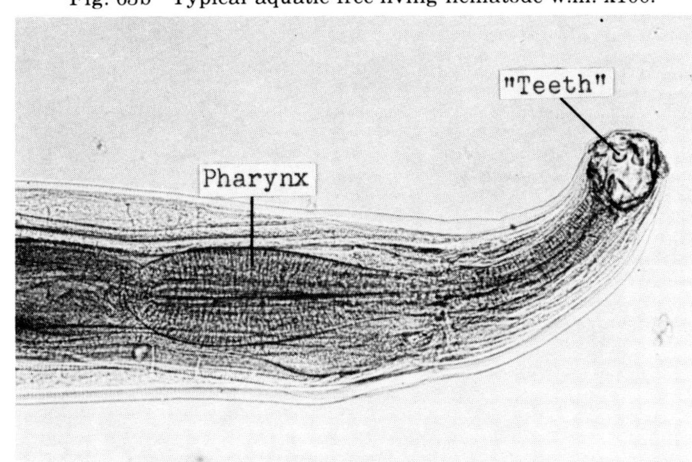

Fig. 63d Hookworm (Necator americanus) male, anterior end w.m. x40. Notice the teeth for drawing blood and the muscular pharynx for sucking the host's blood.

Fig. 63e Trichinella spiralis (trichina) x.s. x100. The larvae are encysted in muscle. This organism causes trichinosis.

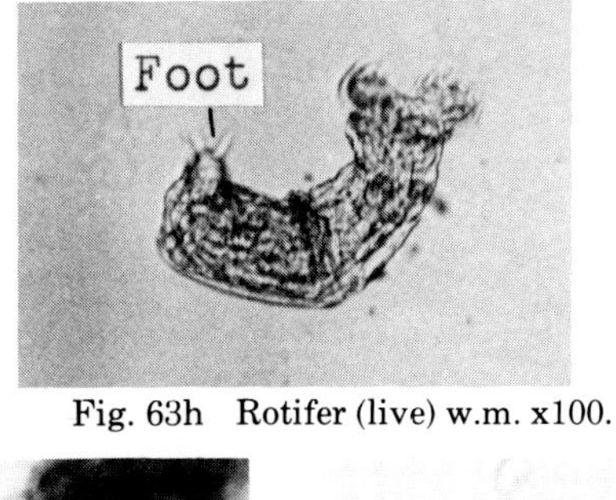

Fig. 63h Rotifer (live) w.m. x100.

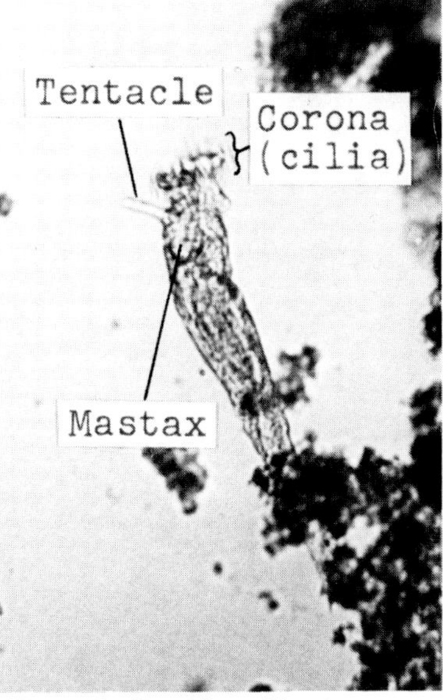

Fig. 63f Rotifer (live) w.m. x100.

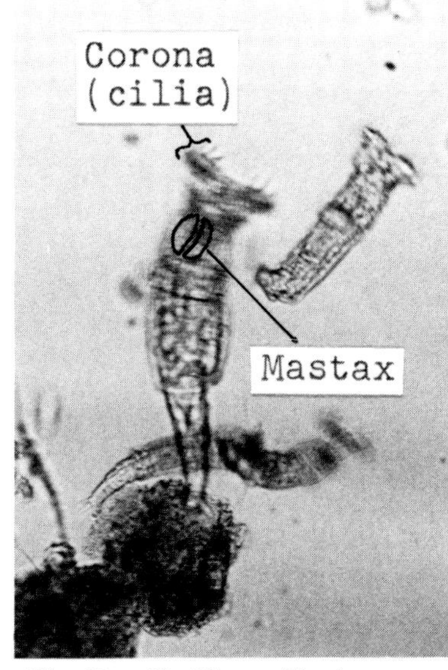

Fig. 63g Rotifers (live) w.m. x100. The mastax acts like jaws.

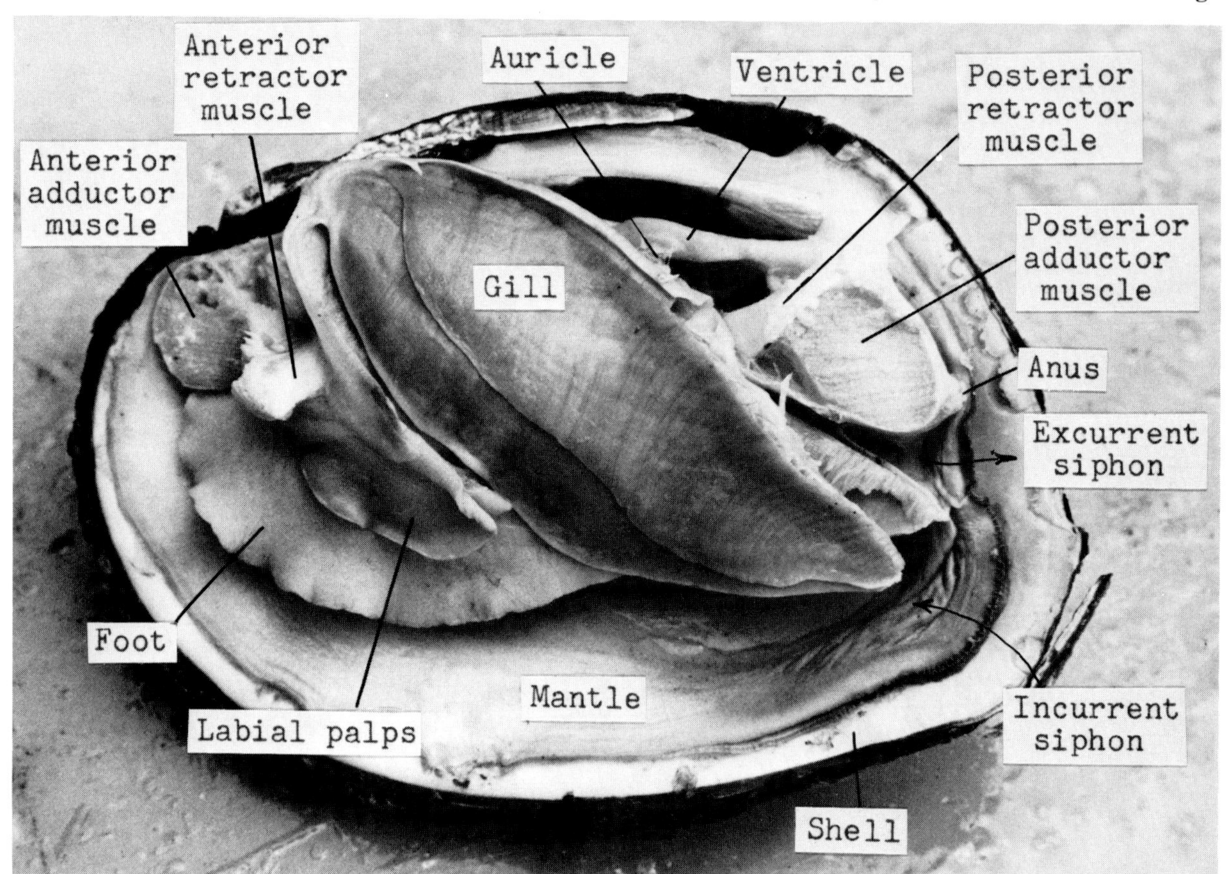

Fig. 64a Clam dissection x1. Only the upper shell (valve) and upper mantle have been removed.

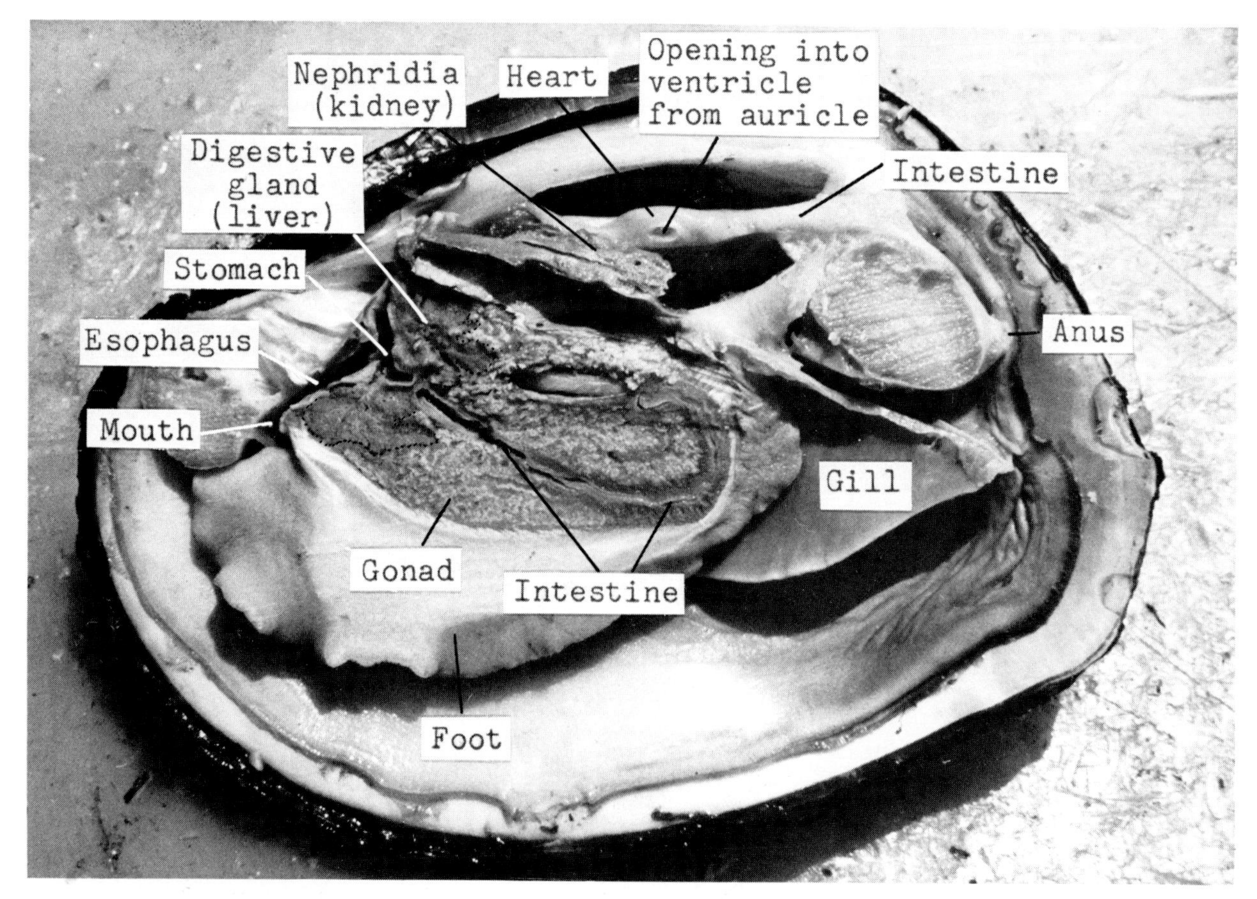

Fig. 64b Clam dissection x1. The upper gills and the top half of the foot have been cut away.

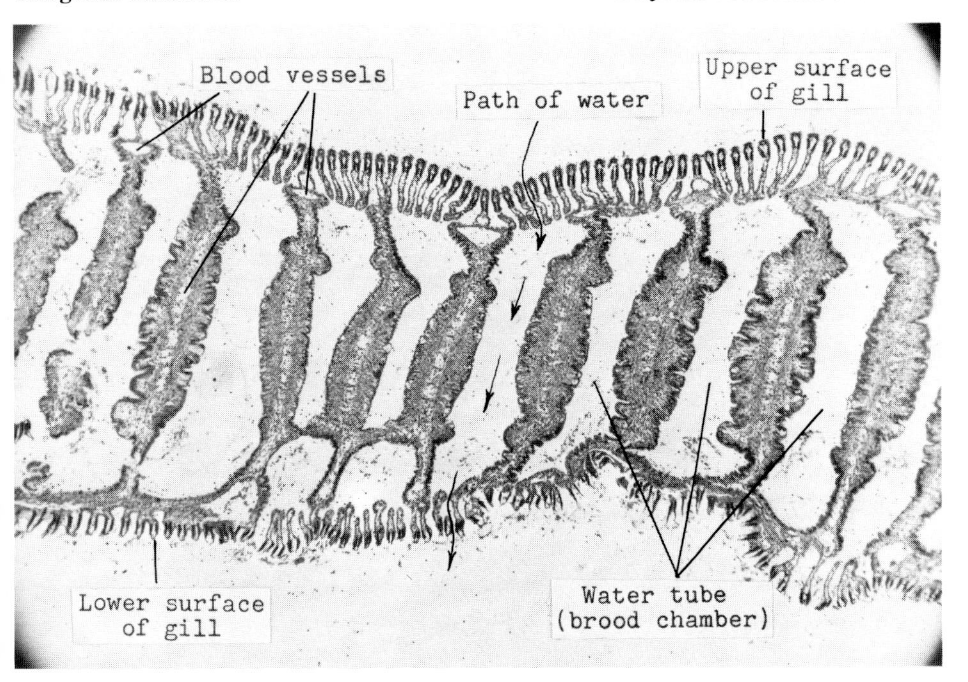

Fig. 65a Clam gill x.s. x40. The brood chambers are where the zygotes are incubated and become glochidia.

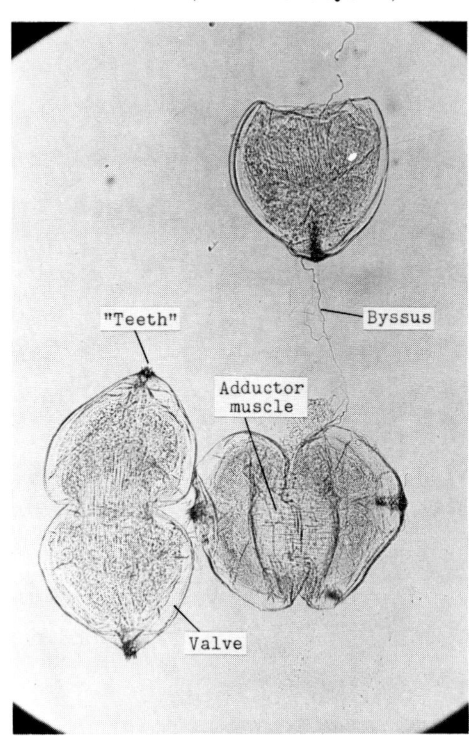

Fig. 65b Glochidia w.m. x40. These are a larval stage in the life cycle of a clam. The teeth at the apex of the valve (shell) and byssus thread are used to attach the glochidium to the gill filaments of a host fish.

Class Cephalopoda

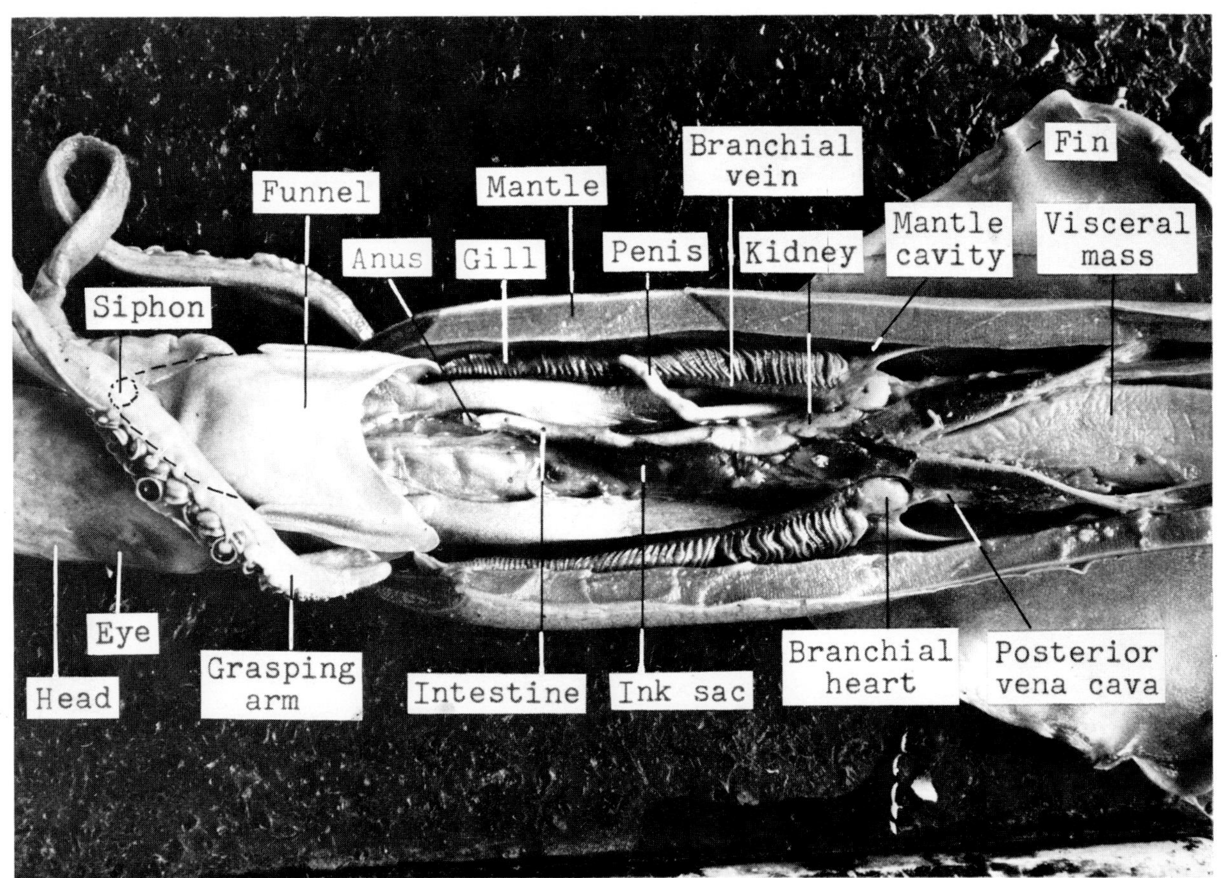

Fig. 65c Loligo (squid) dissection x1.

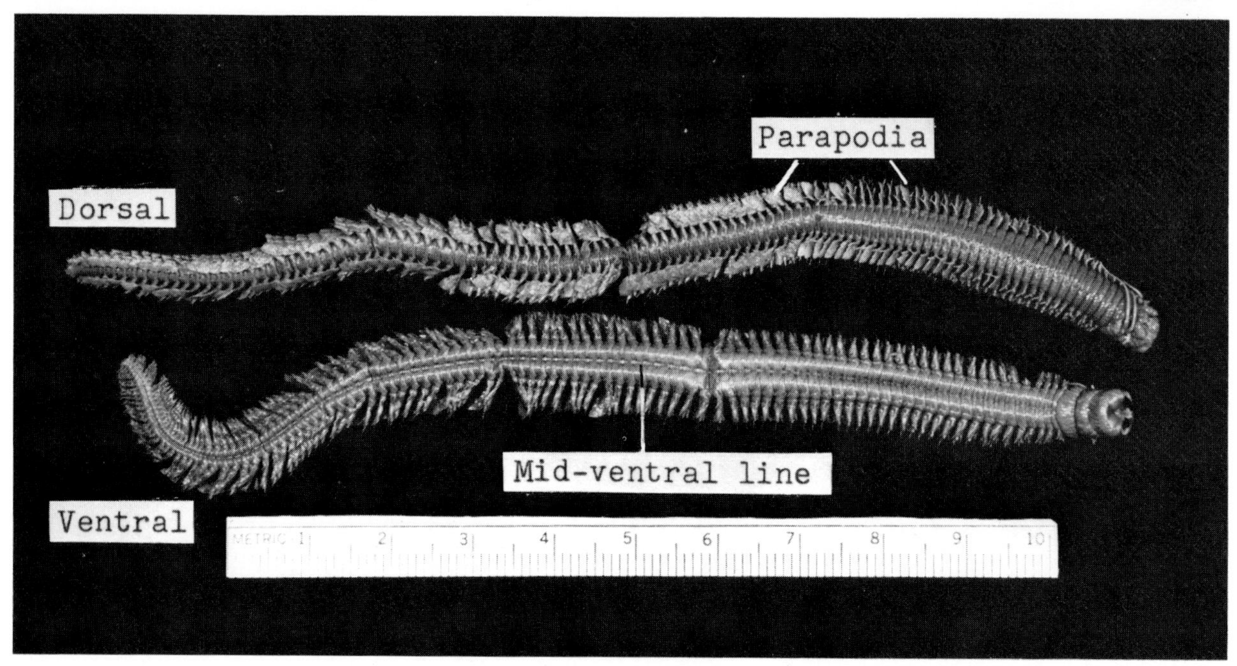

Fig. 66a Neanthes (clam worms) formerly known as Nereis x1. These annelids are marine.

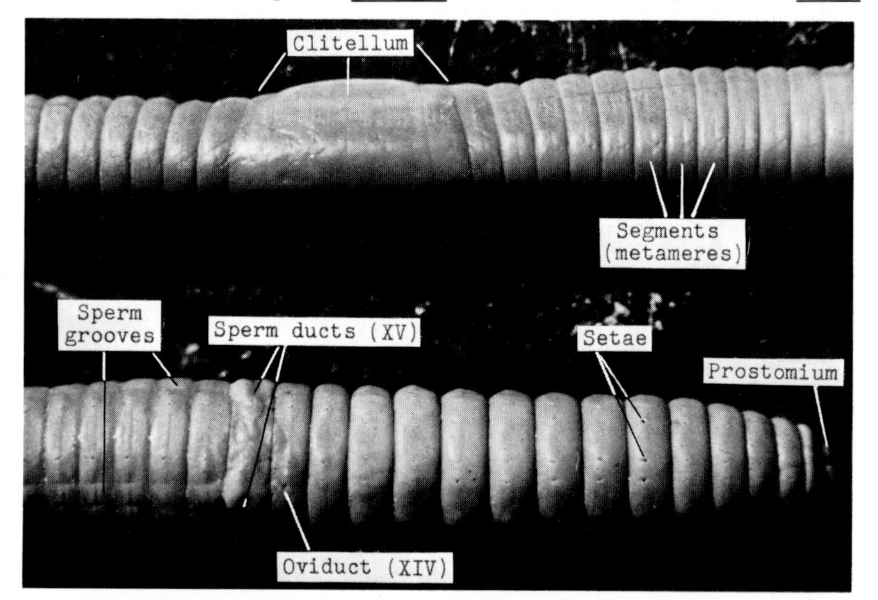

Fig. 66b Lumbricus (earthworm) External anatomy x3.

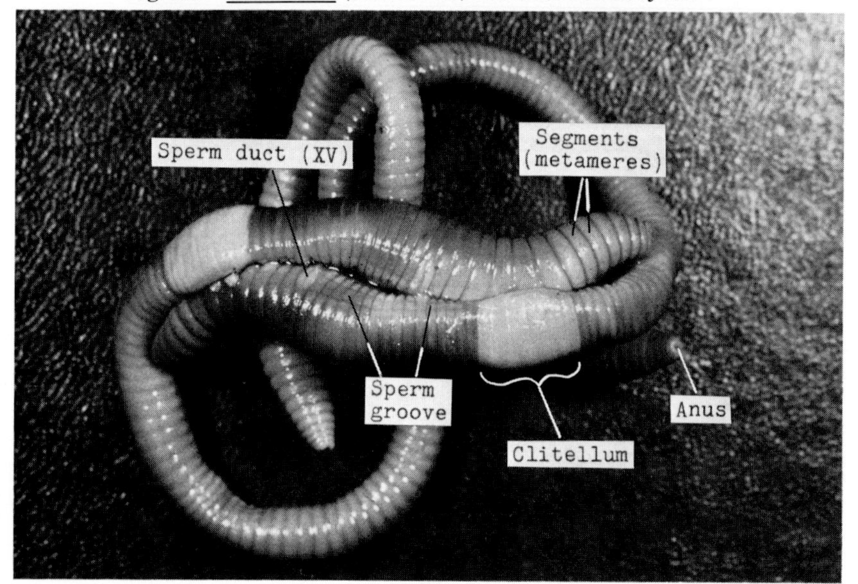

Fig. 66c Copulation between two Lumbricus (earthworms) x1. Sperm is being passed from the sperm duct of each worm into the seminal receptacles of the other.

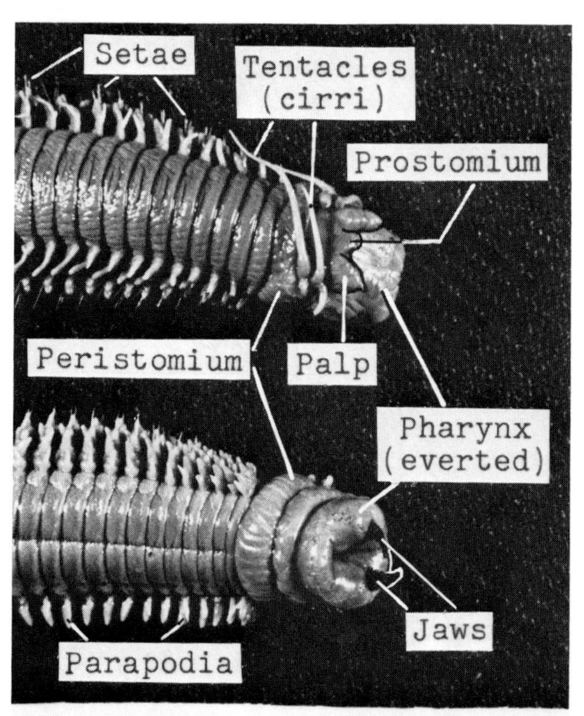

Fig. 66d Neanthes anterior end x2. Dorsal and ventral views.

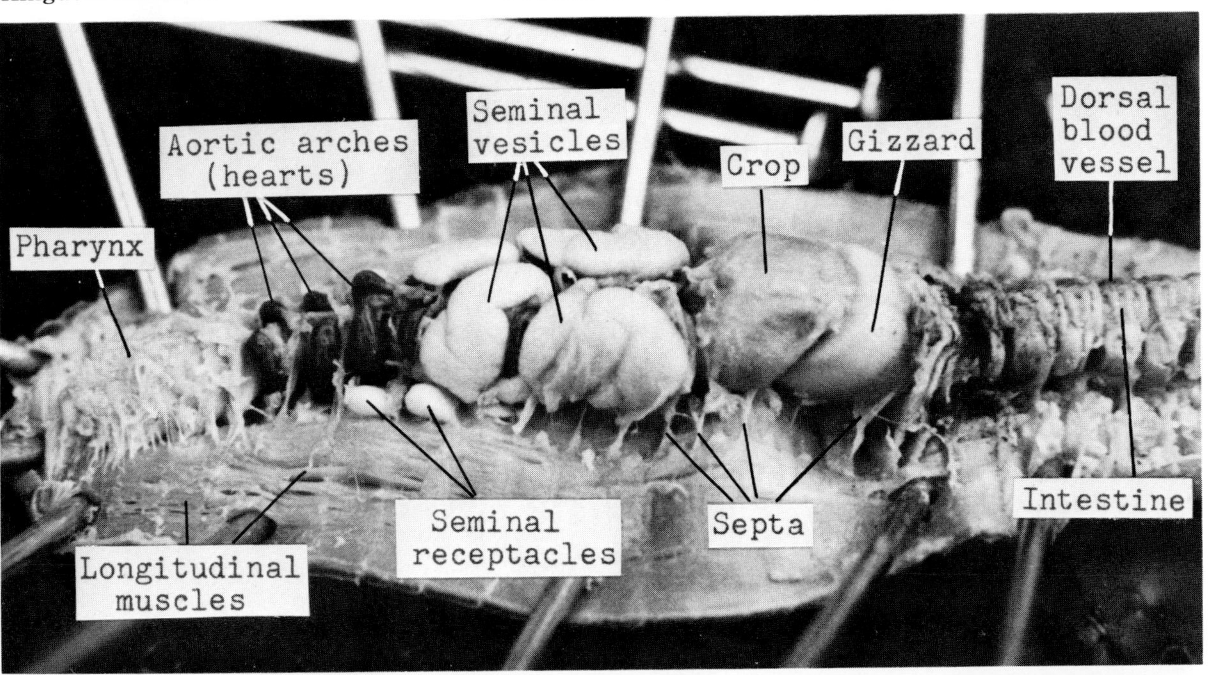

Fig. 67a Lumbricus (earthworm) dissection of anterior end x5.

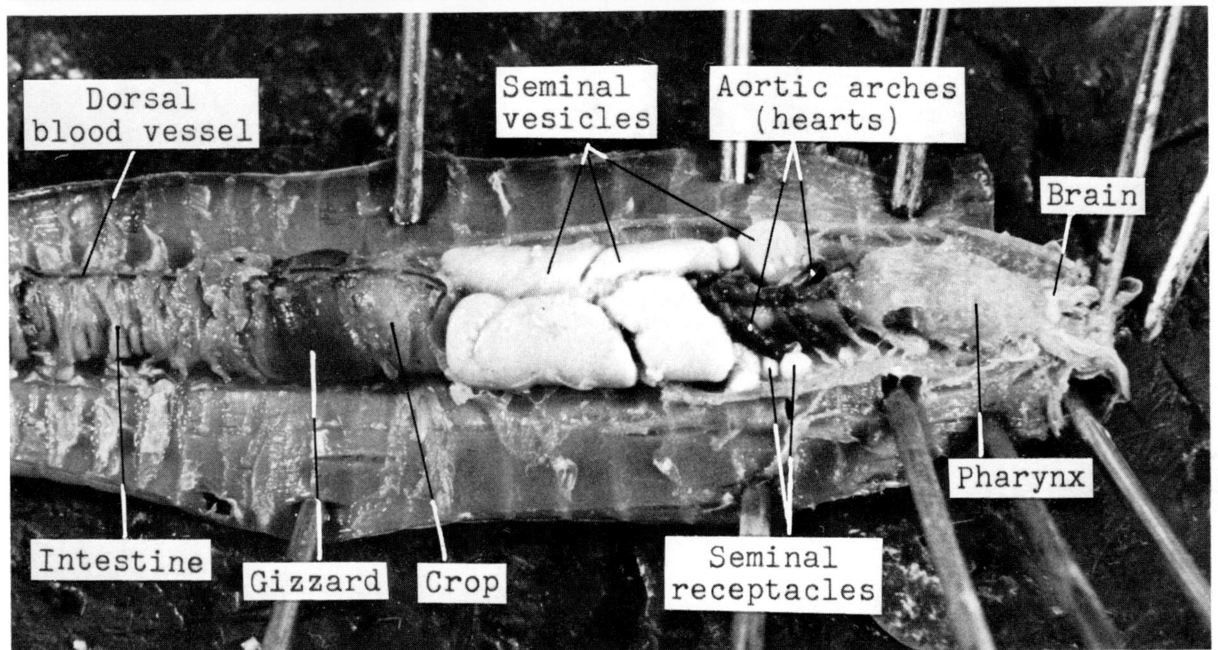

Fig. 67b Lumbricus dissection of anterior end x5. For a better view of the brain see Fig. 68d.

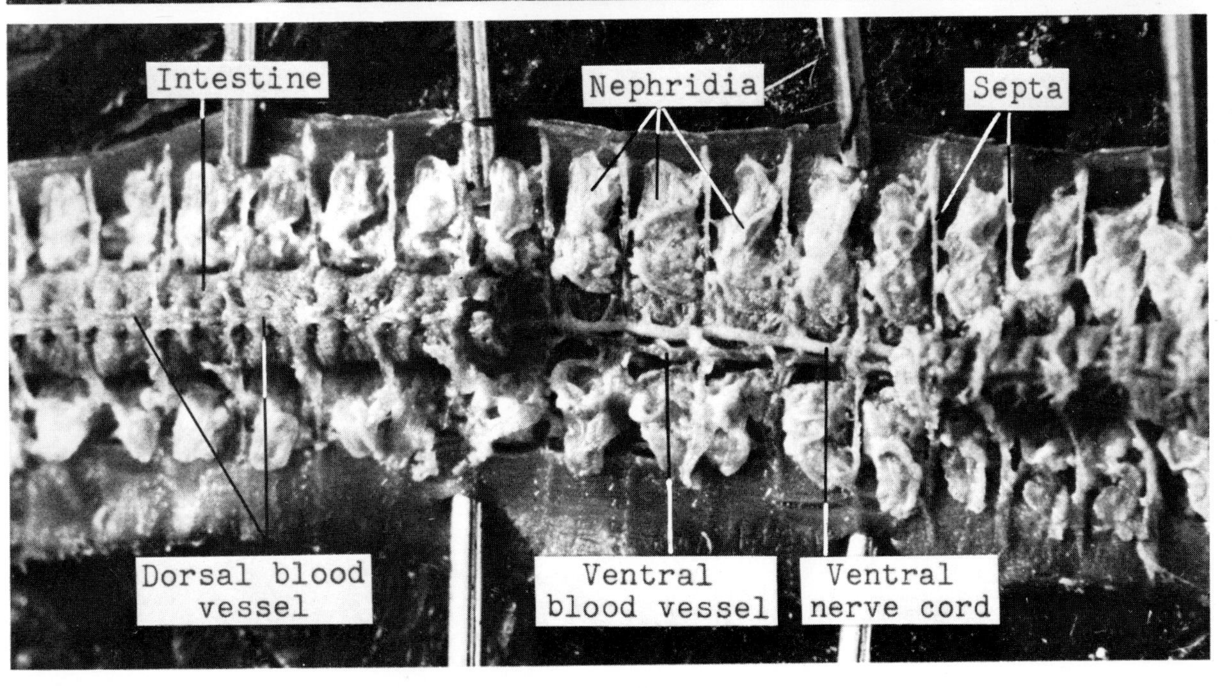

Fig. 67c Lumbricus dissection of posterior region x5. A section of the intestine has been removed in middle to expose the ventral nerve cord and ventral blood vessel.

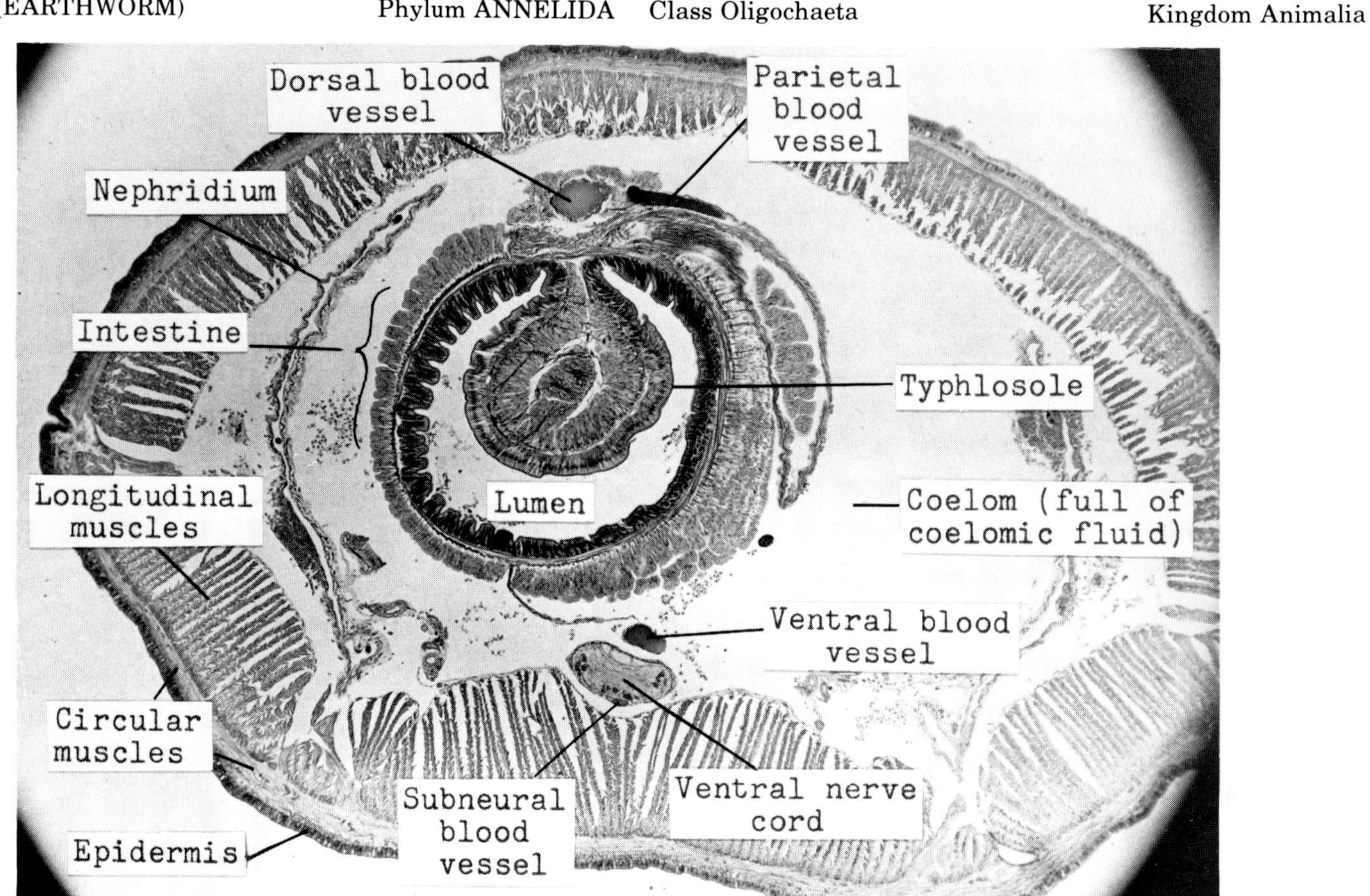

Fig. 68a Lumbricus (earthworm) x.s. intestinal region x40.

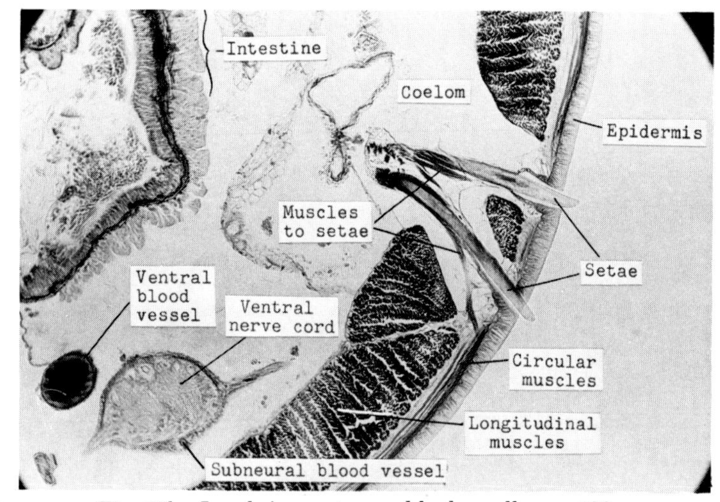

Fig. 68b Lumbricus setae and body wall x.s. x100.

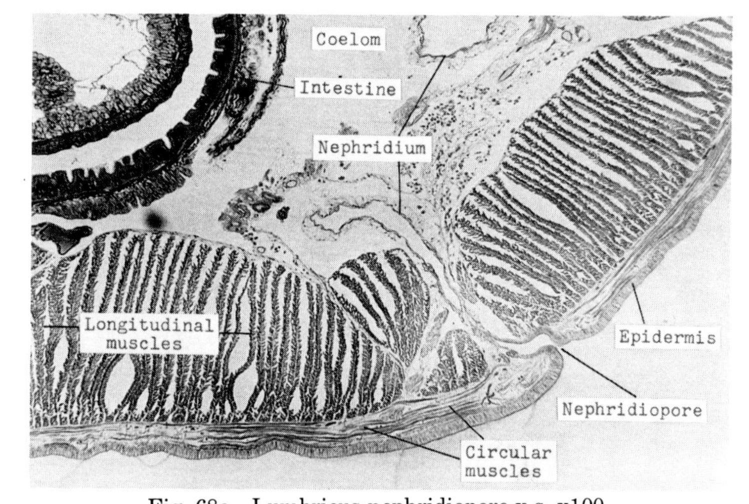

Fig. 68c Lumbricus nephridiopore x.s. x100.

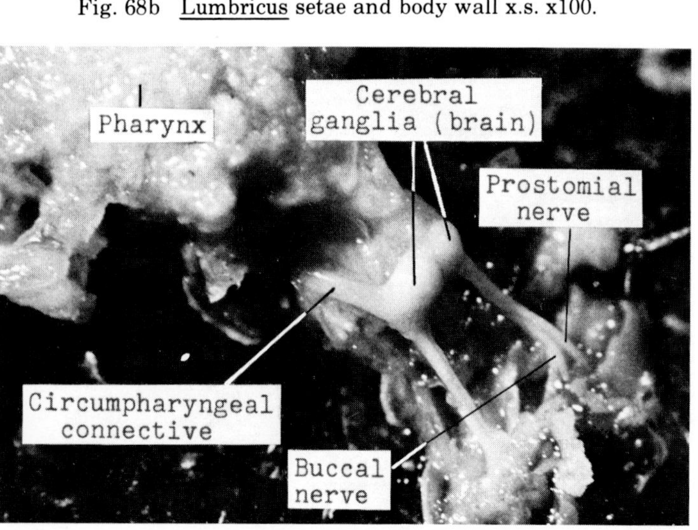

Fig. 68d Lumbricus brain dissection x20.

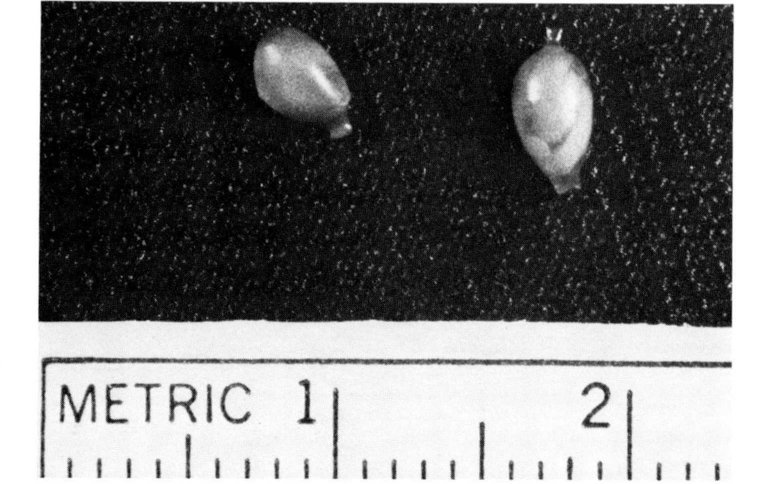

Fig. 68e Lumbricus cocoons x3.

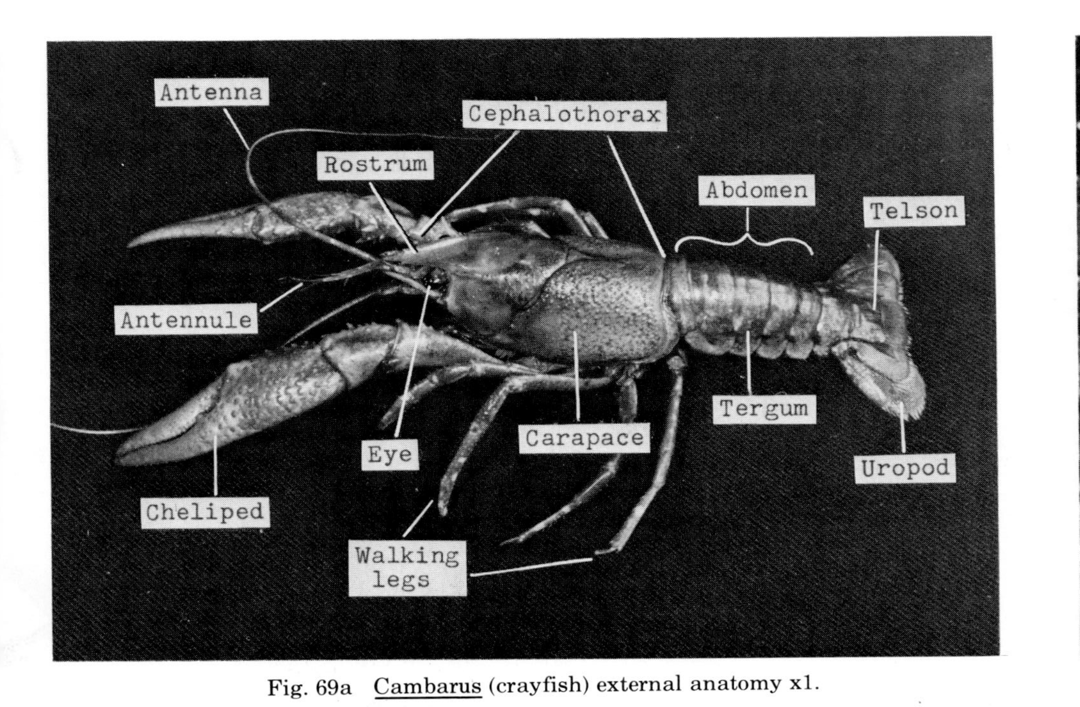

Fig. 69a Cambarus (crayfish) external anatomy x1.

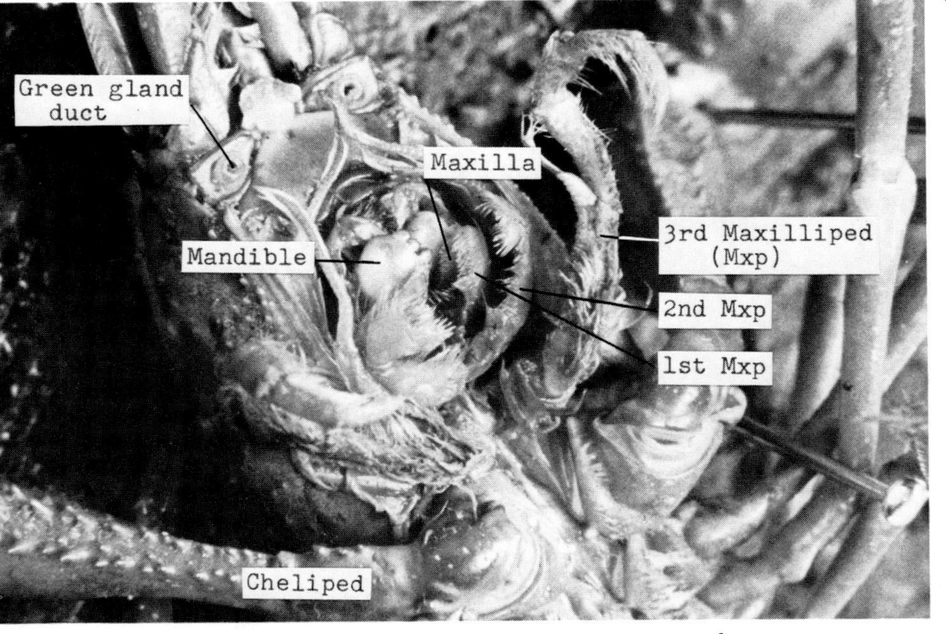

Fig. 69b Cambarus x2. Ventral side of head showing mouth parts.

Disc covering oviduct
1st pair swimmerets (pleopods)
Sperm ducts (genital pores)
Seminal receptacle
Abdomen
Swimmerets (pleopods)

Fig. 69c Cambarus male and female x2. Ventral side of abdomen. Notice that in the male the first two pairs of swimmerets are greatly enlarged (they act as copulatory organs for the transfer of sperm from the sperm duct opening at the base of the last pair of walking legs to the female's seminal receptacle).

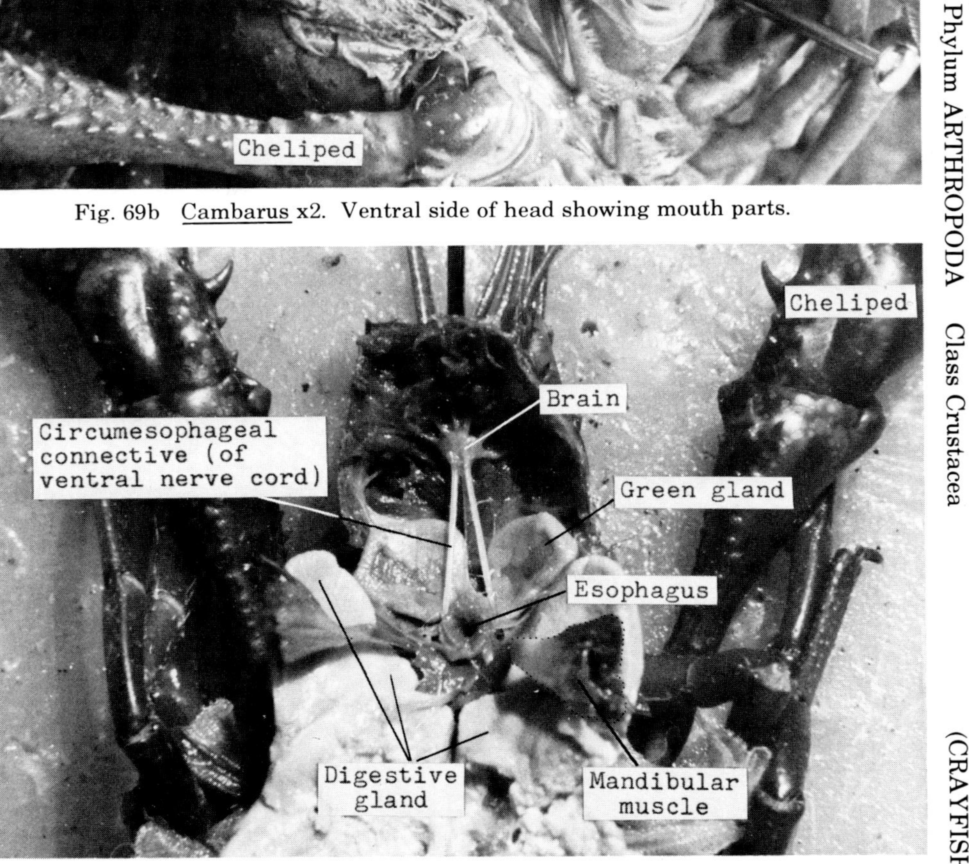

Fig. 69d Cambarus x2. Head region (stomach removed).

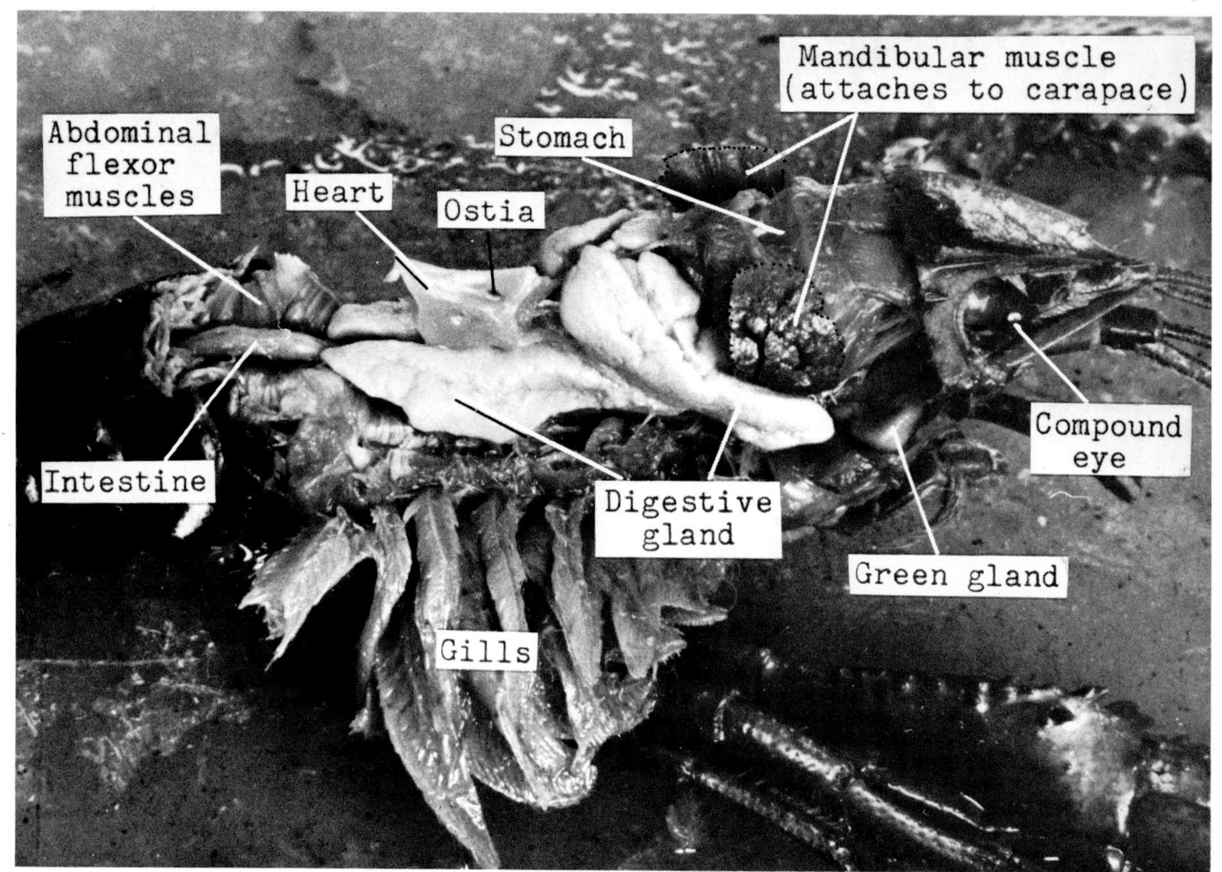

Fig. 70a Cambarus dissection x2. The mandibular muscles (from the mandibles to the carapace) are in place.

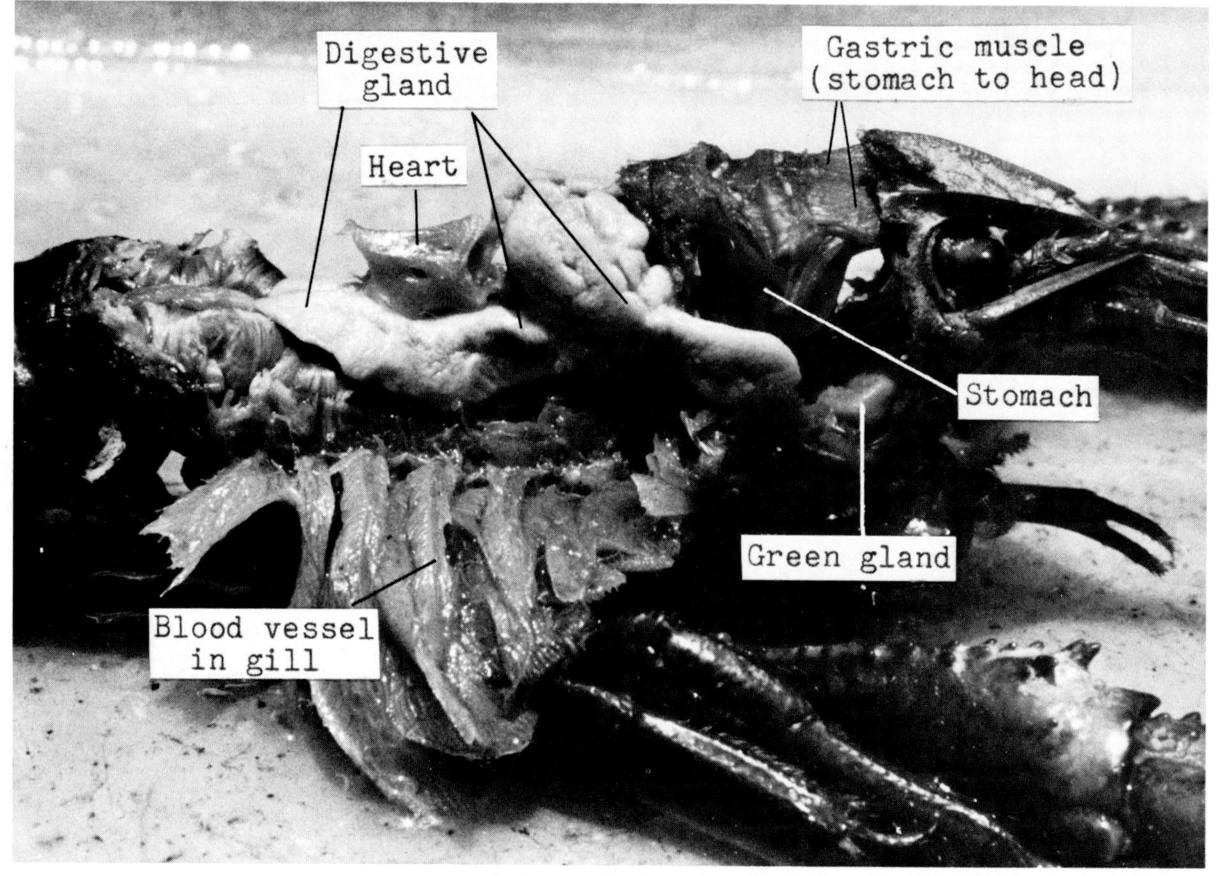

Fig. 70b Cambarus dissection x2. The mandibular muscles have been lowered to show the separation between the stomach and head.

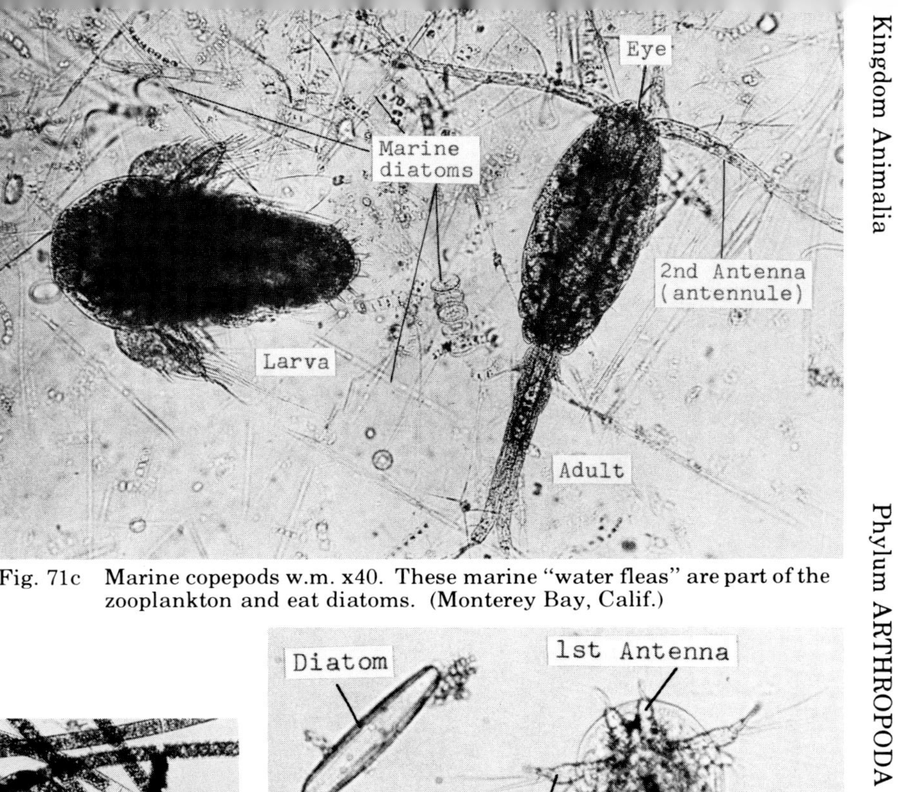

Fig. 71a Cambarus male x2. The heart has been removed to show the testes and vas deferens leading to the genital pore on the last walking leg.

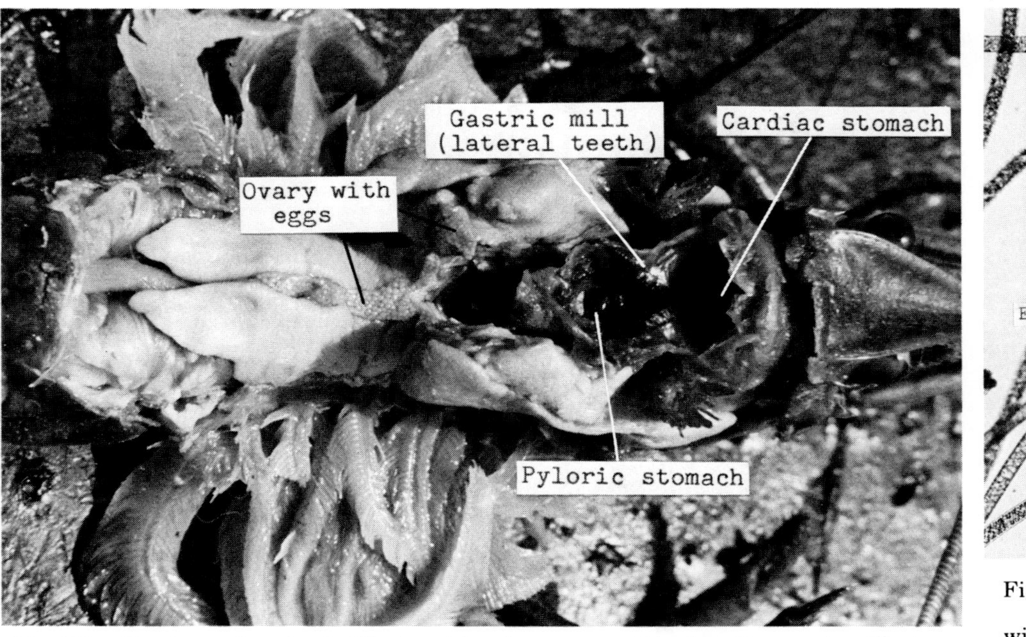

Fig. 71b Cambarus female x2. The heart has been removed to show the ovaries. The top of the stomach has been removed to show the gastric mill within.

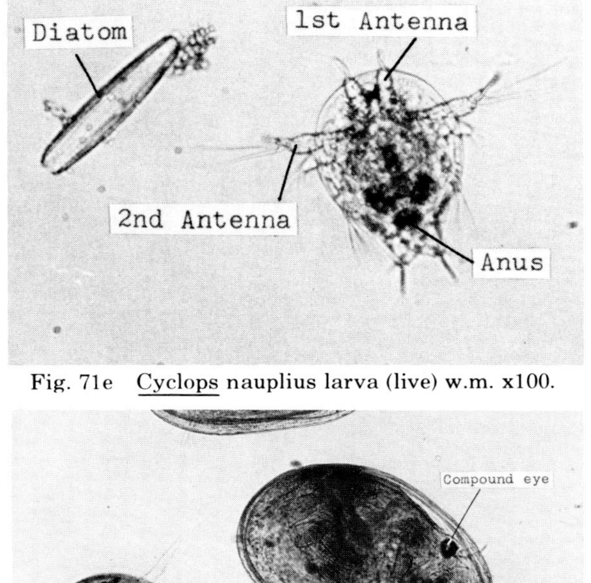

Fig. 71c Marine copepods w.m. x40. These marine "water fleas" are part of the zooplankton and eat diatoms. (Monterey Bay, Calif.)

Fig. 71d Cyclops (live) w.m. x40. A fresh-water copepod with two egg sacs attached.

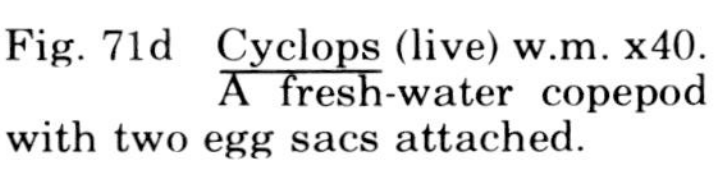

Fig. 71e Cyclops nauplius larva (live) w.m. x100.

Fig. 71f Ostracods w.m. x40. Free swimming crustaceans with a bivalve carapace.

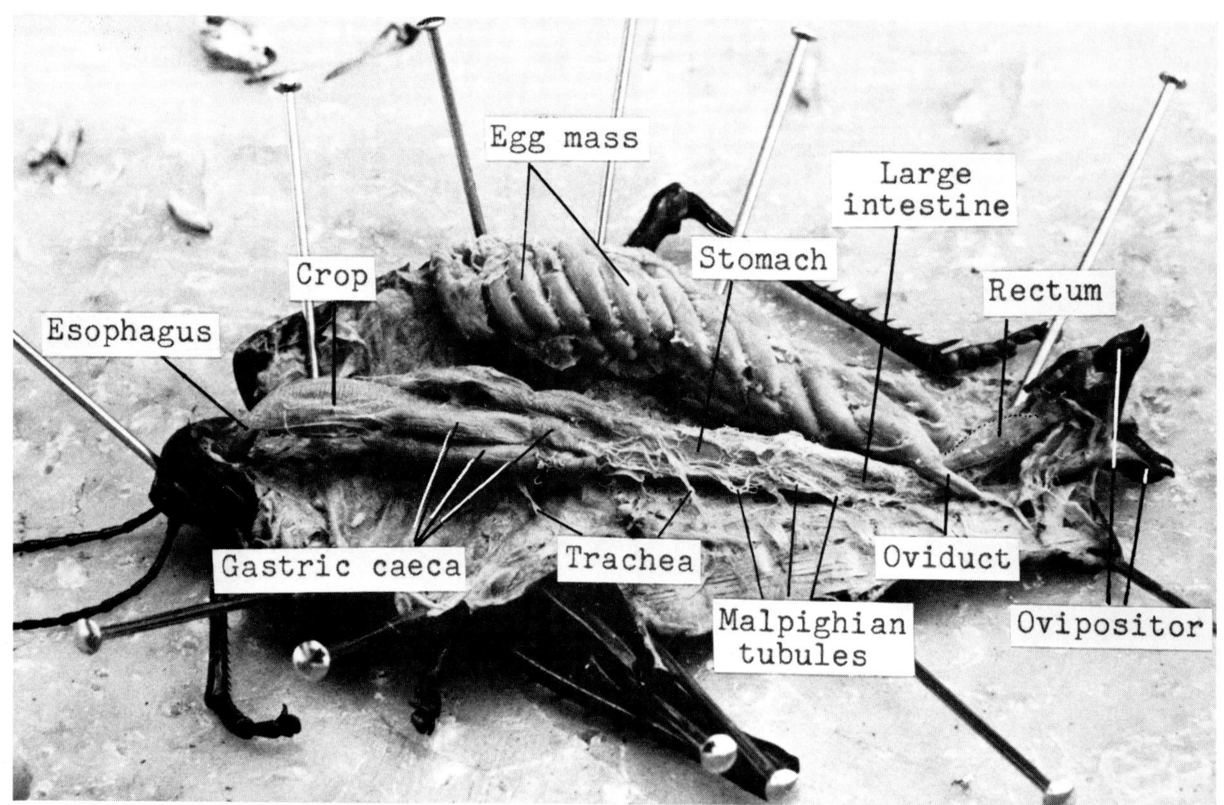

Fig. 72a Romalea (grasshopper) female dissection x2.

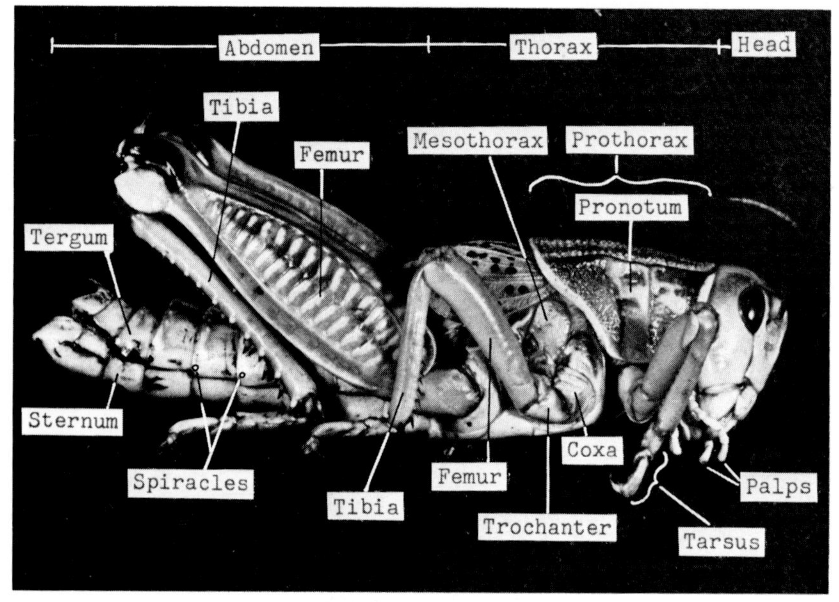

Fig. 72b Grasshopper external anatomy x1.

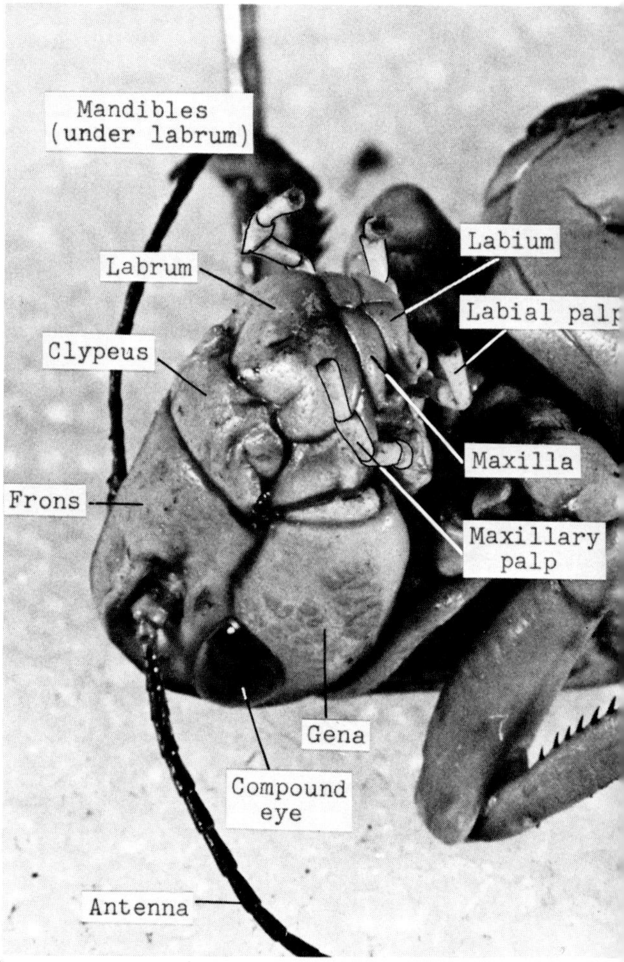

Fig. 72c Grasshopper head anatomy x10.

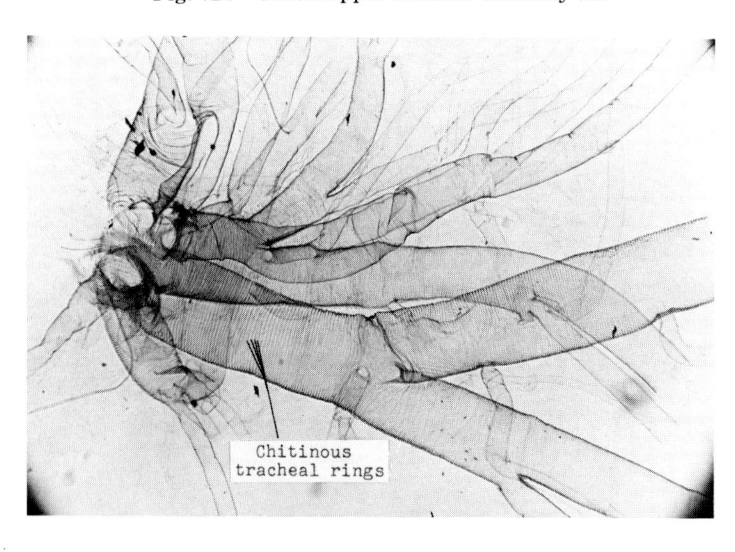

Fig. 72d Insect trachea w.m. x40. The trachea carry air directly to the cells in all regions of the insect. The chitinous rings help prevent the collapse of the tubes.

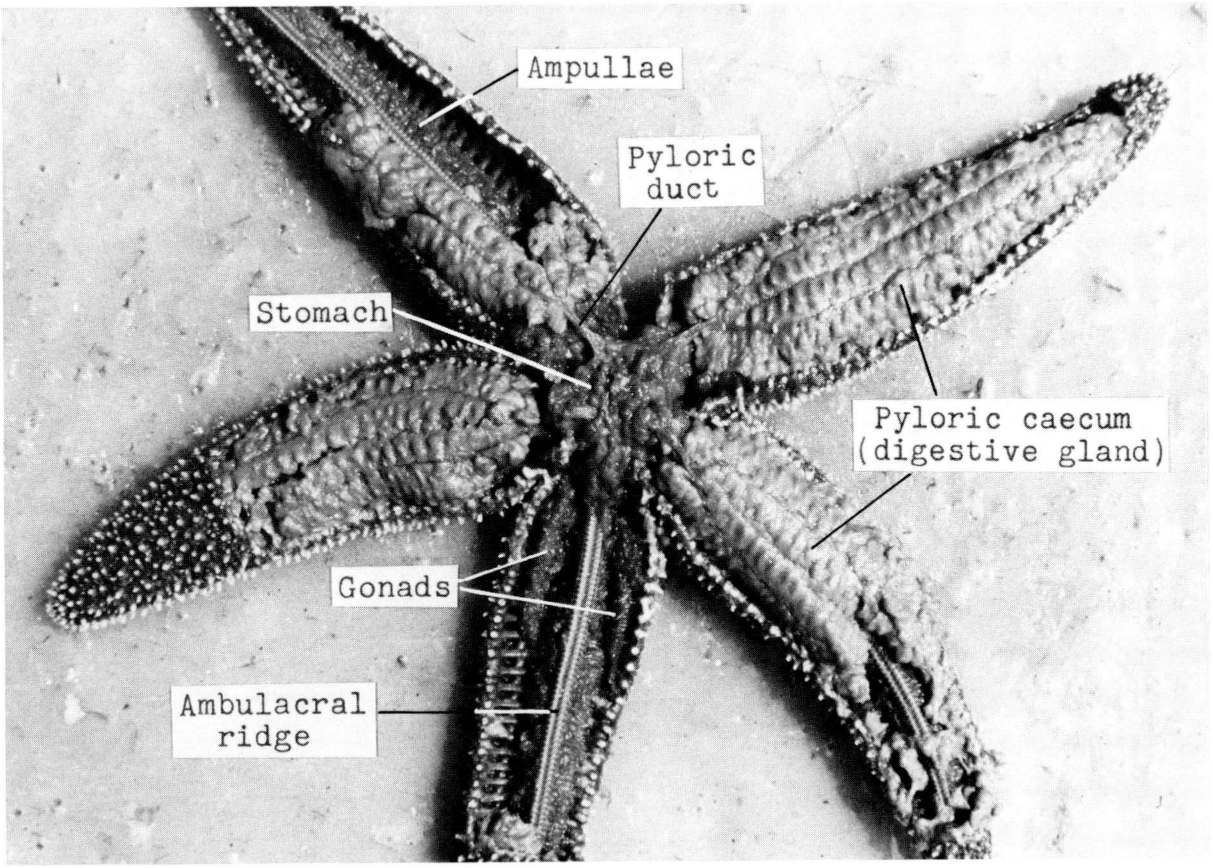

Fig. 73a Starfish dissection, dorsal surface x1.

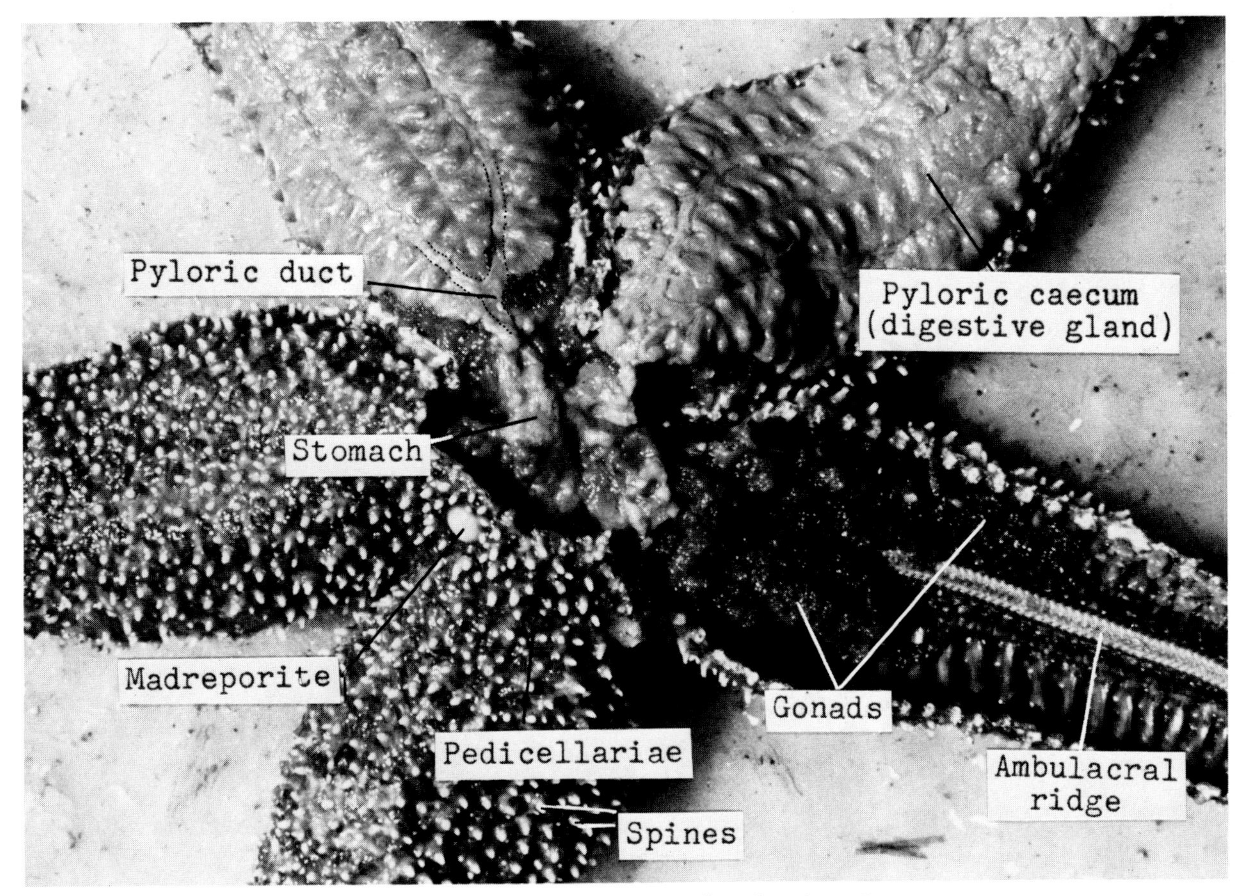

Fig. 73b Starfish dissection, dorsal surface x2.

Fig. 74a Starfish dissection, dorsal surface x2. The stomach has been removed to expose the stone canal and peristome. The pyloric caecum has also been removed on the right and lower arms to expose the gonads and ampullae.

Fig. 74b Starfish, ventral surface x2.

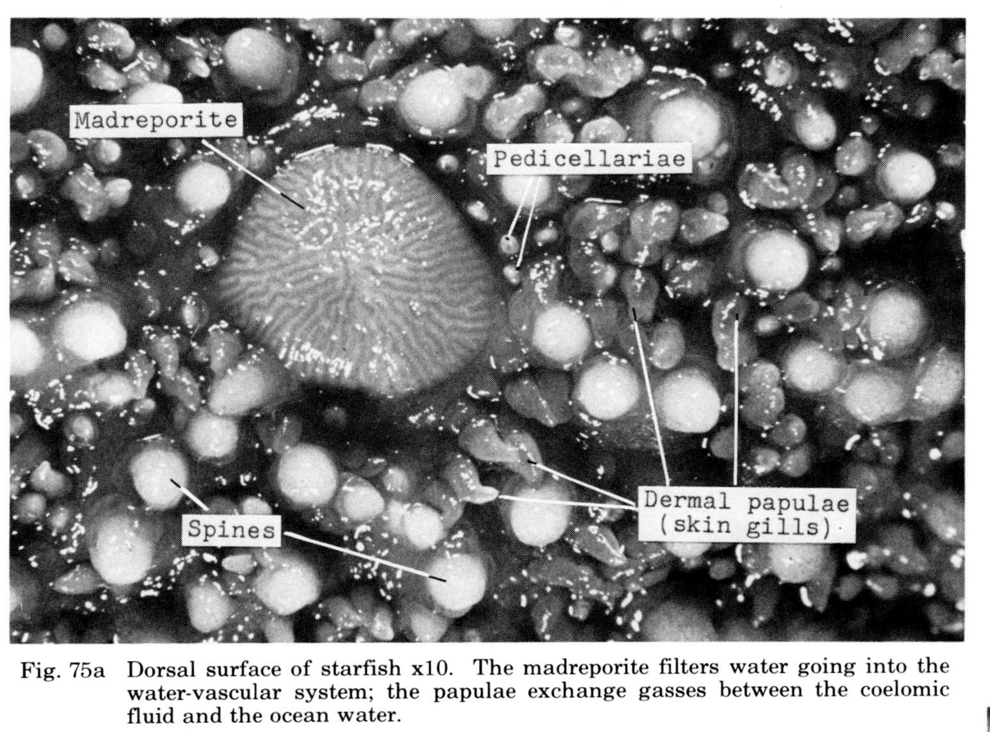

Fig. 75a Dorsal surface of starfish x10. The madreporite filters water going into the water-vascular system; the papulae exchange gasses between the coelomic fluid and the ocean water.

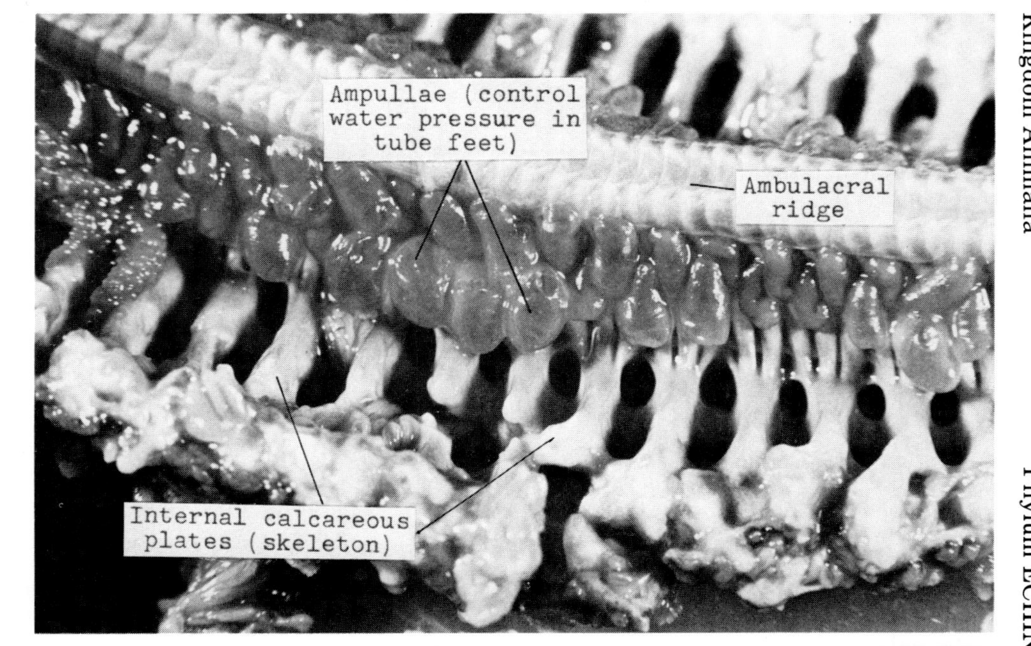

Fig. 75b Ampullae x5.

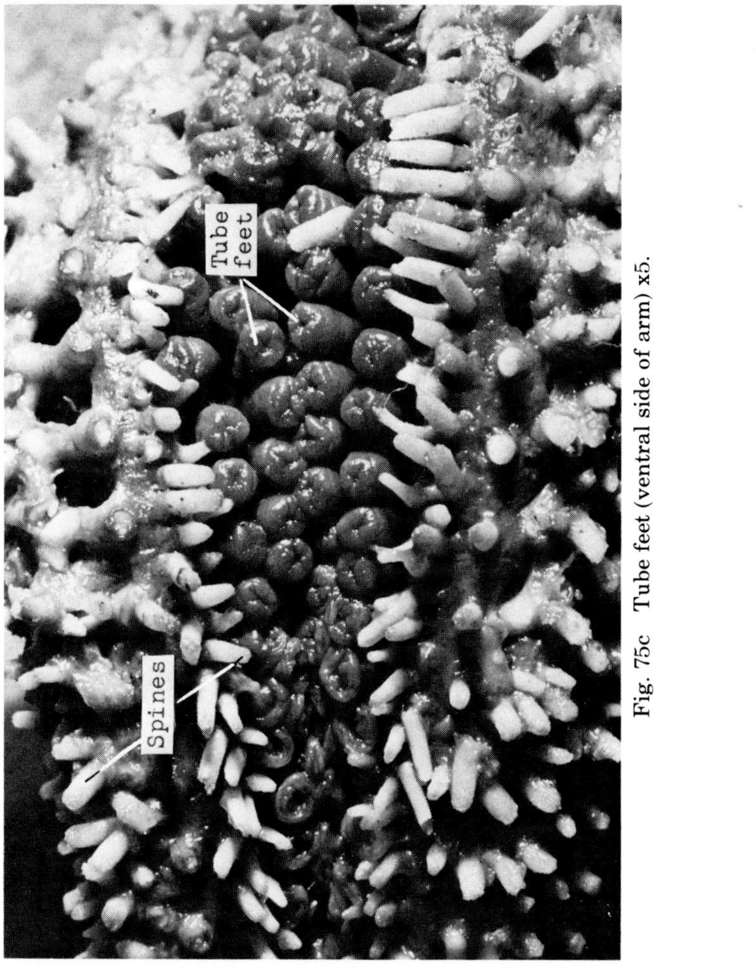

Fig. 75c Tube feet (ventral side of arm) x5.

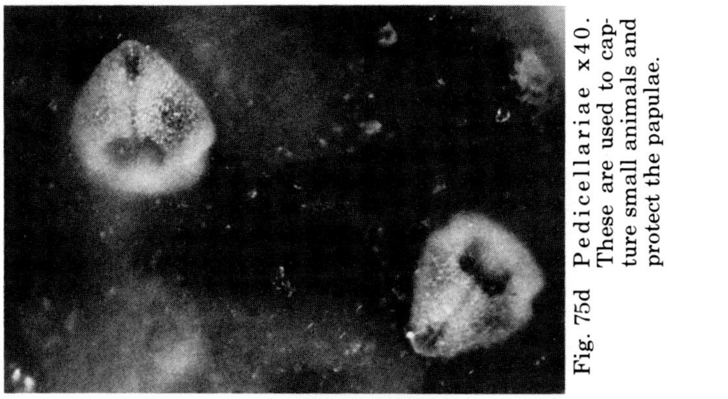

Fig. 75d Pedicellariae x40. These are used to capture small animals and protect the papulae.

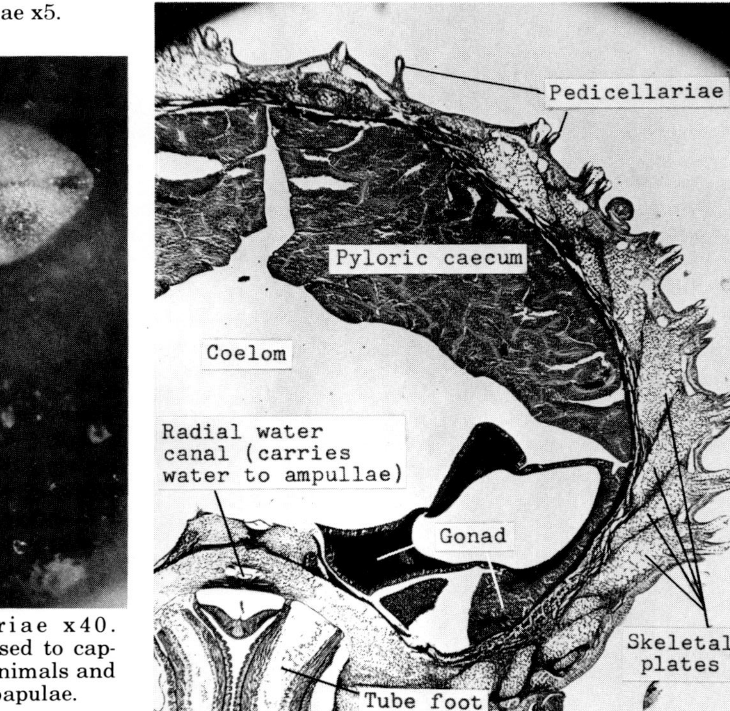

Fig. 75e Starfish arm x.s. x40.

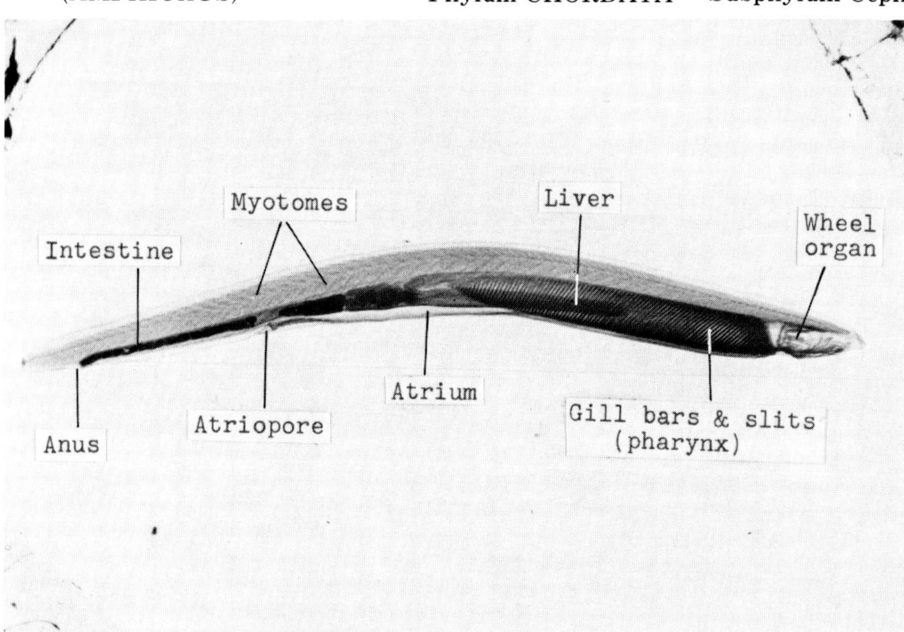

Fig. 76a Amphioxus w.m. x10.

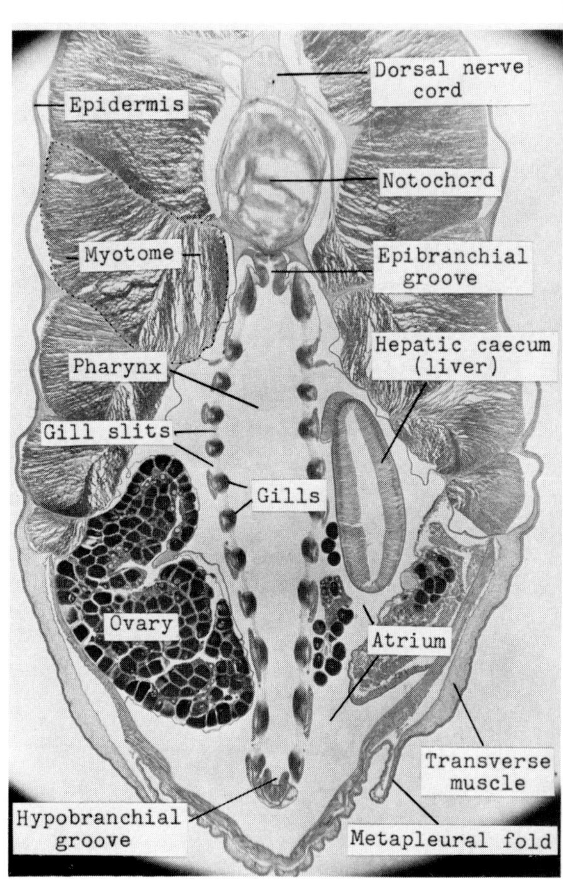

Fig. 76d Amphioxus pharynx x.s. x40. Section taken at point "A" in Fig. 76c.

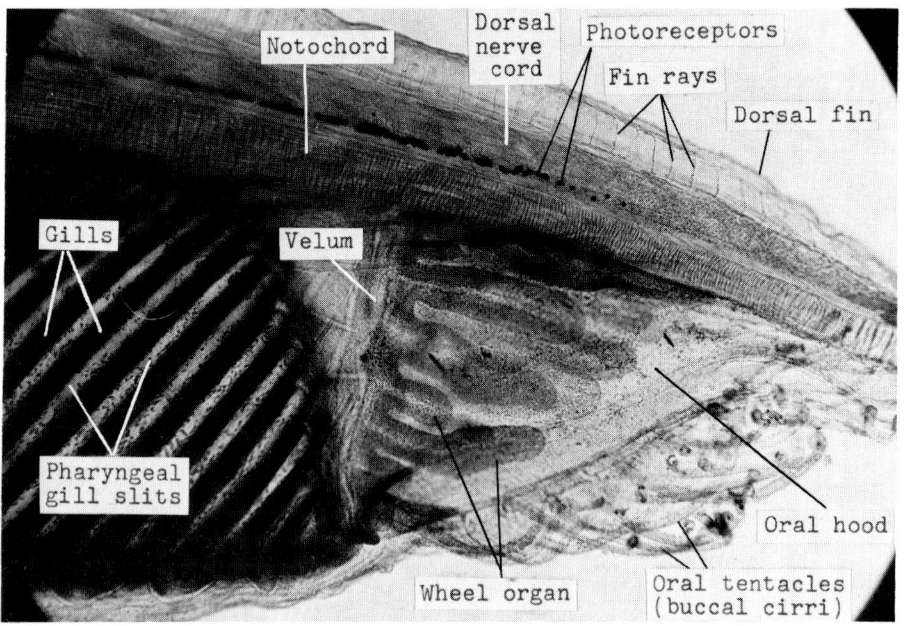

Fig. 76b Amphioxus anterior end w.m. x100.

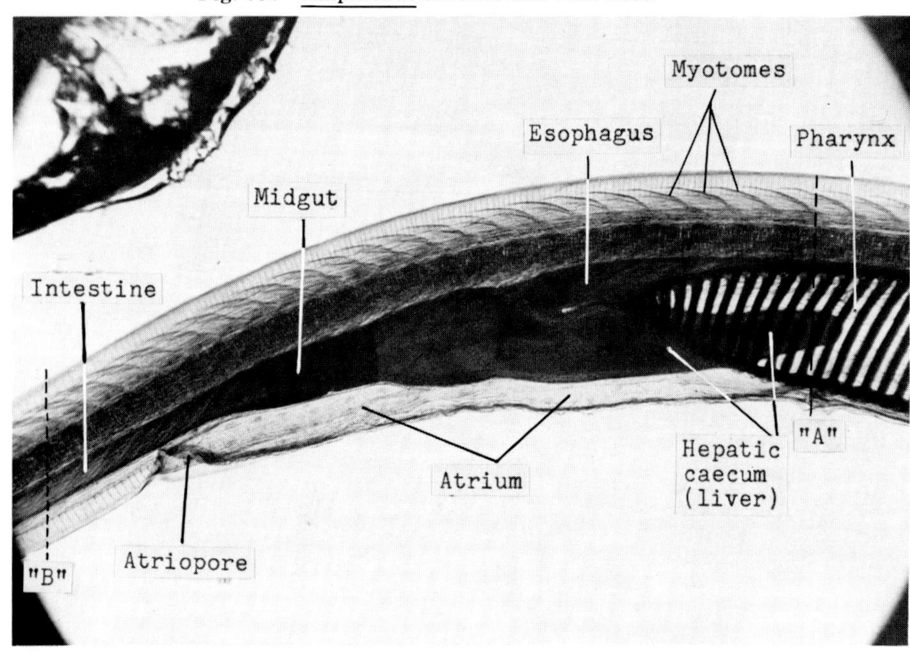

Fig. 76c Amphioxus mid-region w.m. x40.

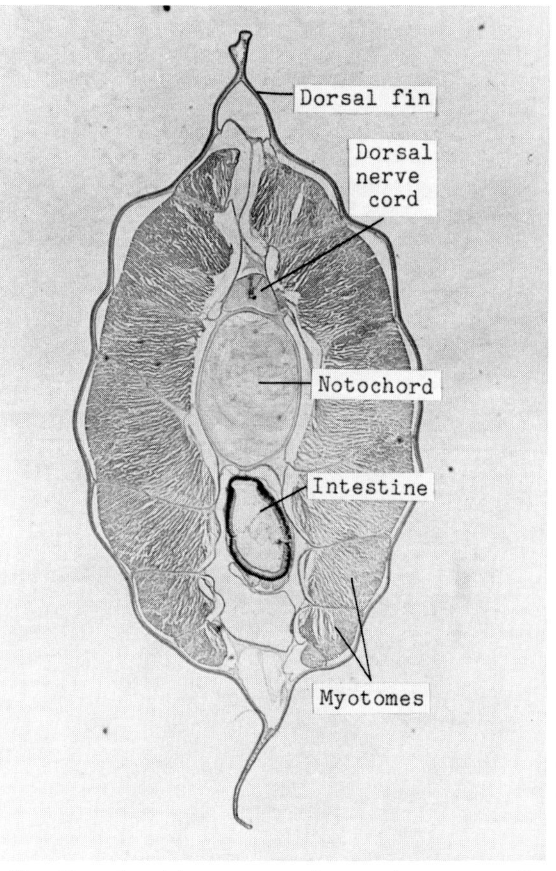

Fig. 76e Amphioxus posterior region x.s. x40. Section taken at point "B" in Fig. 76c.

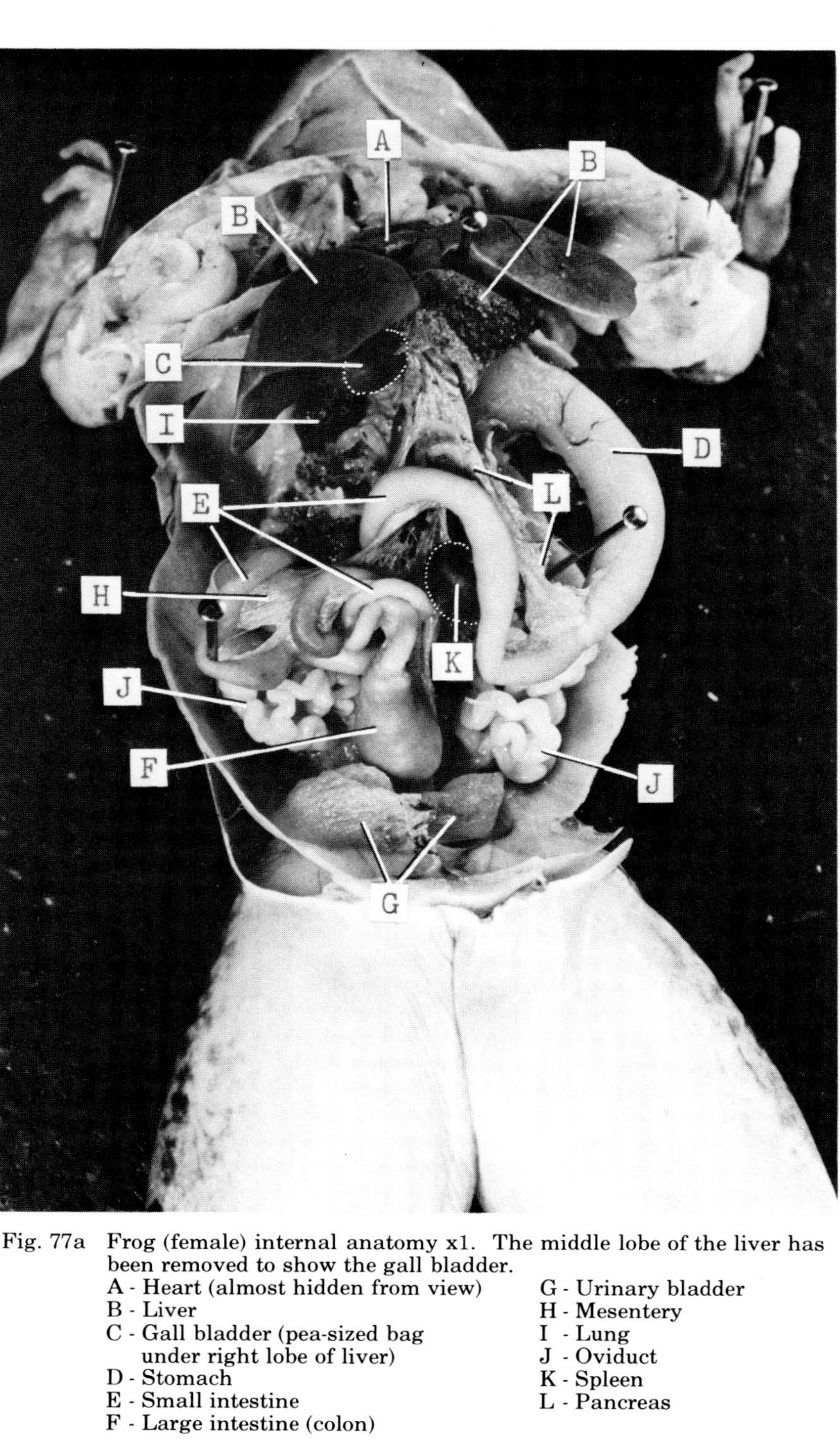

Fig. 77a Frog (female) internal anatomy x1. The middle lobe of the liver has been removed to show the gall bladder.
A - Heart (almost hidden from view)
B - Liver
C - Gall bladder (pea-sized bag under right lobe of liver)
D - Stomach
E - Small intestine
F - Large intestine (colon)
G - Urinary bladder
H - Mesentery
I - Lung
J - Oviduct
K - Spleen
L - Pancreas

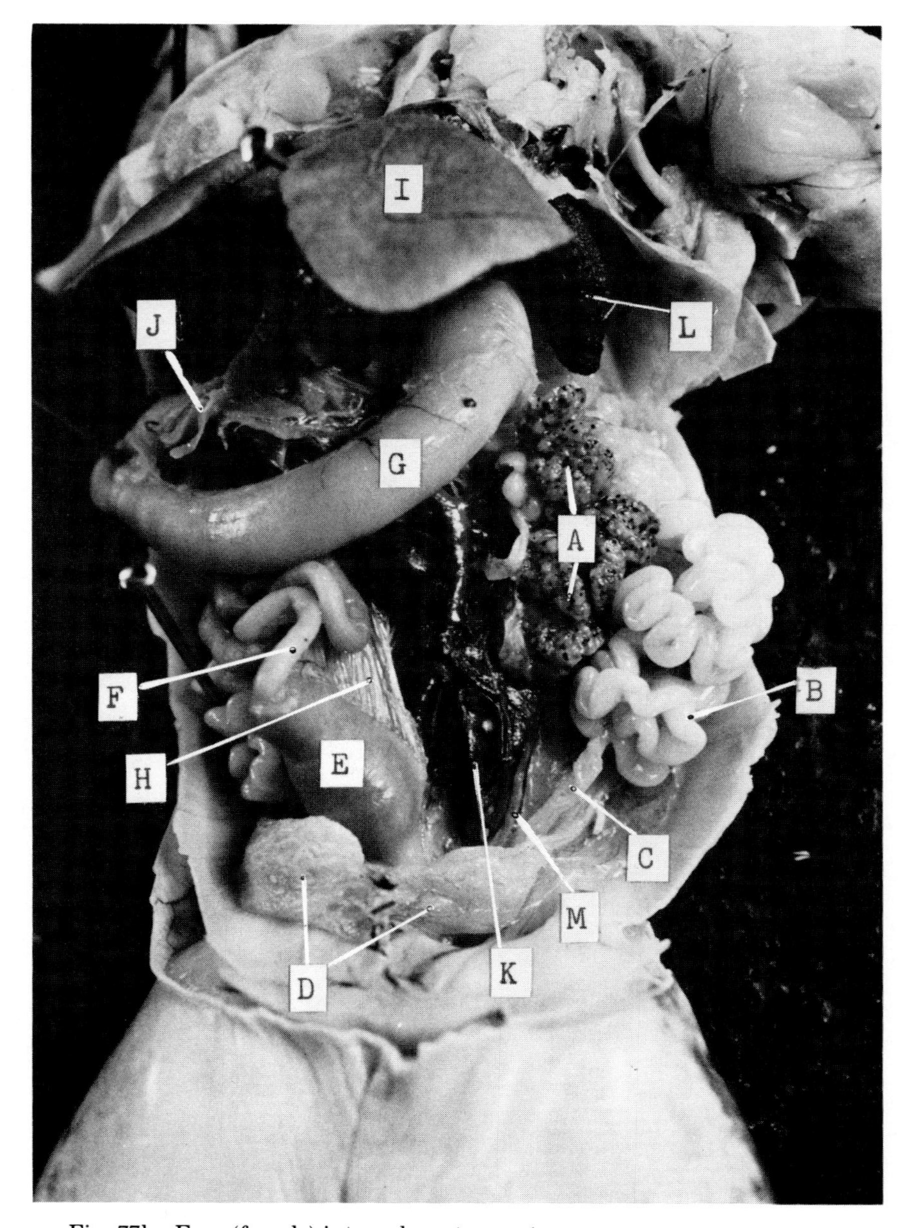

Fig. 77b Frog (female) internal anatomy x1.
A - Ovaries containing immature eggs
B - Oviducts
C - Uterus
D - Urinary bladder
E - Large intestine (colon)
F - Small intestine
G - Stomach
H - Mesentery
I - Liver
J - Pancreas
K - Kidney
L - Lung
M - Ureter

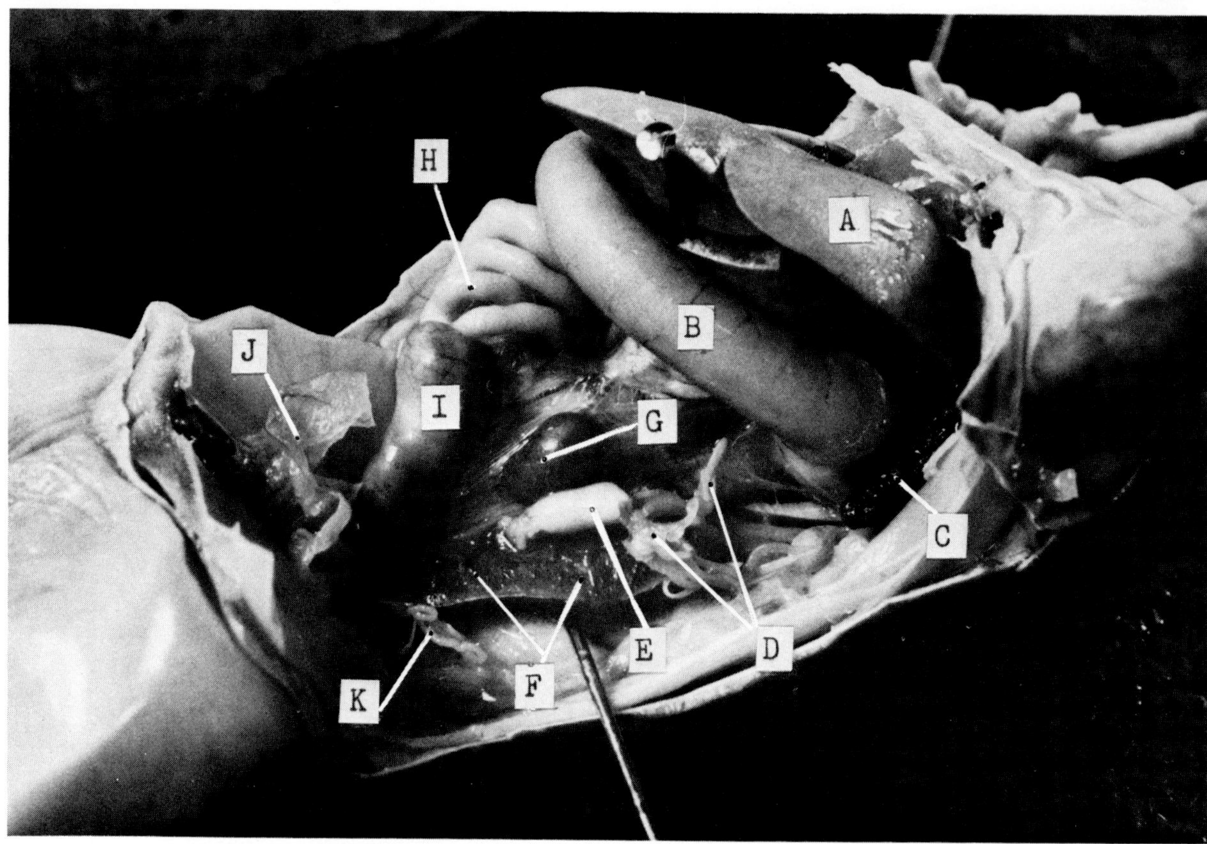

Fig. 78a Frog (male) internal anatomy.
A - Liver
B - Stomach
C - Lung
D - Fat bodies
E - Testis
F - Kidney
G - Spleen
H - Small intestine
I - Large intestine (colon)
J - Urinary bladder
K - Vestigial oviduct

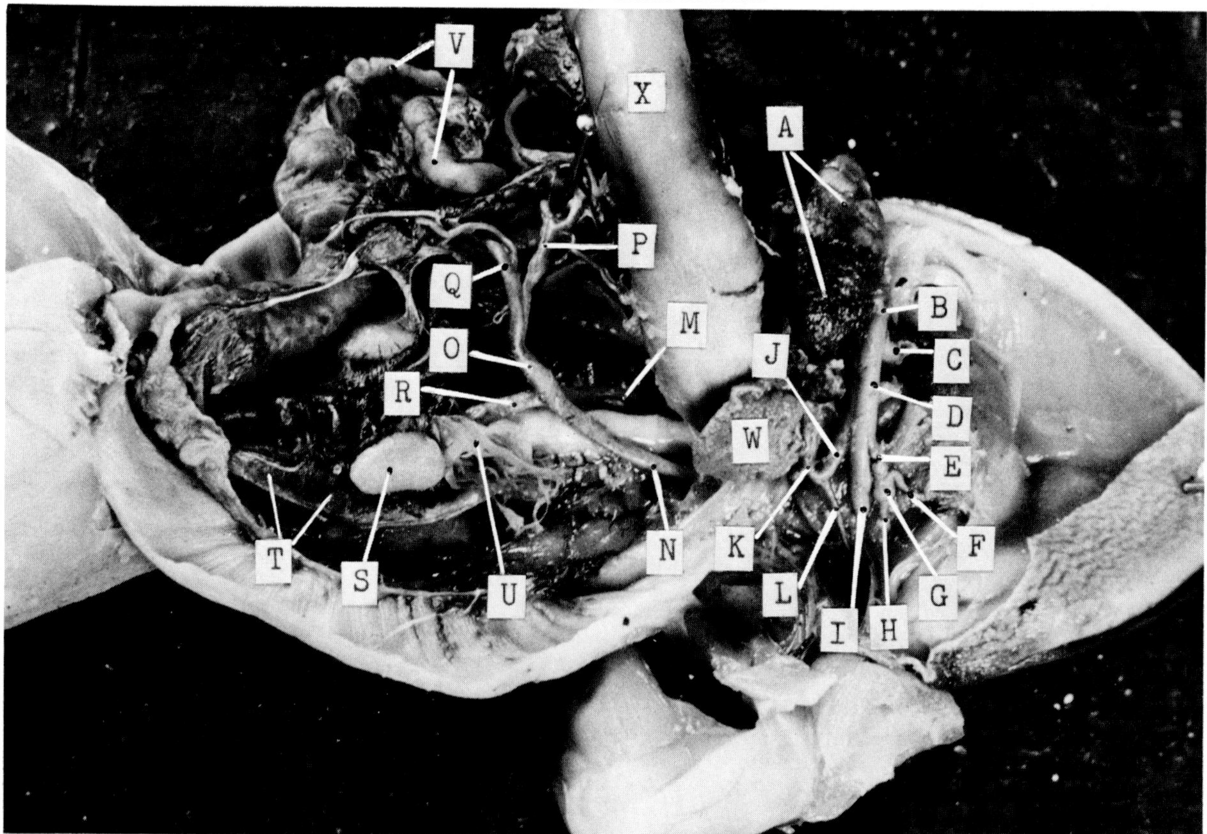

Fig. 78b Frog (male) arterial system.
A - Heart
B - Conus arteriosus
C - Right truncus arteriosus
D - Left truncus arteriosus
E - Carotid arch
F - External carotid
G - Carotid gland
H - Internal carotid
I - Systemic arch (left)
J - Pulmocutaneous arch
K - Pulmonary
L - Cutaneous
M - Systemic arch (right)
N - Systemic arch (left)
O - Coeliaco-mesenteric
P - Coeliac
Q - Mesenteric (anterior)
R - Dorsal aorta
S - Testis
T - Kidney
U - Fat bodies
V - Small intestine
W - Lung
X - Stomach

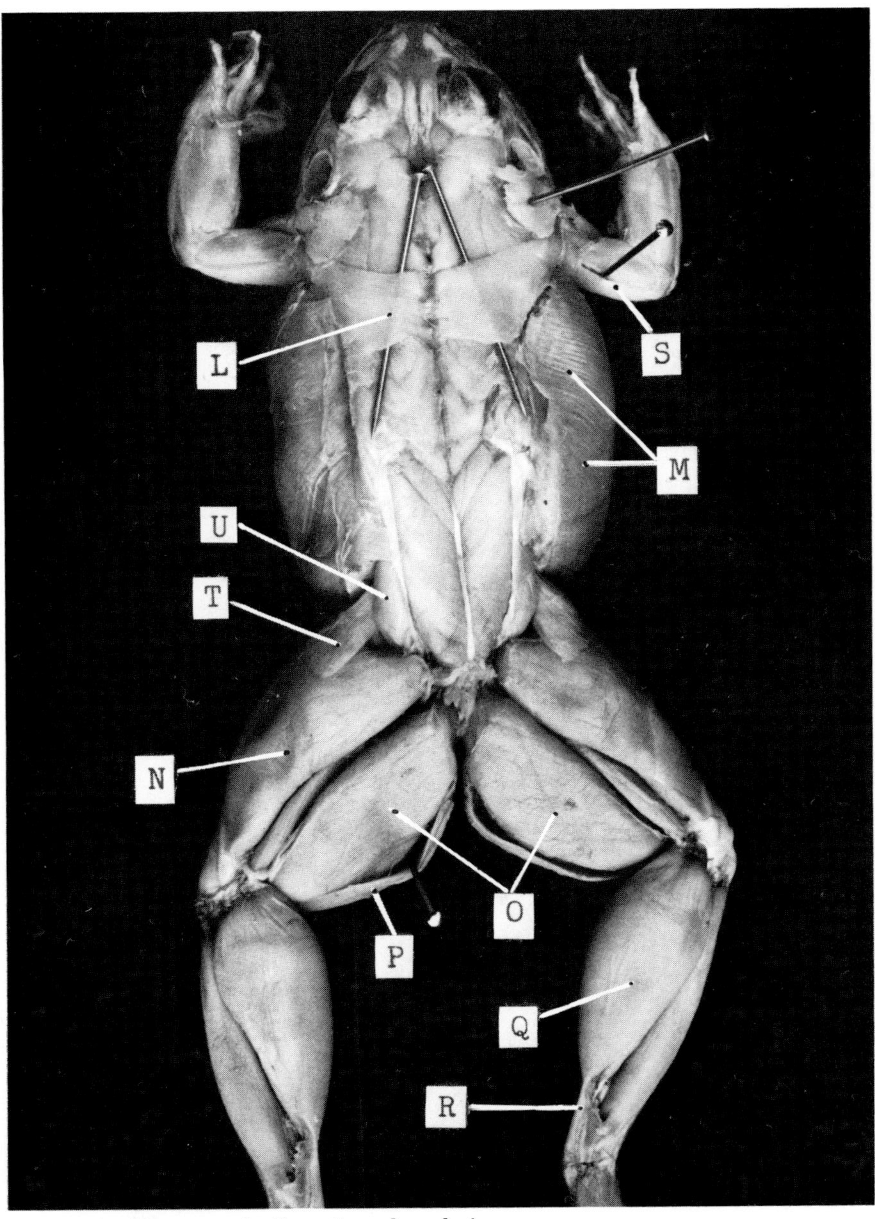

Fig. 79a Frog muscle dissection, ventral view.
A - Deltoid
B - Pectoralis
C - Rectus abdominis
D - Sartorius (note that this muscle is cut & raised on one leg to better expose the Adductor longus)
E - Triceps femoris
F - Adductor magnus
G - Adductor longus (a very flat & thin muscle just under the Sartorius)
H - Gracilis major
I - Gracilis minor
J - External oblique
K - Gastrocnemius

Fig. 79b Frog muscle dissection, dorsal view.
L - Latissimus dorsi
M - External oblique
N - Triceps femoris
O - Semimembranosus
P - Gracilis minor
Q - Gastrocnemius
R - Tendon of Achilles
S - Triceps brachii
T - Cutaneous abdominis
U - Gluteus

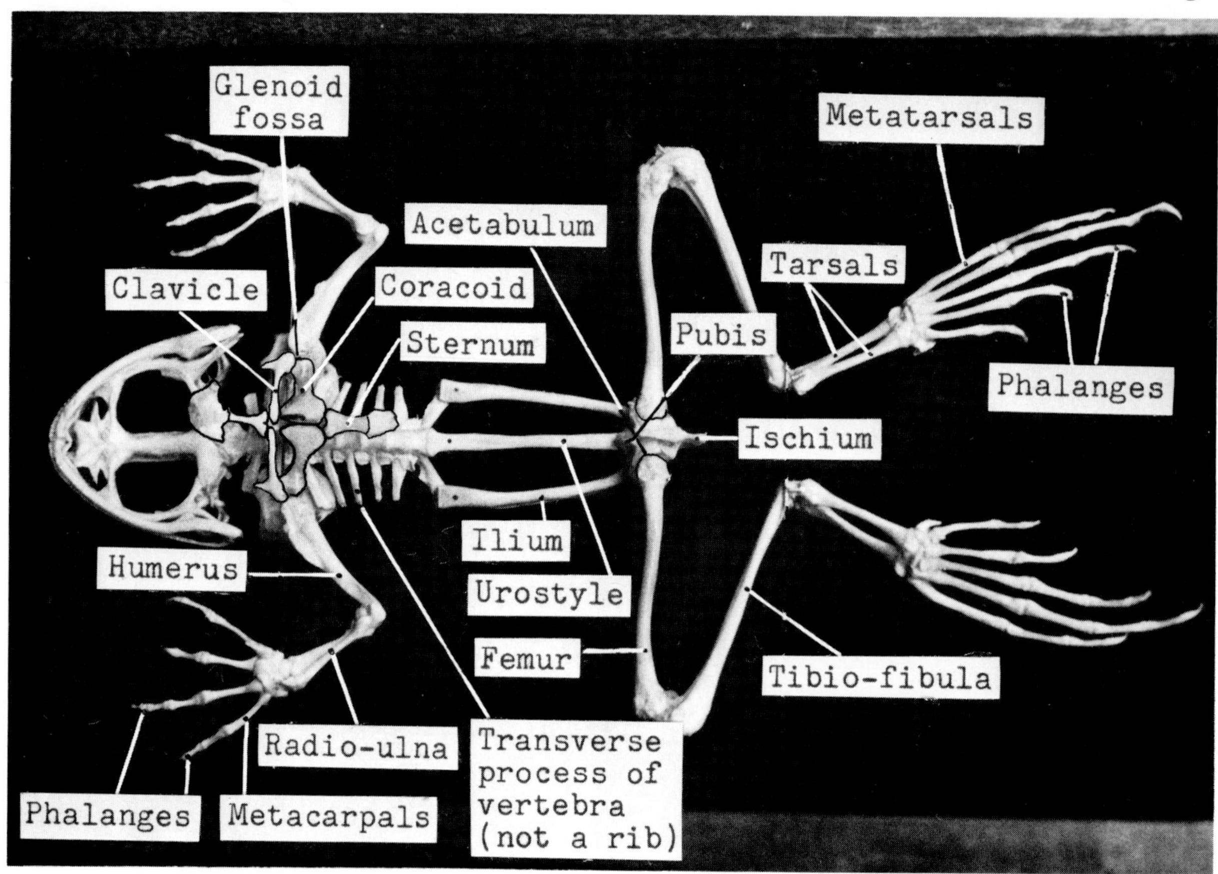

Fig. 80a Frog skeleton, ventral view x1.

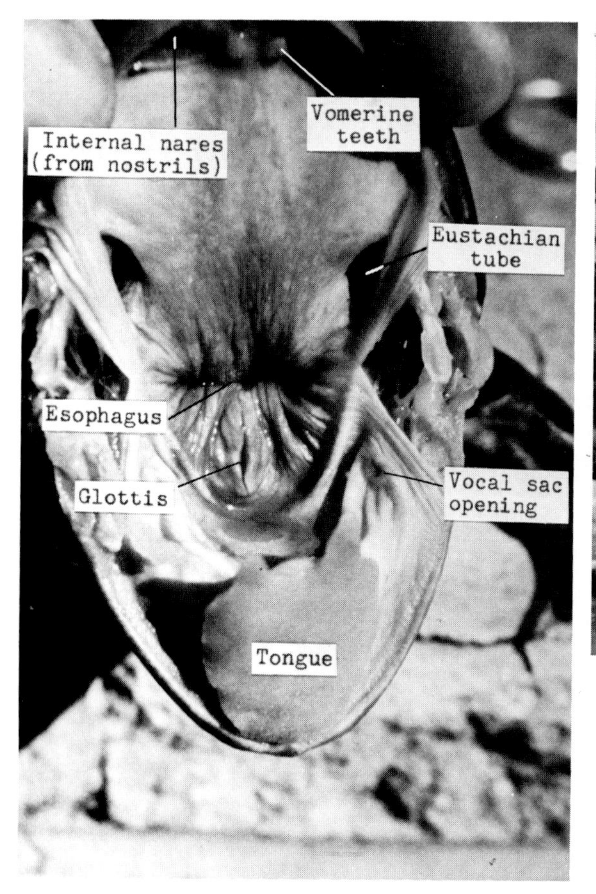

Fig. 80b Frog mouth interior anatomy x2.

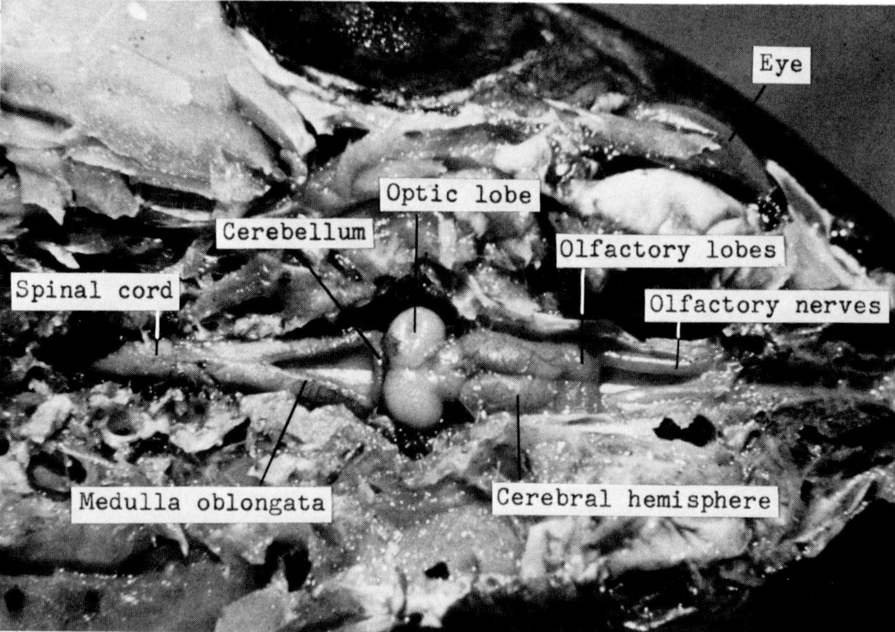

Fig. 80c Frog brain dissection x4.

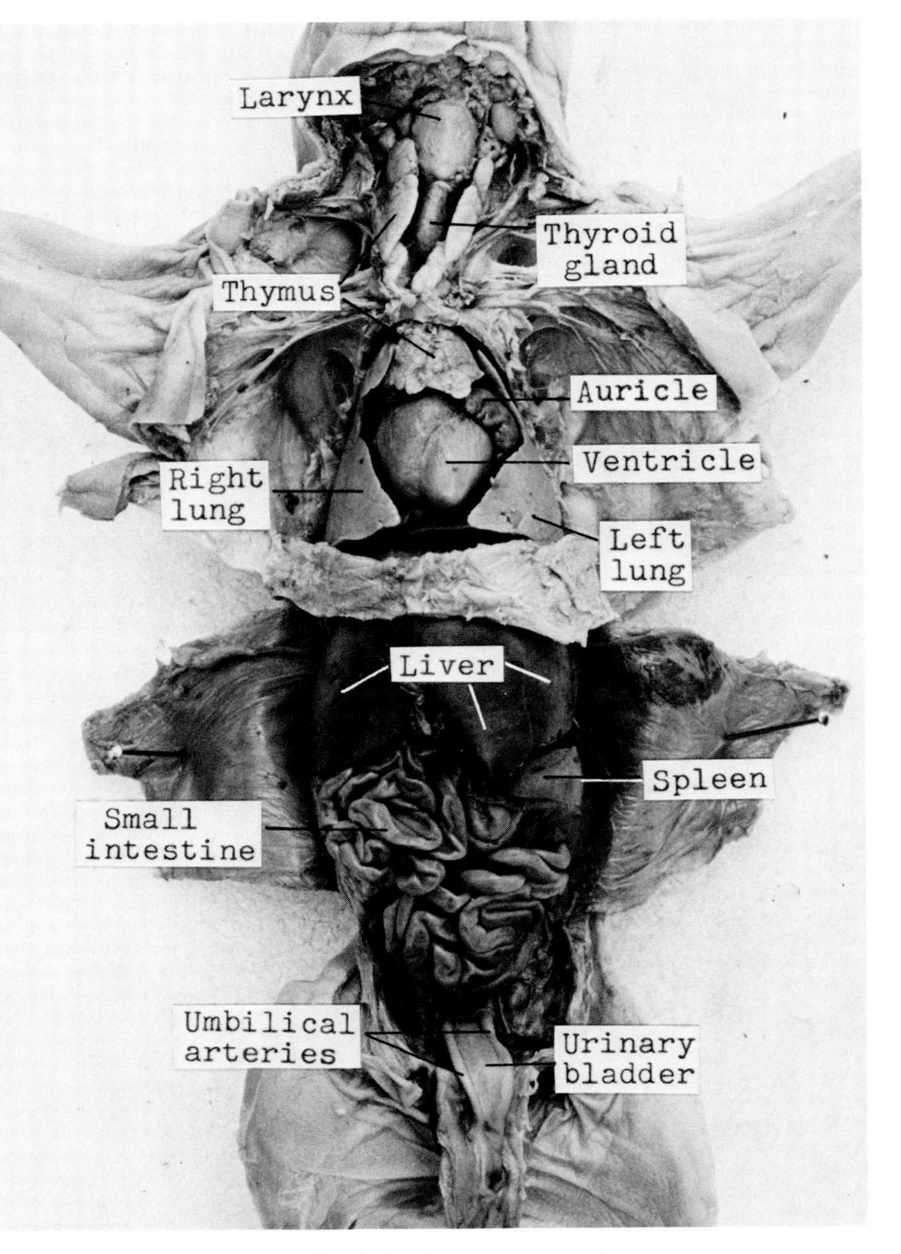

Fig. 81a Fetal pig dissection, ventral view.

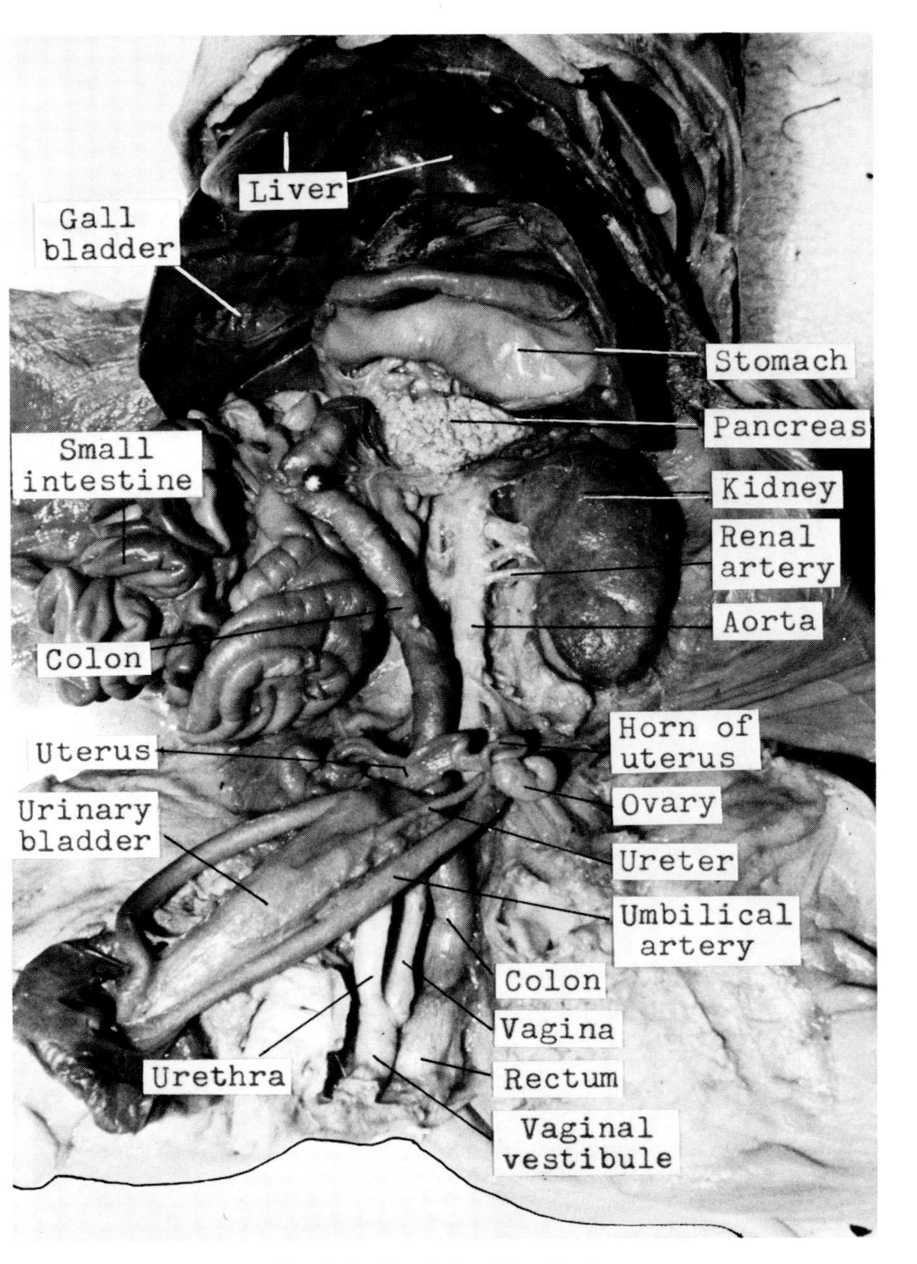

Fig. 81b Fetal pig abdominal organs.

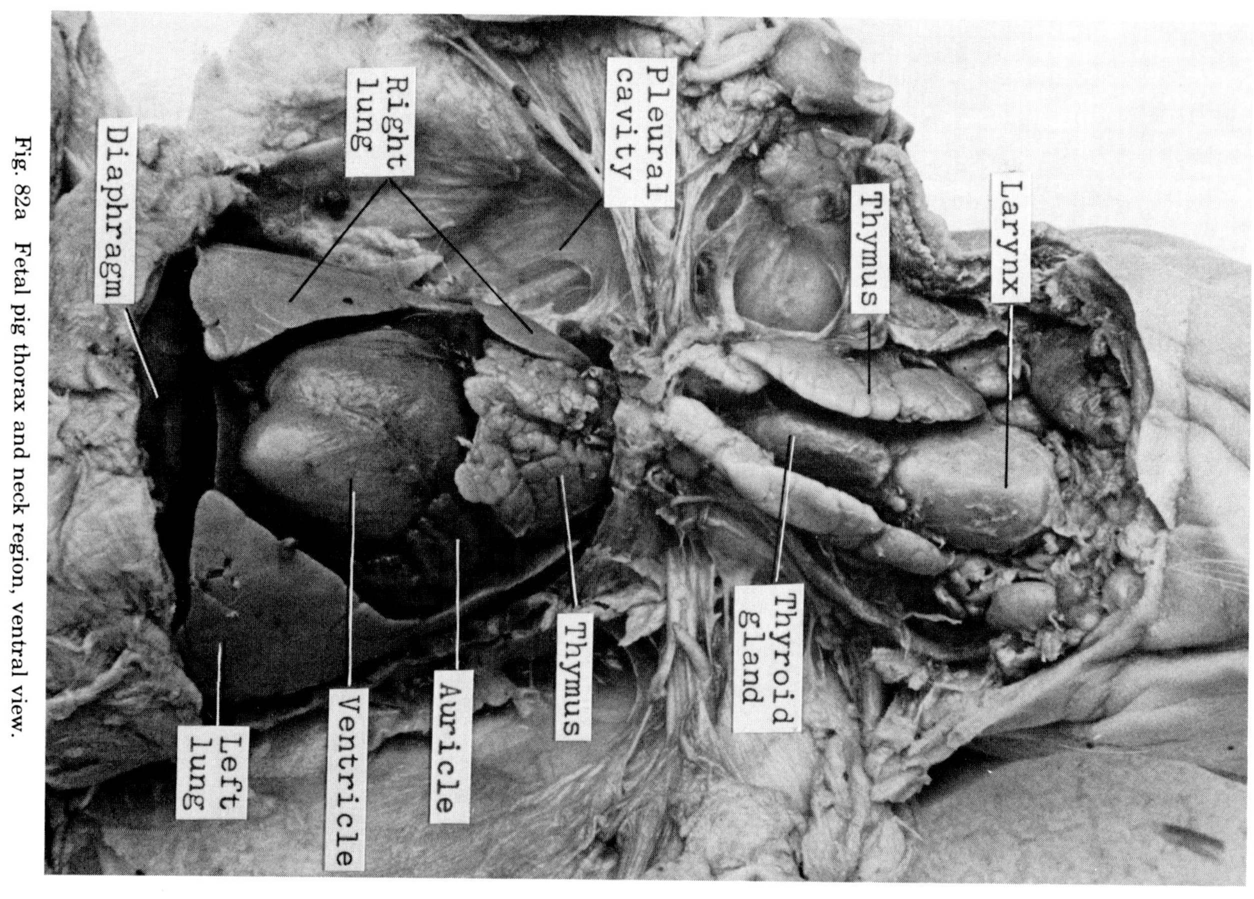

Fig. 82a Fetal pig thorax and neck region, ventral view.

Fig. 82b Fetal pig urinary system. (The digestive system has been removed.)

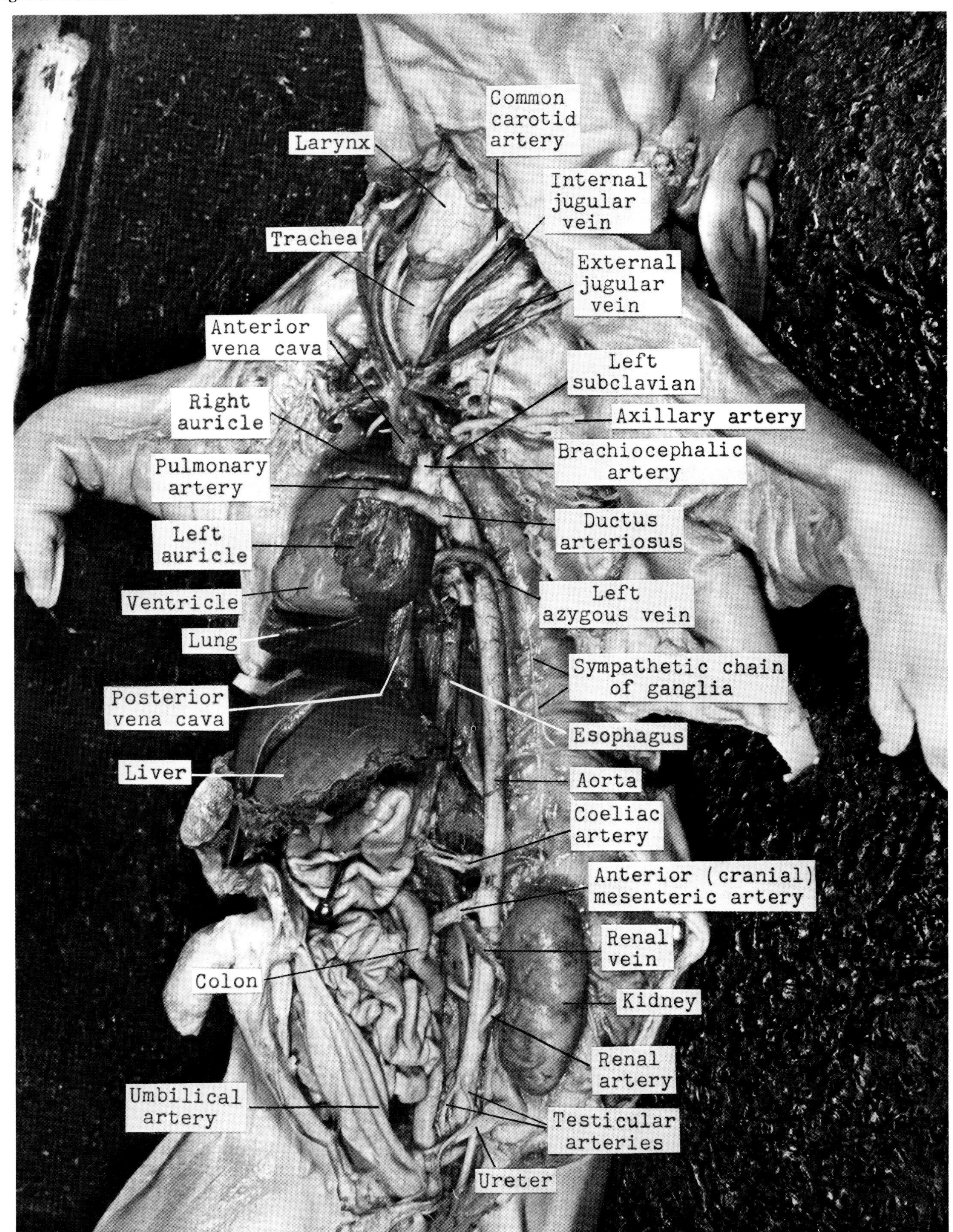

Fig. 83 Fetal pig, ventral view. (Dissection courtesy of Stephen Davenport.)

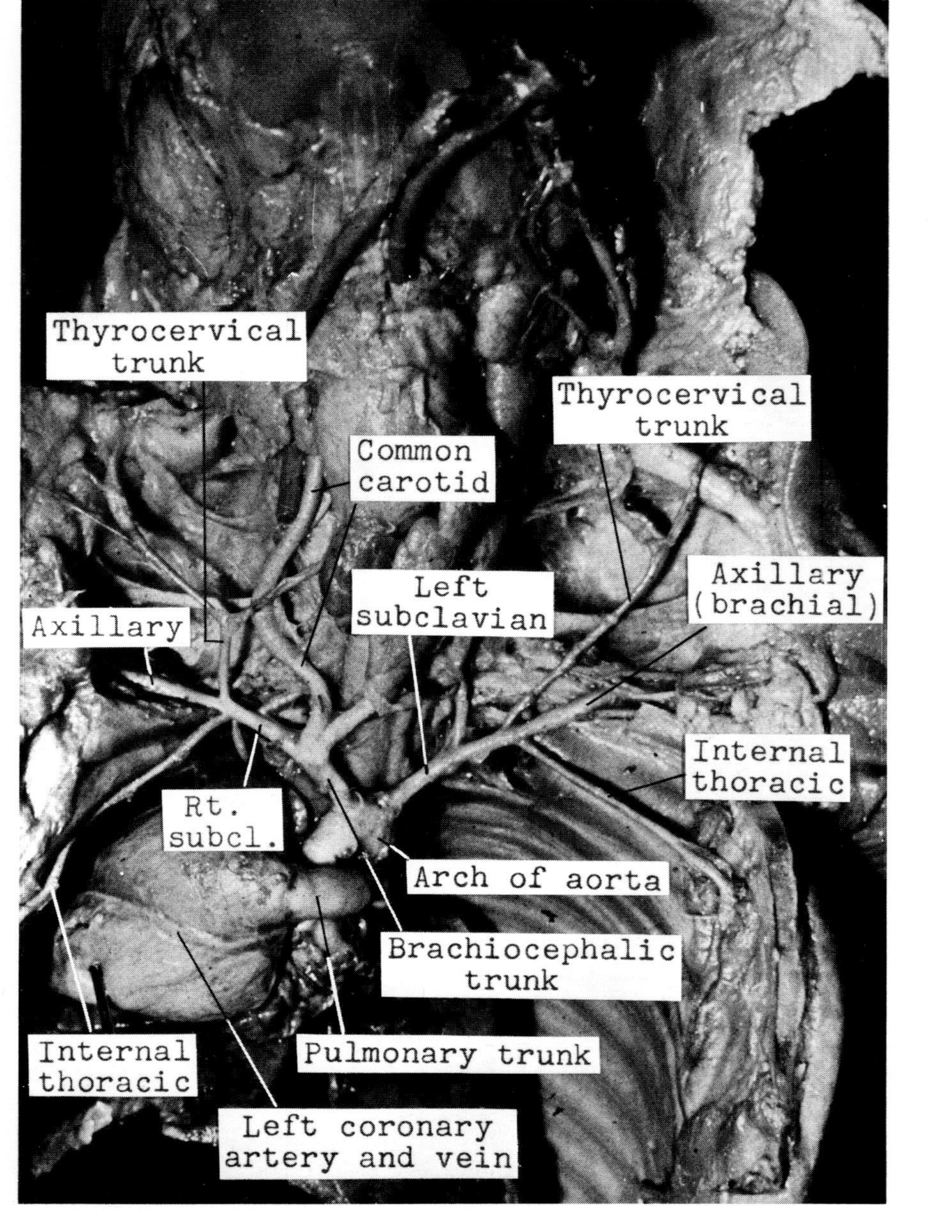

Fig. 84a Fetal pig. Arteries of the thoracic and neck regions. (Photo courtesy of Stephen Davenport.)

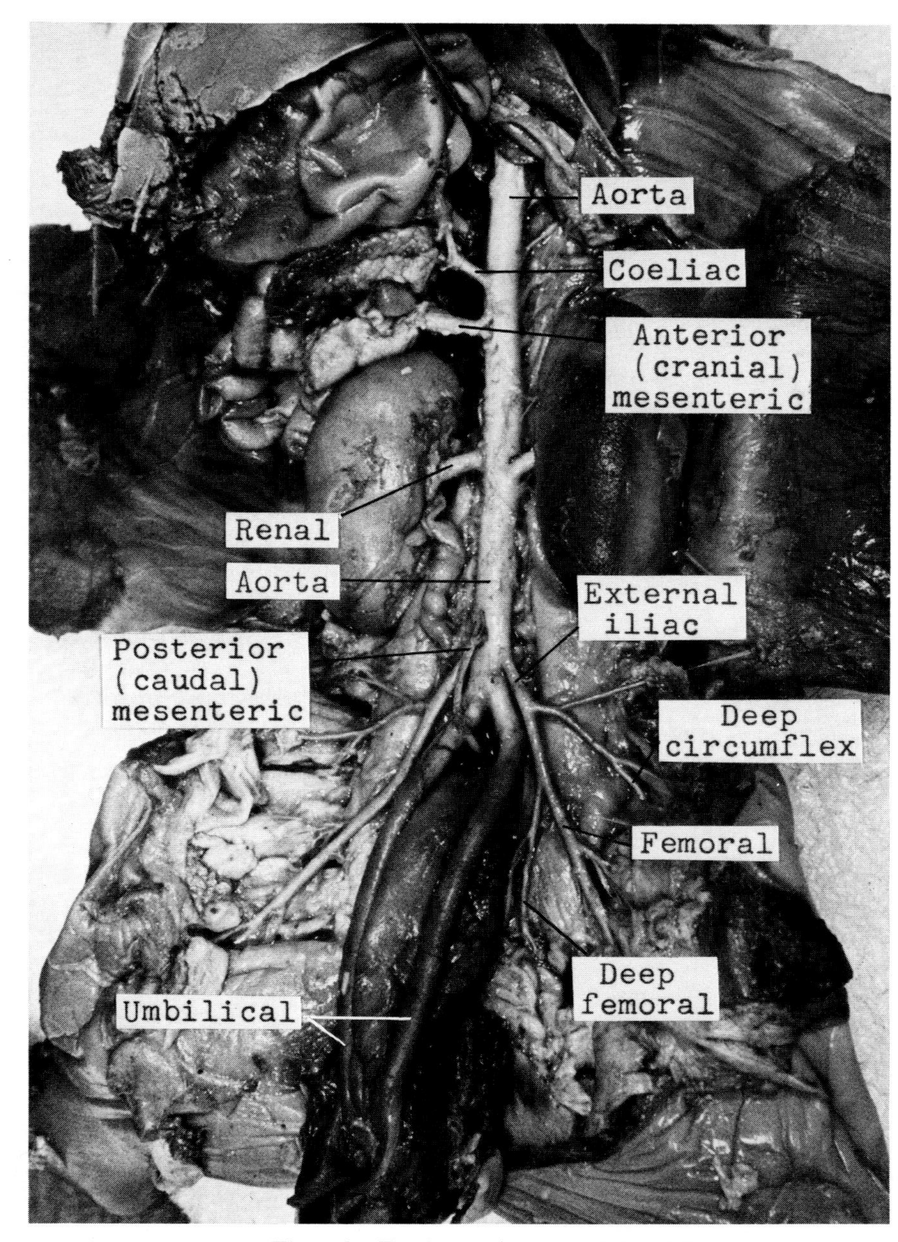

Fig. 84b Fetal pig. Arteries of the abdomen.

Common carotid artery
Internal jugular
External jugular
Cephalic
Internal thoracic
Axillary (brachial)
Anterior (cranial) vena cava
Left brachiocephalic
Posterior (caudal) vena cava

Fig. 85a Fetal pig. Veins of the thoracic and neck regions. (Photo courtesy of Stephen Davenport.)

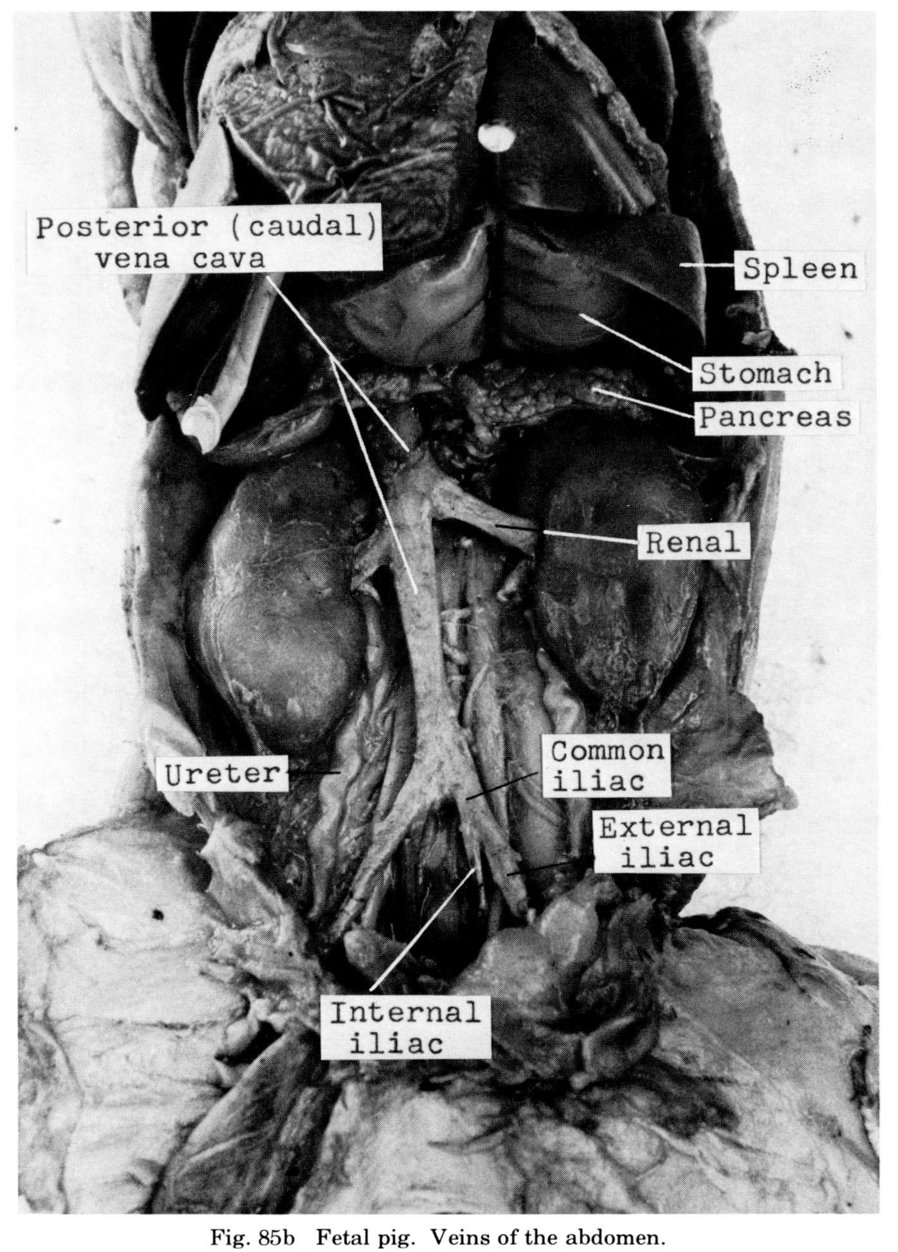

Fig. 85b Fetal pig. Veins of the abdomen.

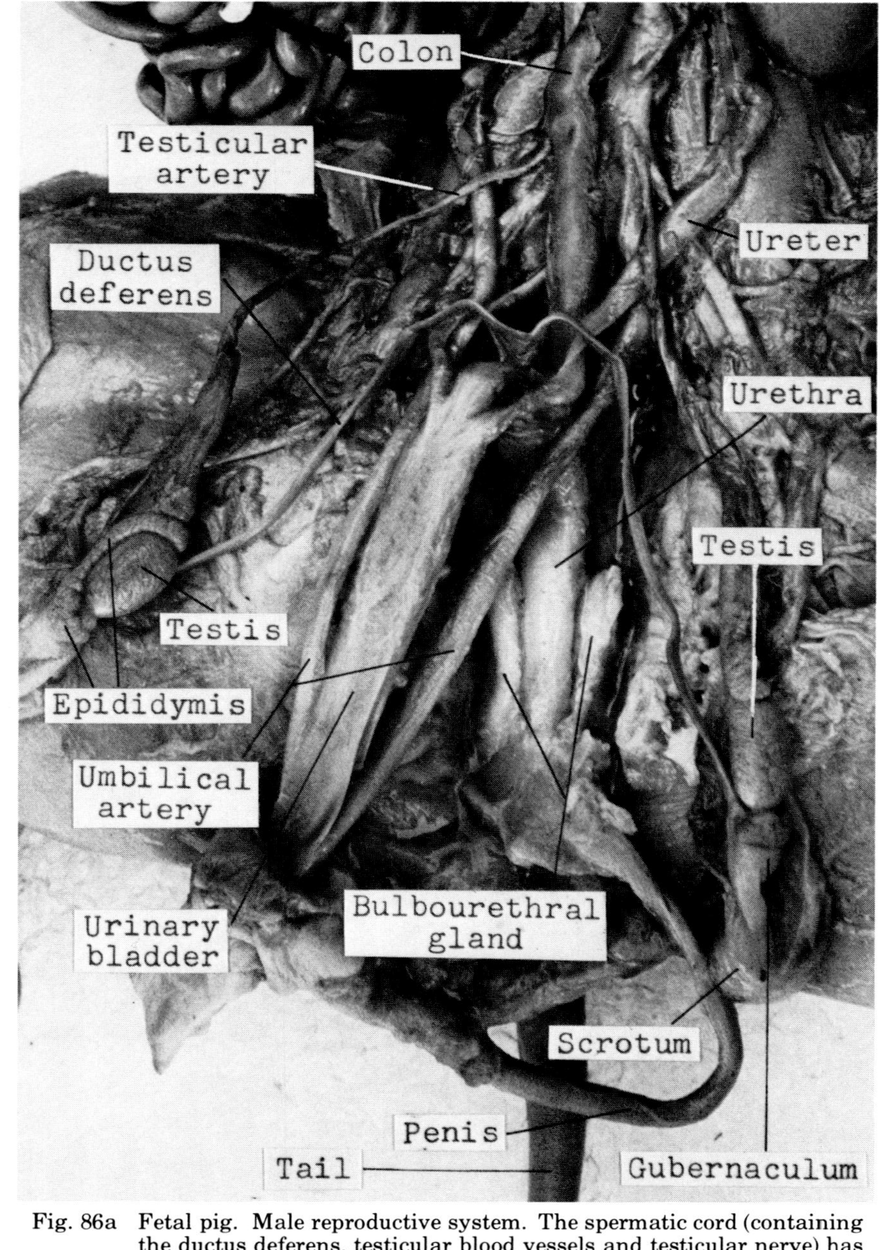

Fig. 86a Fetal pig. Male reproductive system. The spermatic cord (containing the ductus deferens, testicular blood vessels and testicular nerve) has been separated to show the individual components.

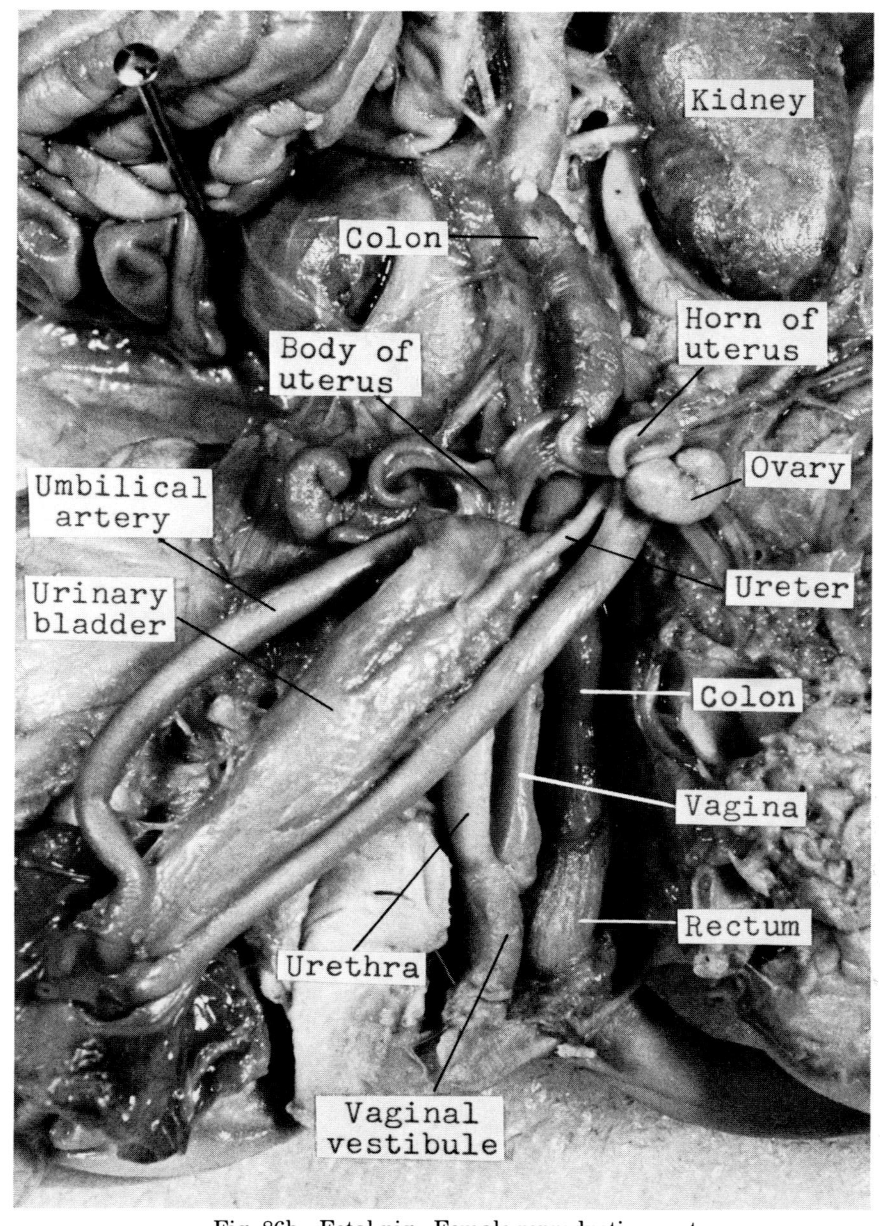

Fig. 86b Fetal pig. Female reproductive system.

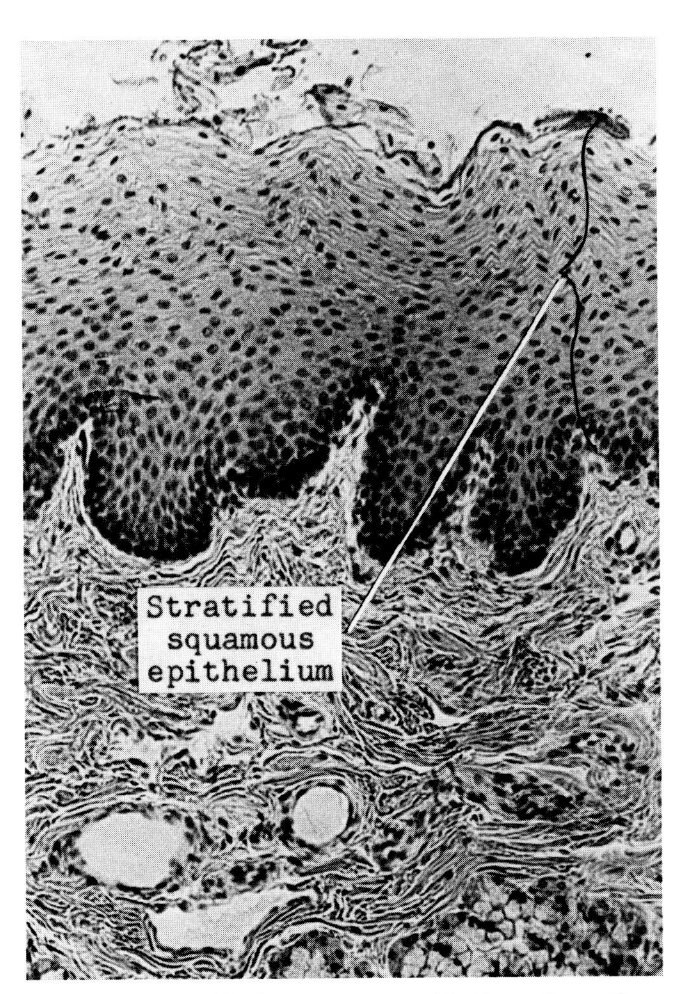

Fig. 87a Stratified squamous epithelium from the esophagus x.s. x100.

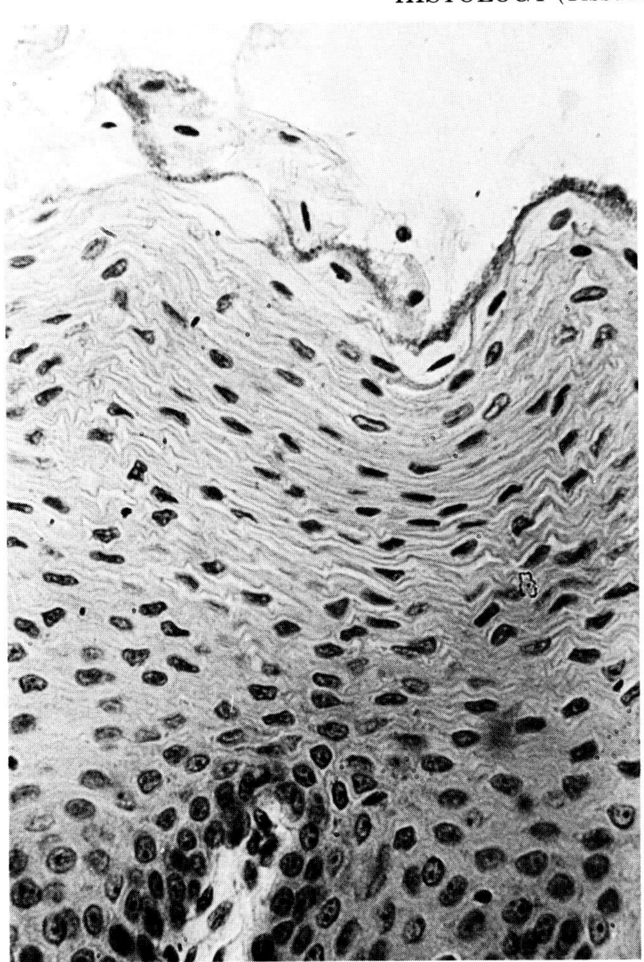

Fig. 87b Stratified squamous epithelium x.s. x430. Also see Fig. 1a & 1b for individual cells.

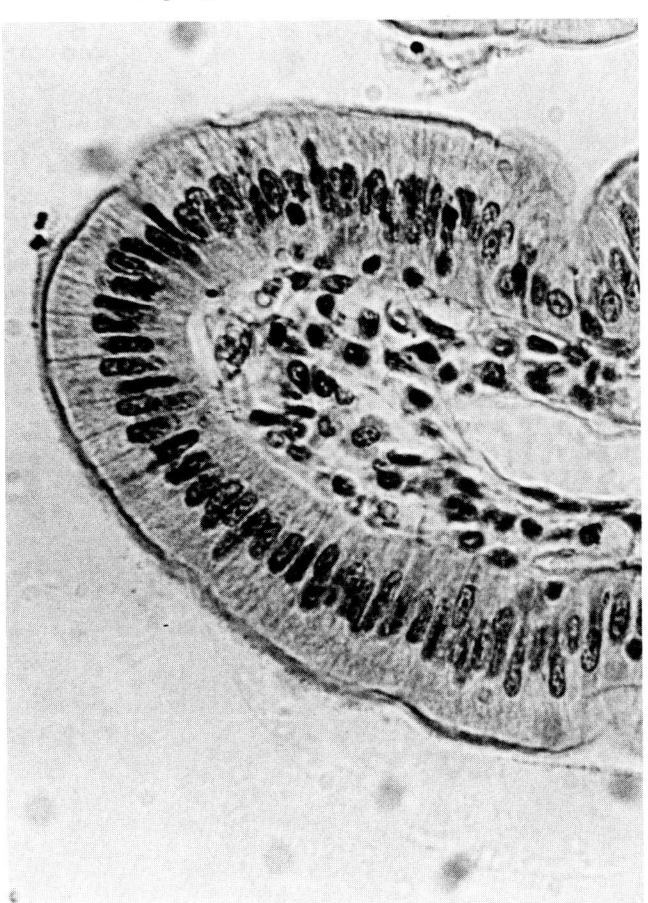

Fig. 87c Simple columnar epithelium from a villus of the small intestine x.s. x430.

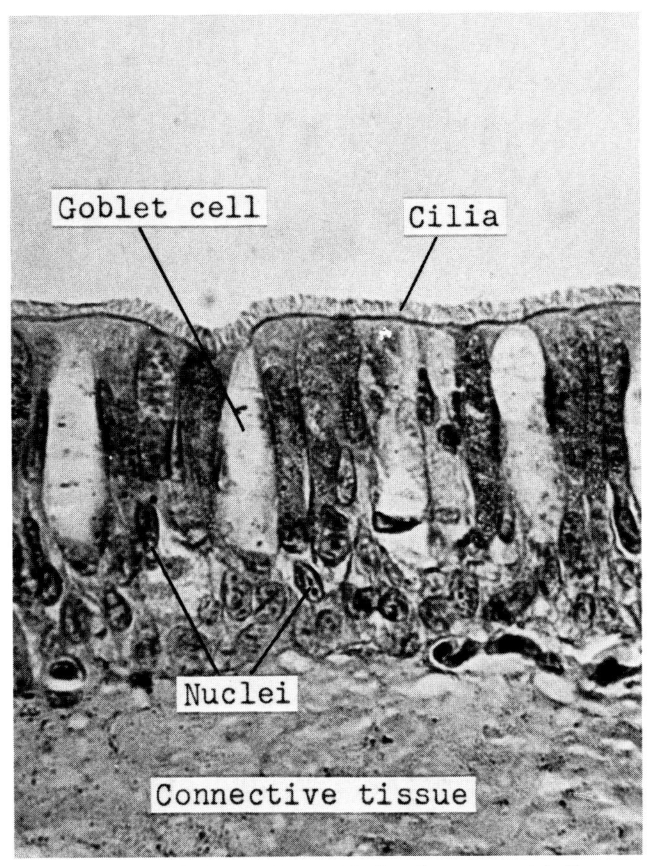

Fig. 87d Ciliated pseudostratified columnar epithelium from the trachea x.s. x430. Also see Fig. 104a for a discussion of function.

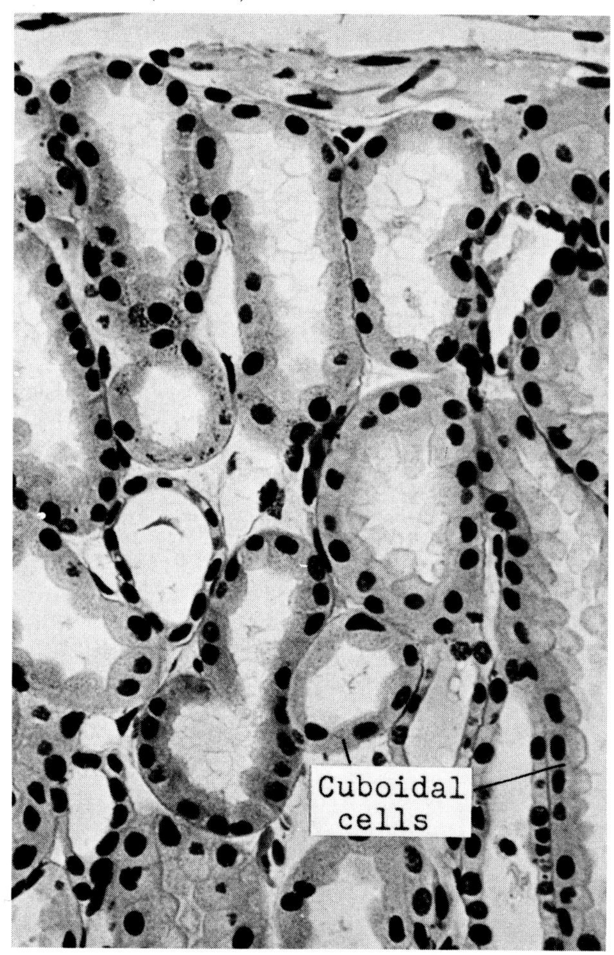

Fig. 88a Cuboidal epithelium lining the distal convoluted tubules of the kidney x.s. x430.

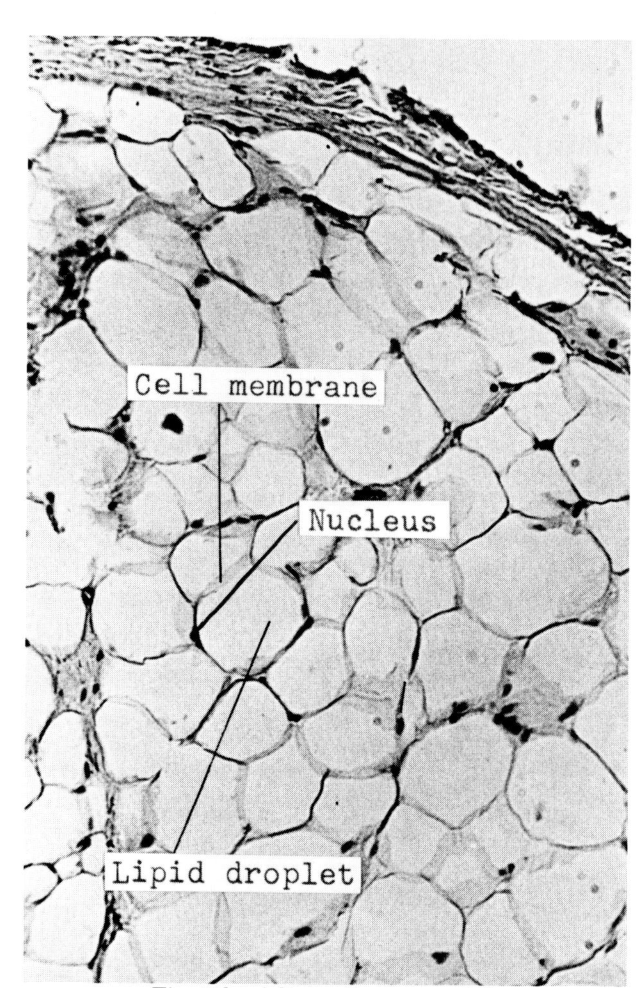

Fig. 88b Adipose tissue x.s. x100.

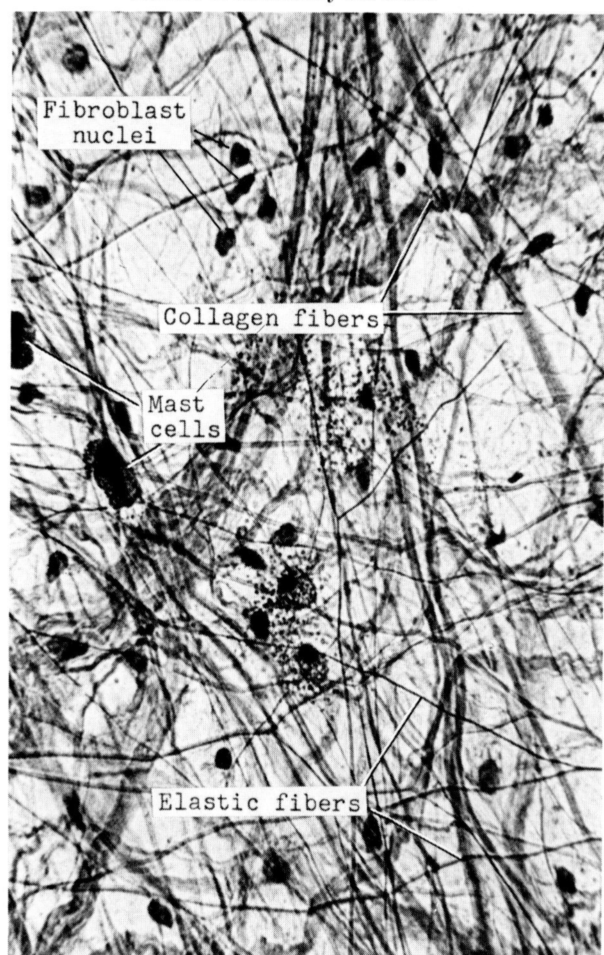

Fig. 88c Loose (areolar) connective tissue w.m. x430.

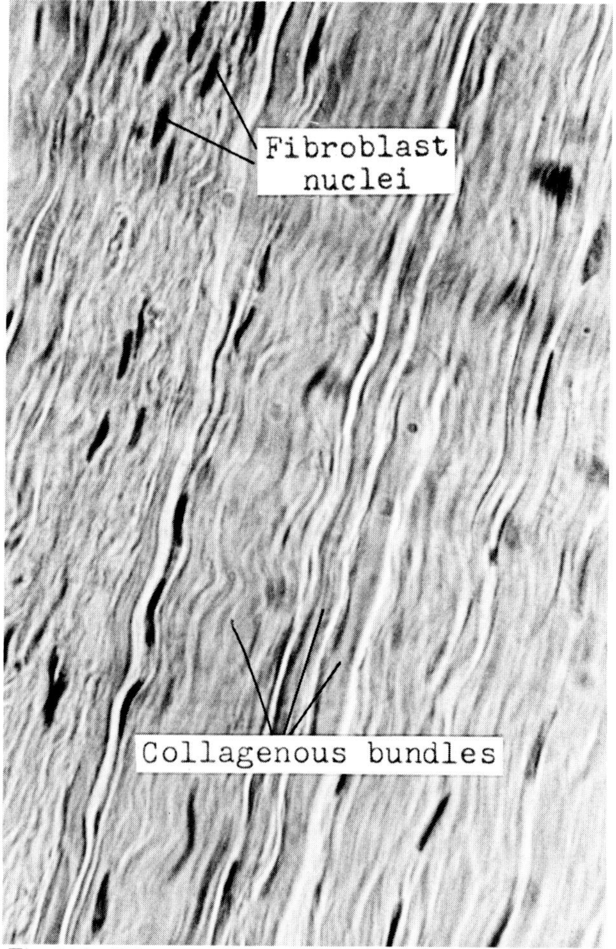

Fig. 88d Fibrous (dense) connective tissue from the tendon l.s. x430.

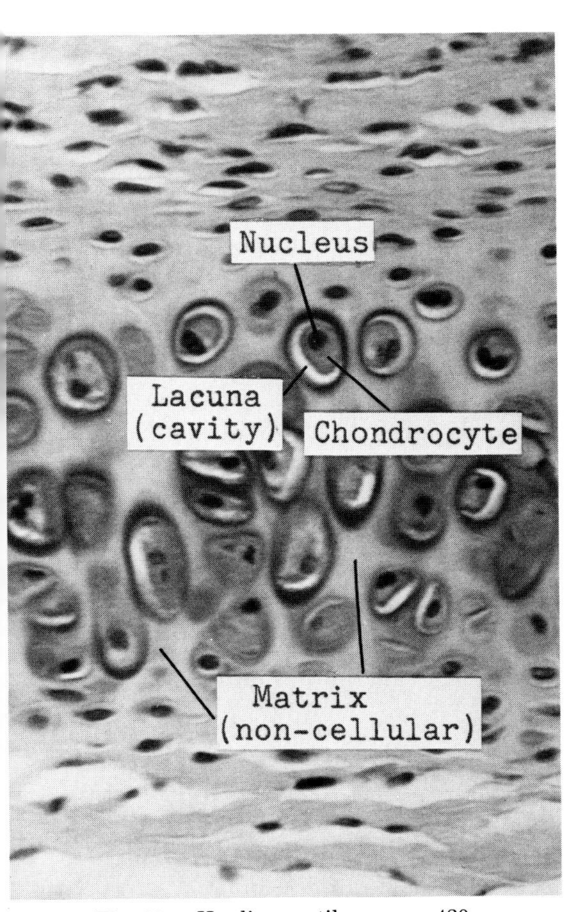

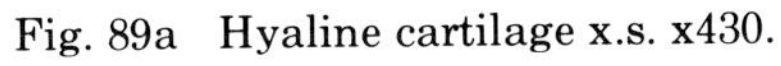
Fig. 89a Hyaline cartilage x.s. x430.

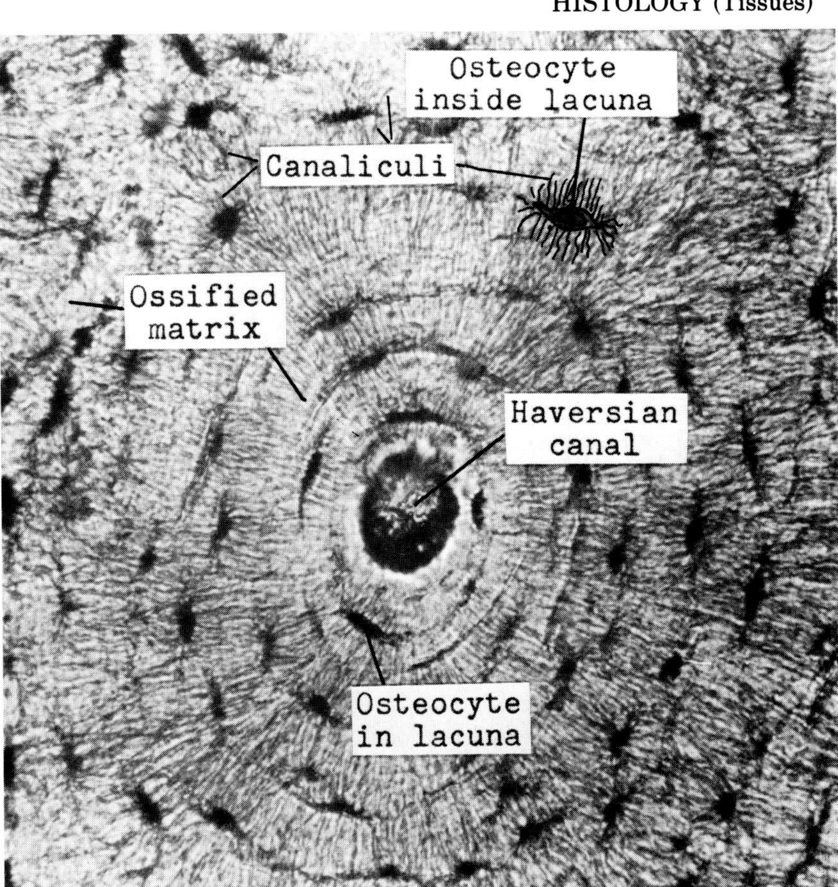

Fig. 89b Haversian System (Osteon) in compact bone x.s. x430.

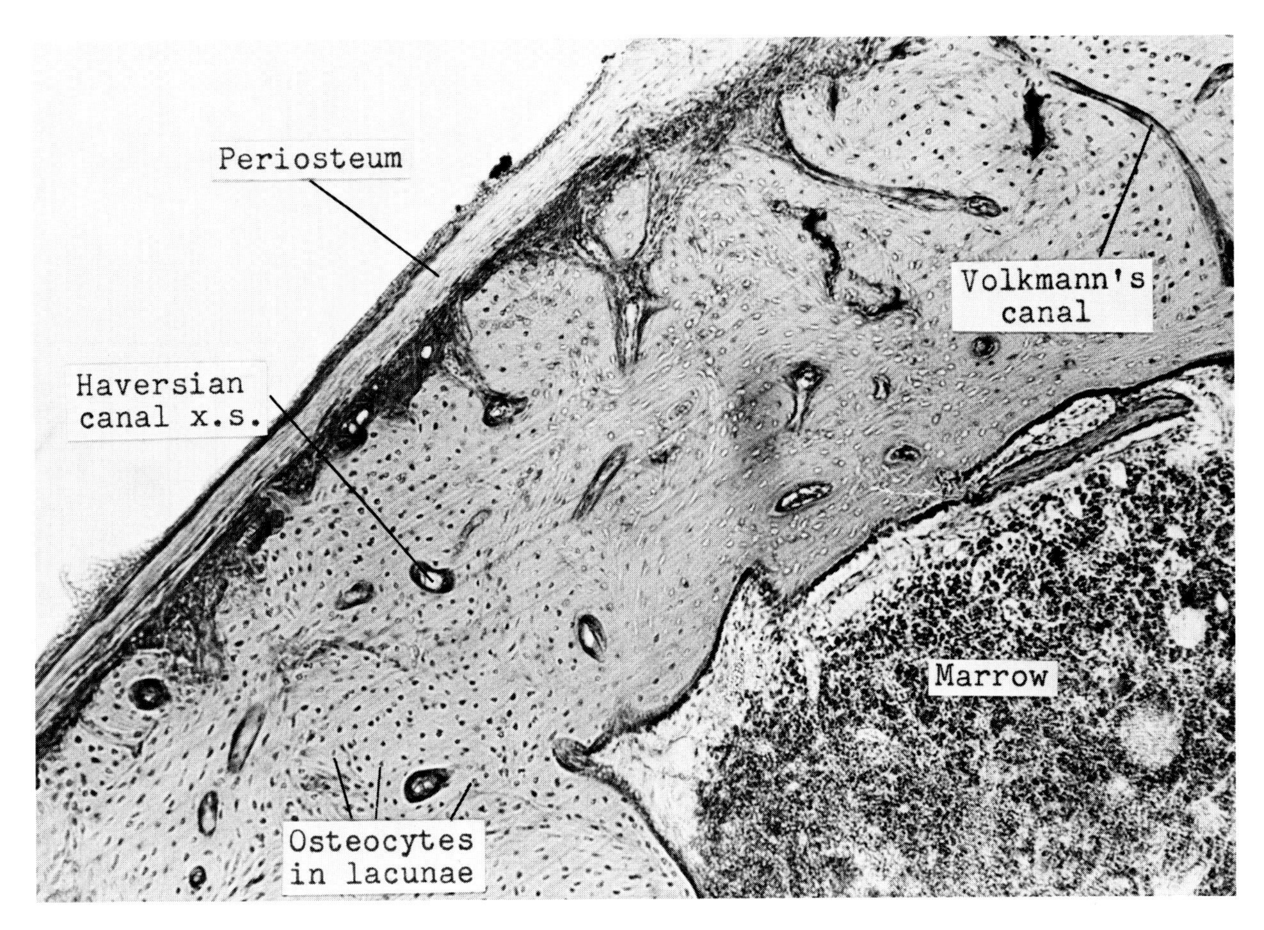

Fig. 89c Bone x.s. x100.

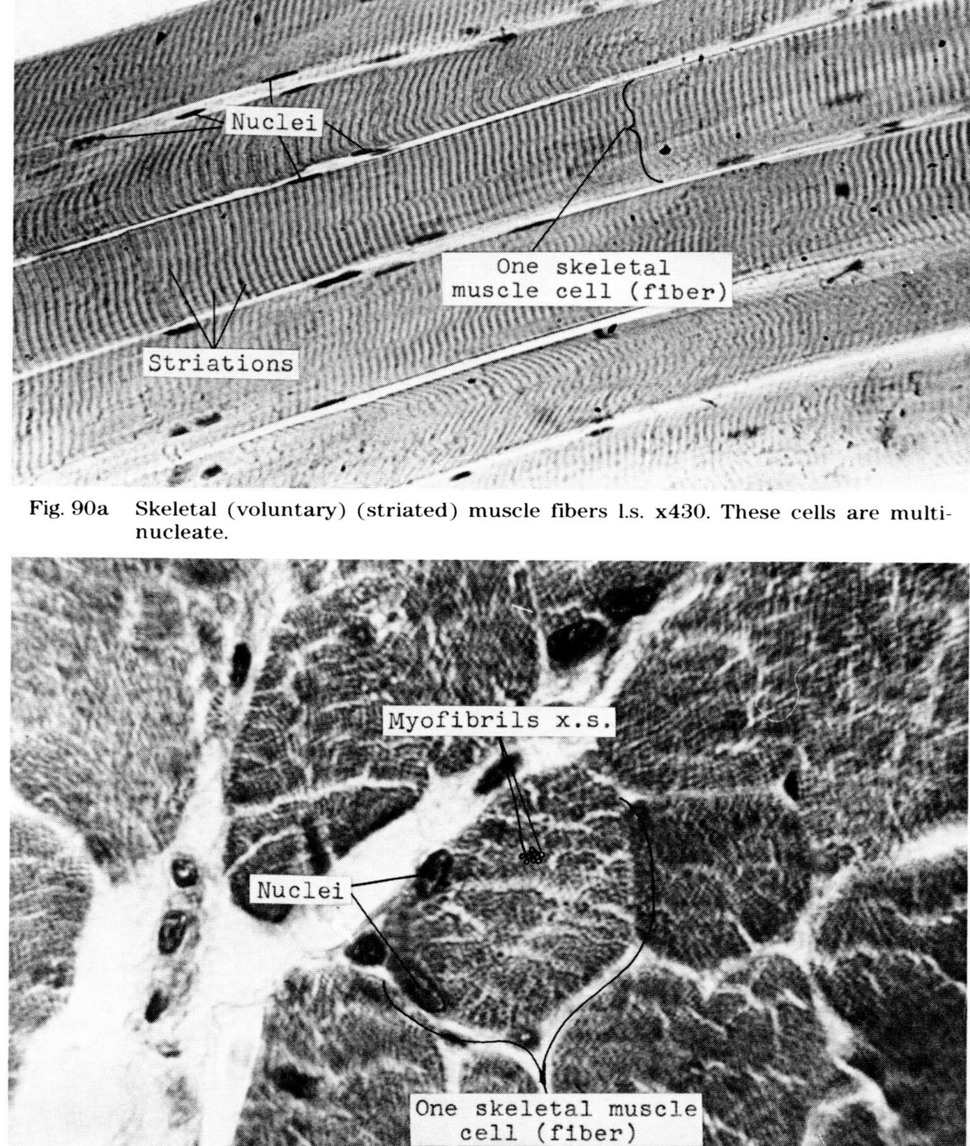

Fig. 90a Skeletal (voluntary) (striated) muscle fibers l.s. x430. These cells are multinucleate.

Fig. 90c Skeletal muscle fibers (cells) x.s. x430. Notice how closely each skeletal muscle cell resembles a coaxial cable of myofibrils. Compare with Fib. 90b.

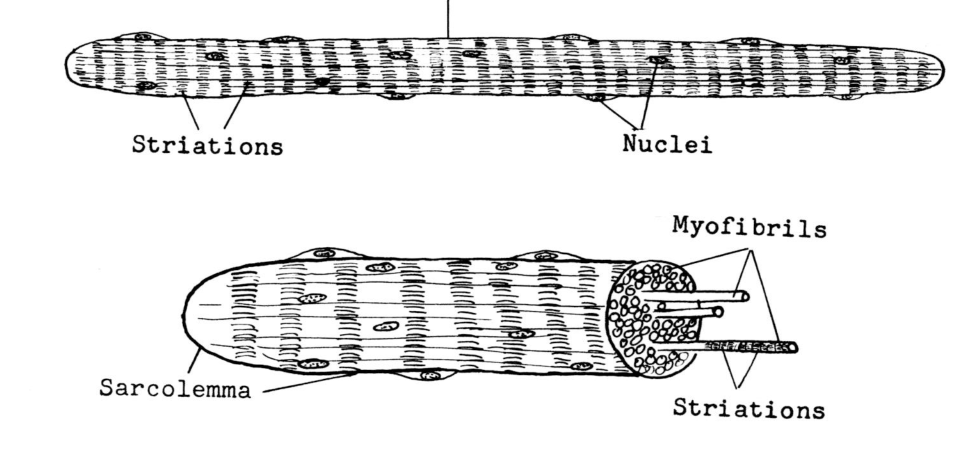

Fig. 90b Diagrams of skeletal muscle cells. The lower cell has been cut in cross section to show features of the cell's interior. The sarcolemma is the cell membrane.

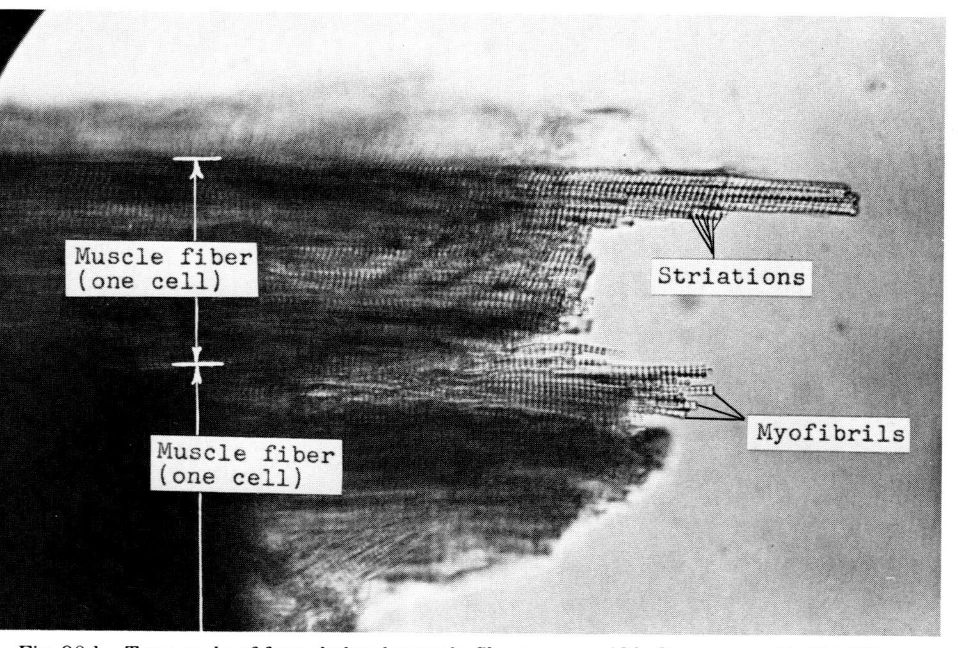

Fig. 90d Torn ends of frog skeletal muscle fibers w.m. x430. Compare with Fig. 90b.

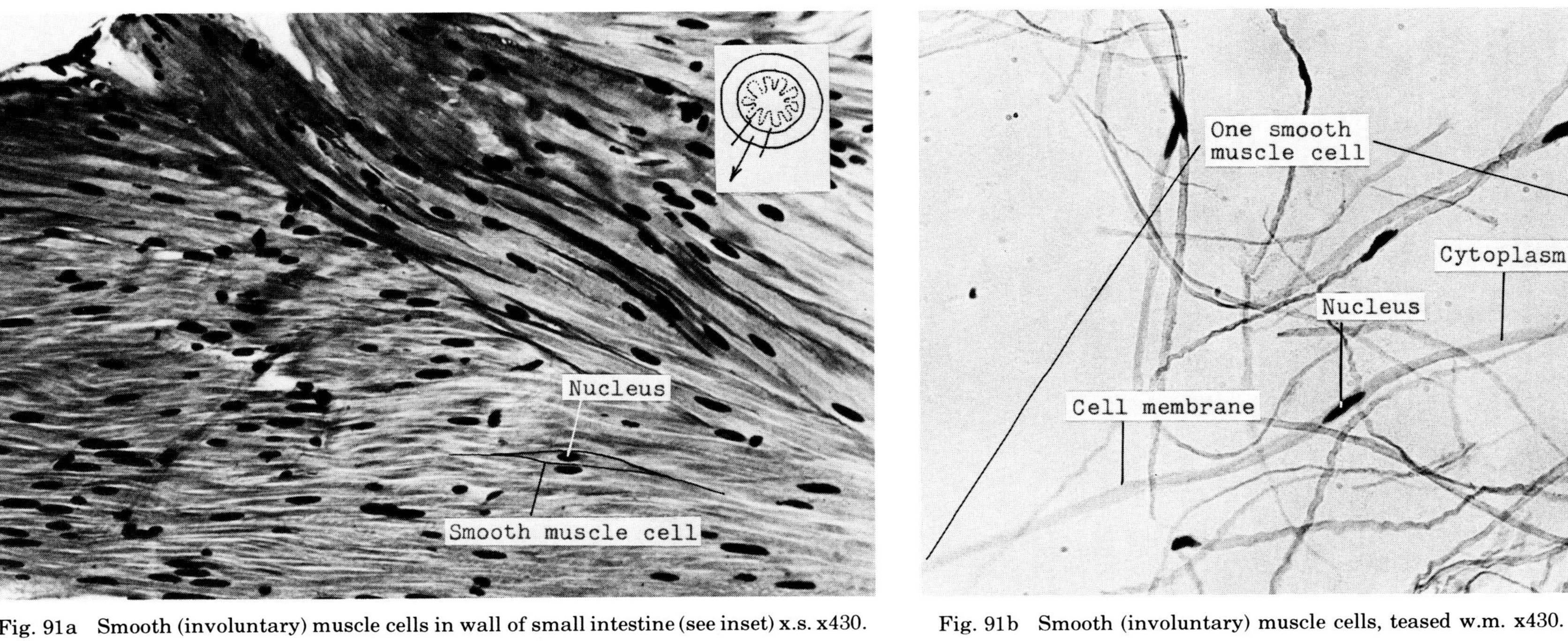

Fig. 91a Smooth (involuntary) muscle cells in wall of small intestine (see inset) x.s. x430.

Fig. 91b Smooth (involuntary) muscle cells, teased w.m. x430. Compare with Fig. 91a.

Fig. 91c Cardiac muscle tissue l.s. x430. The intercalated discs have been shown by the electron microscope to be the abutting cell membranes at the ends of two different cells.

Fig. 91d Neurons from the spinal cord of an ox w.m. x100.

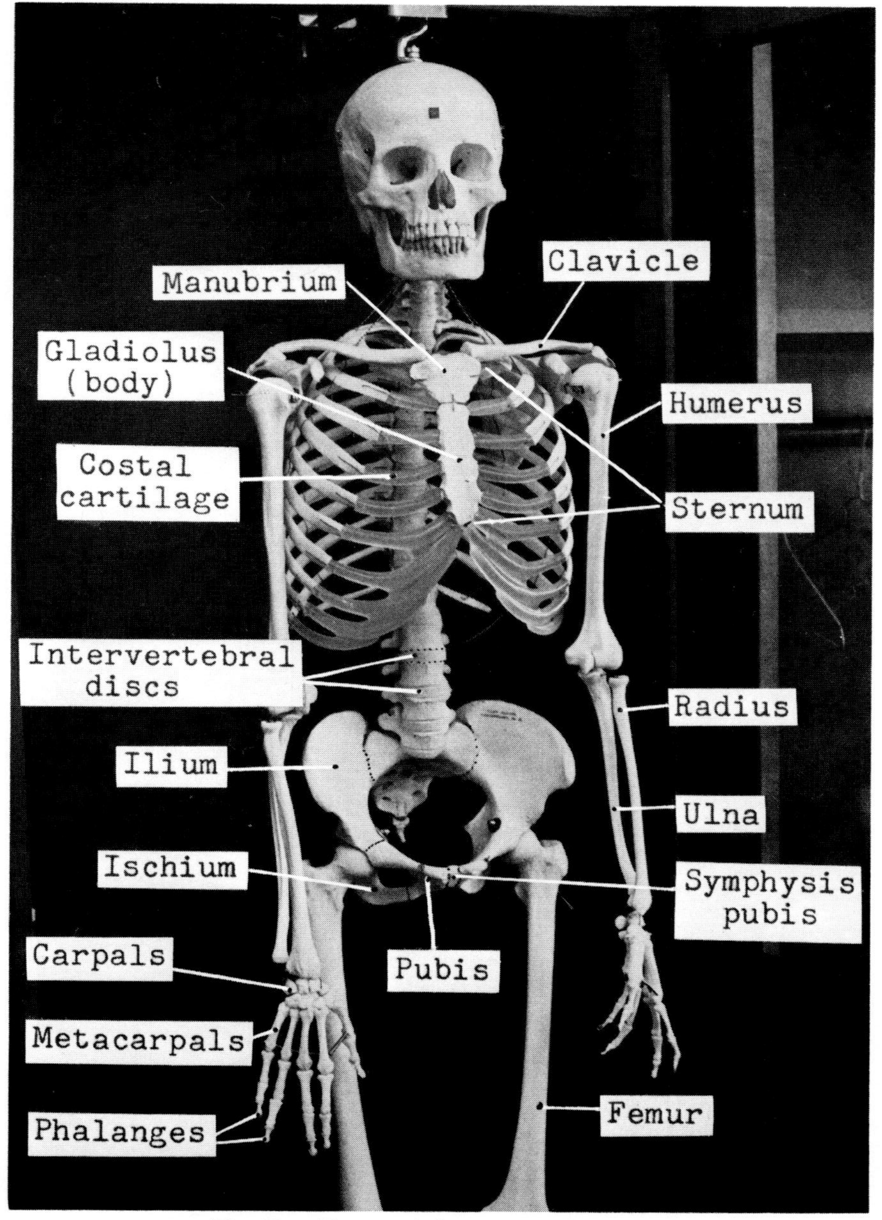

Fig. 92a Human skeleton, ventral view.

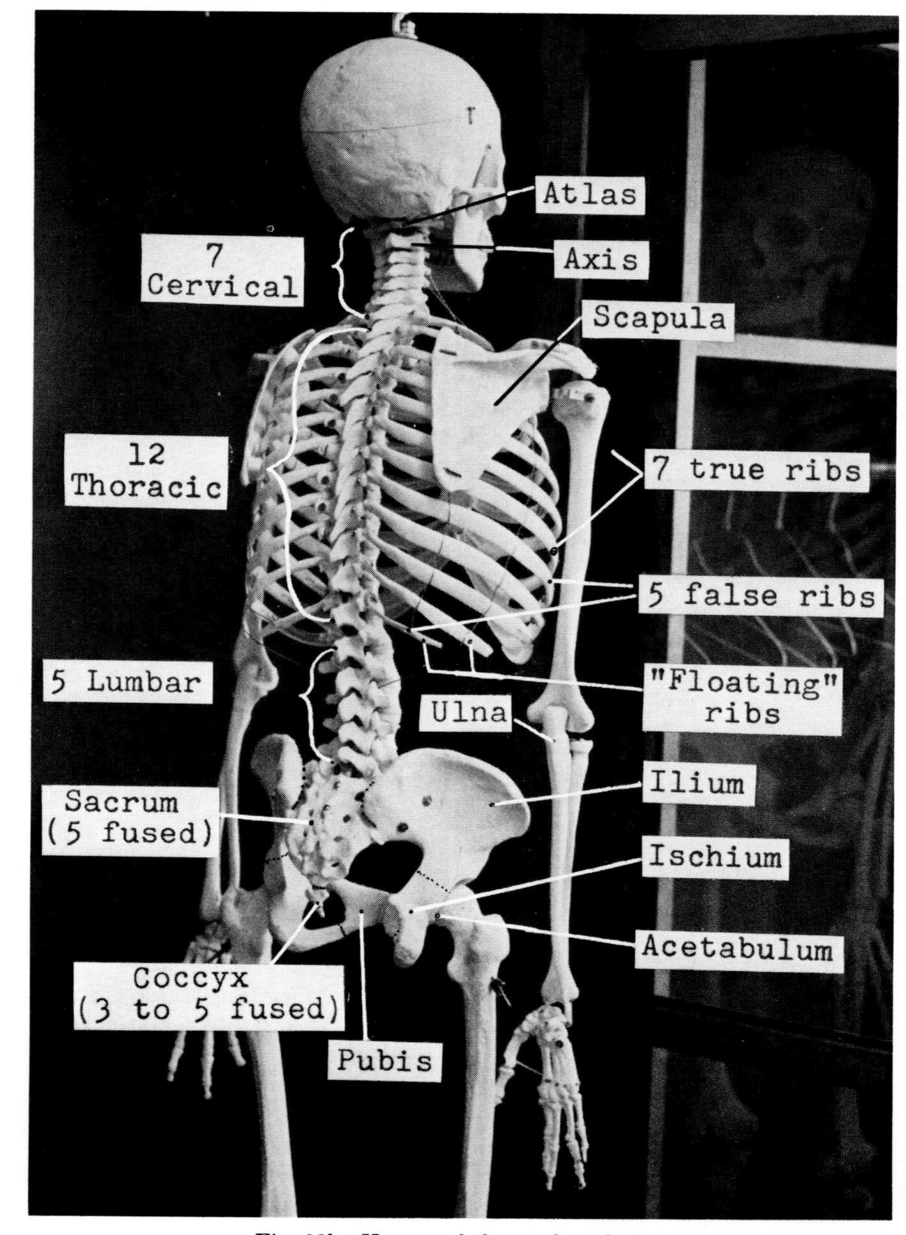

Fig. 92b Human skeleton, dorsal view.

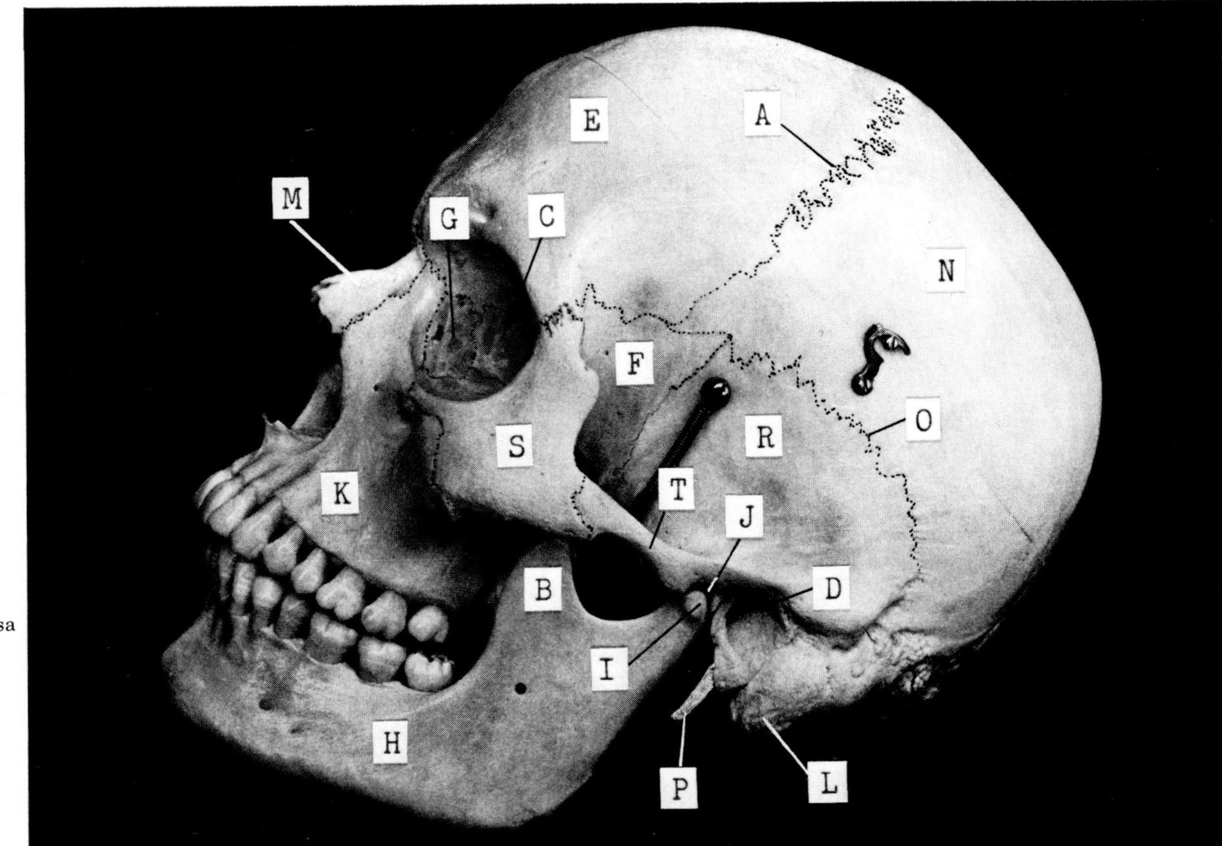

Fig. 93a Human skull.

- Coronal suture
- Coronoid process
- Ethmoid
- External acoustic meatus
- Frontal
- Sphenoid
- Lacrimal
- Mandible
- Mandibular condyle
- Mandibular (Glenoid) fossa
- Maxilla
- Mastoid process
- Nasal
- Parietal
- Squamosal suture
- Styloid process
- Superior orbital fissure
- Temporal
- Zygomatic (Malar)
- Zygomatic process of temporal

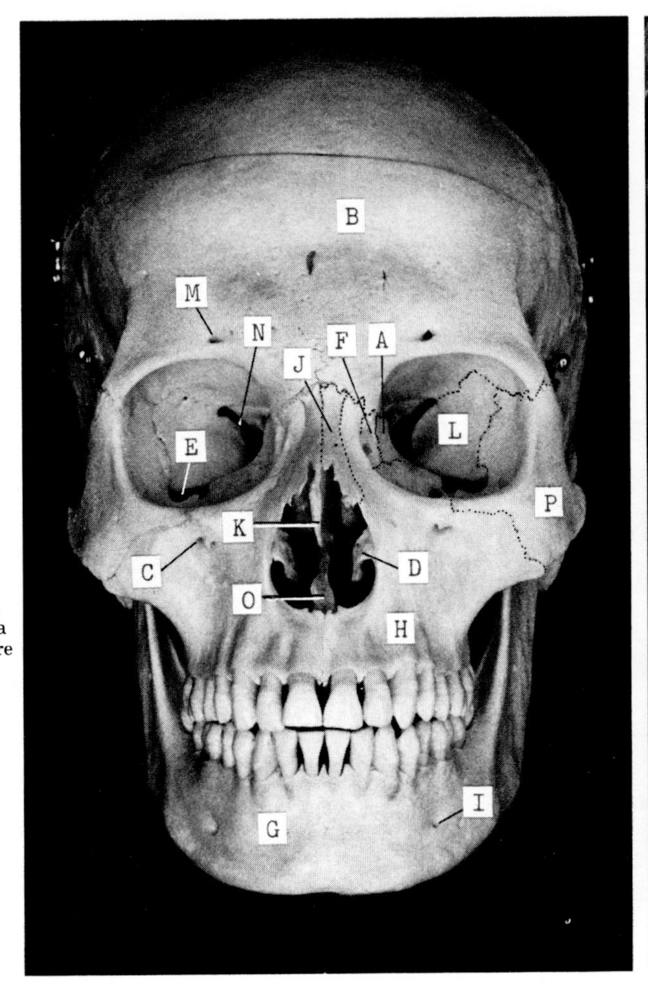

Fig. 93b Human skull.

A - Ethmoid
B - Frontal
C - Infraorbital foramen
D - Inferior nasal concha
E - Inferior orbital fissure
F - Lacrimal
G - Mandible
H - Maxilla
I - Mental foramen
J - Nasal
K - Perpendicular plate of Ethmoid
L - Orbital surface of Sphenoid
M - Supraorbital foramen
N - Superior orbital fissure
O - Vomer
P - Zygomatic (malar)

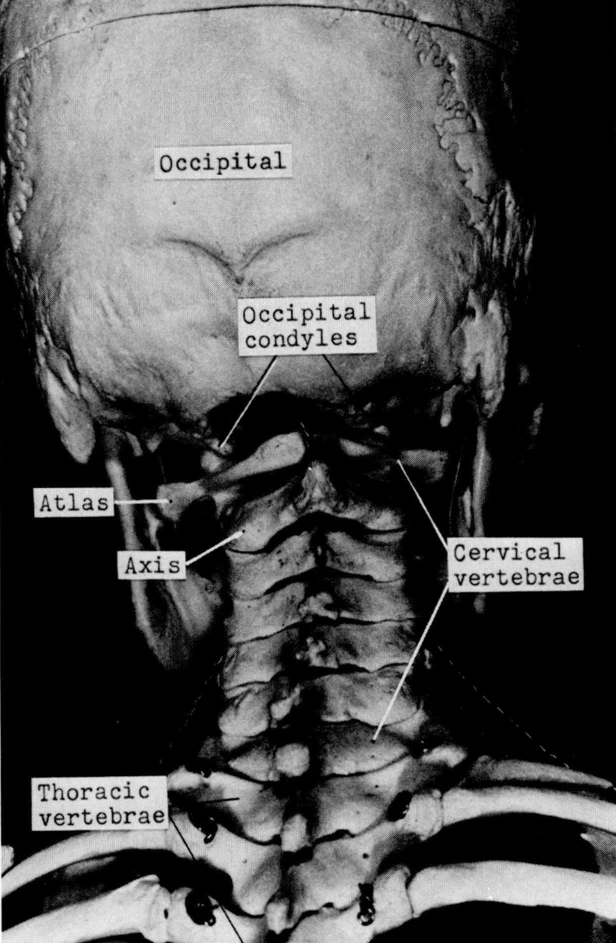

Fig. 93c Cervical vertebrae.

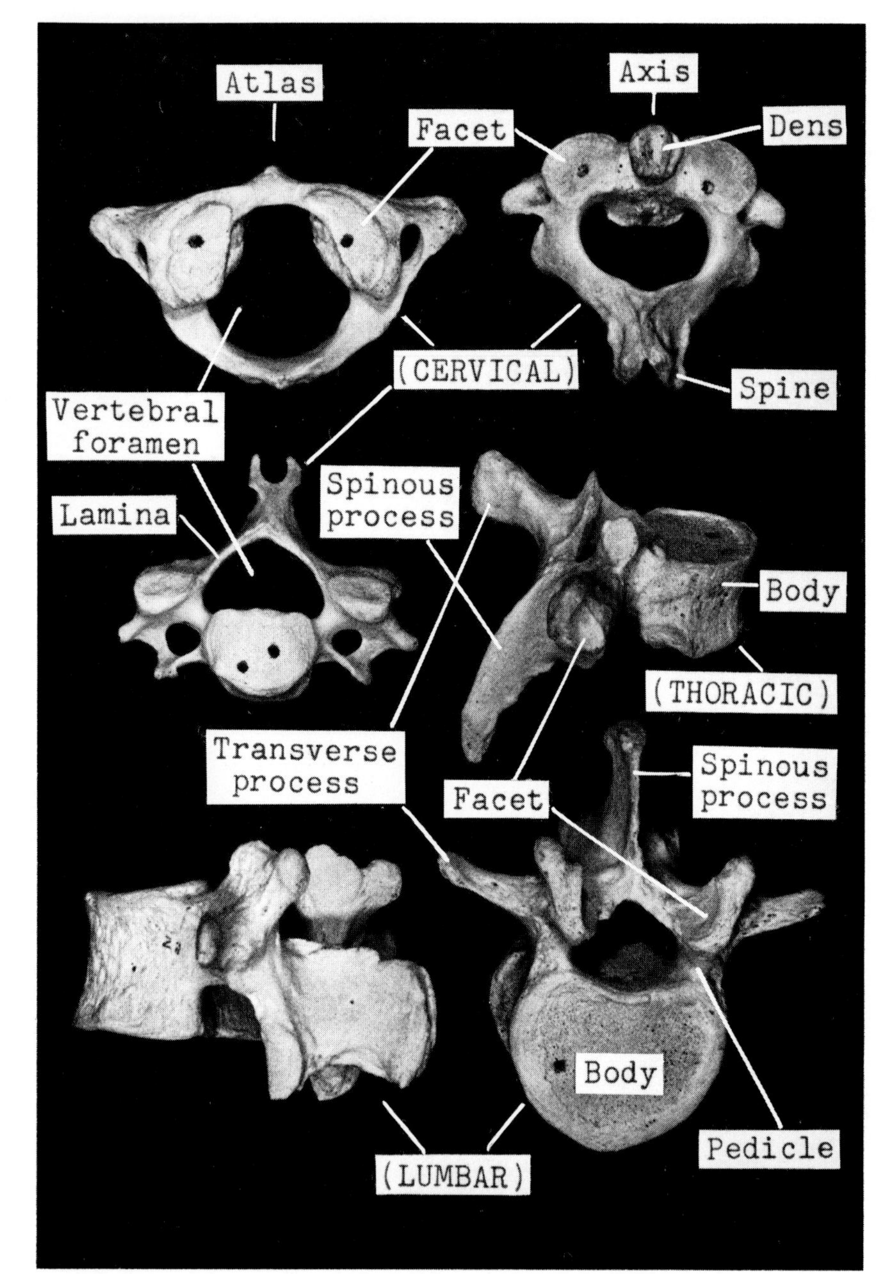

Fig. 94a Human vertebrae, three types.

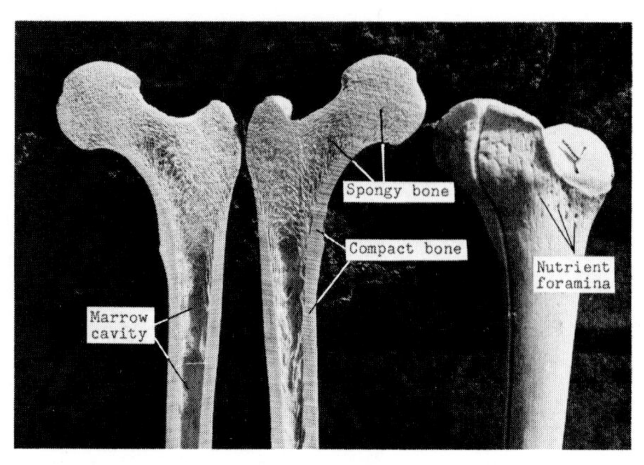

Fig. 94b Sectioned human femur.

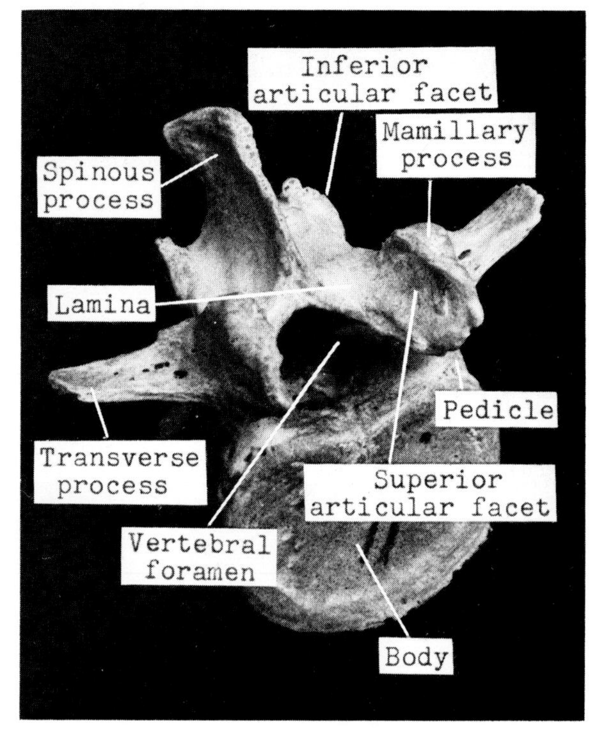

Fig. 94c Lumbar vertebra, superior surface.

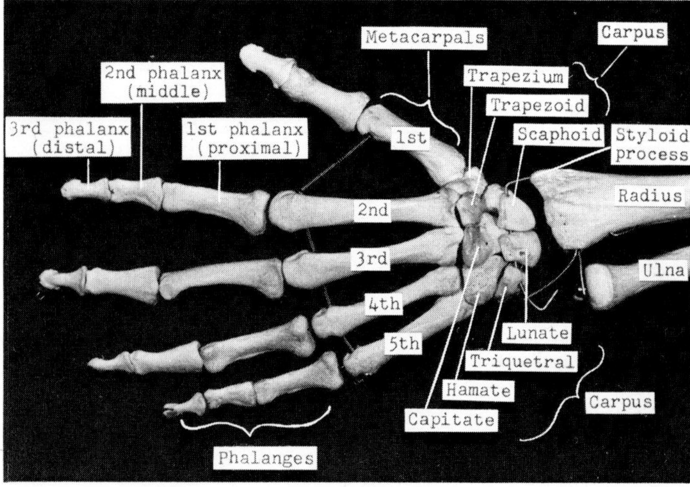

Fig. 94b Left hand, dorsal aspect.

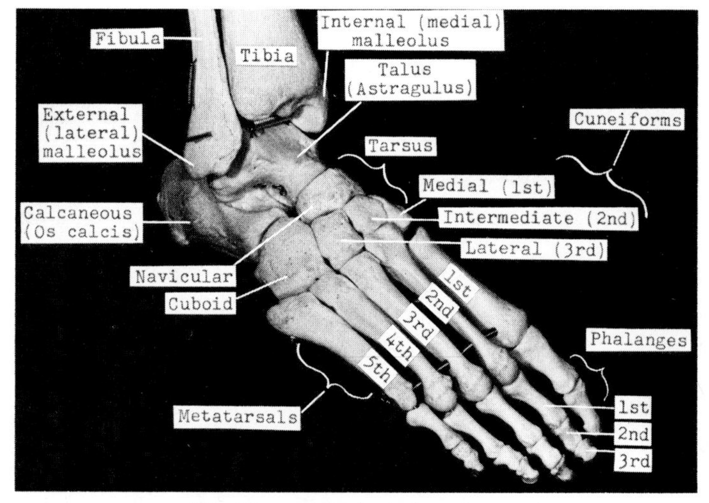

Fig. 94e Right foot, dorsal aspect.

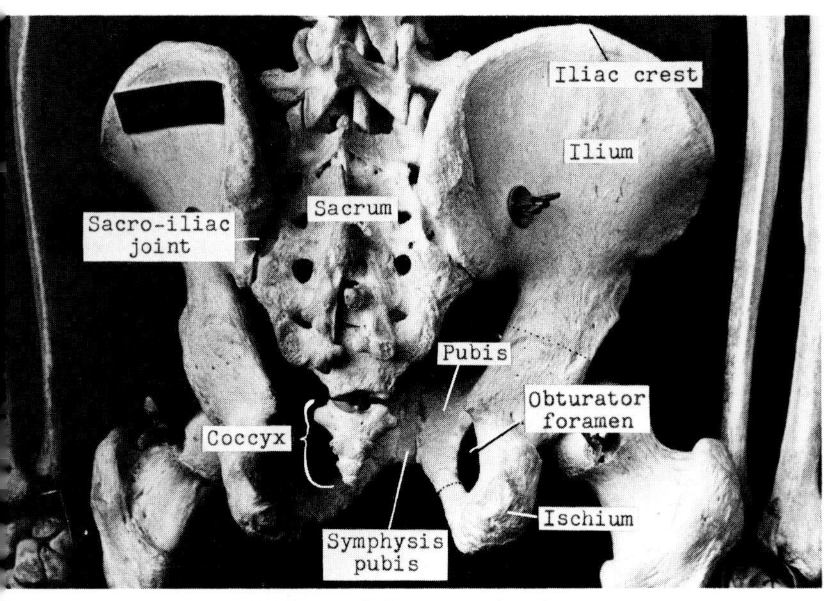

Fig. 95a Pelvis, posterior view.

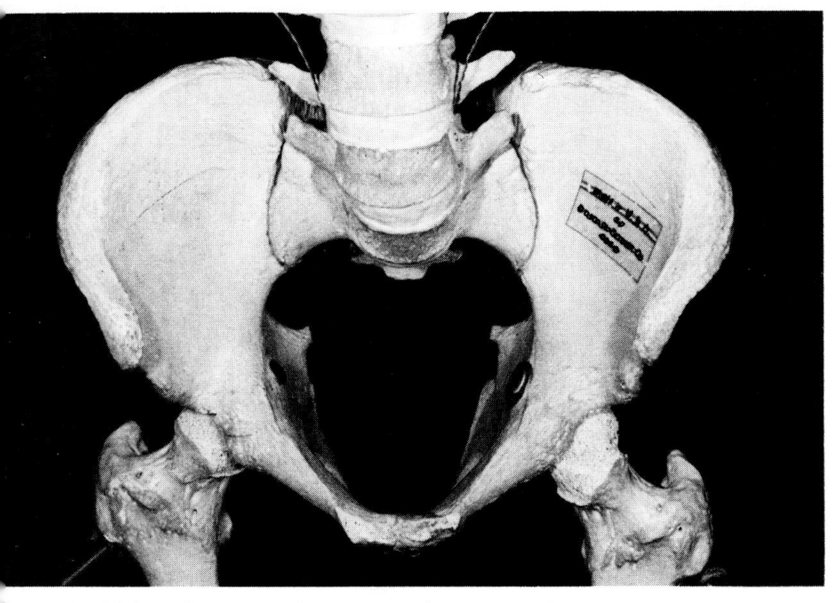

Fig. 95b Male pelvis, anterior view. (Coccyx accidentally missing on this specimen.)

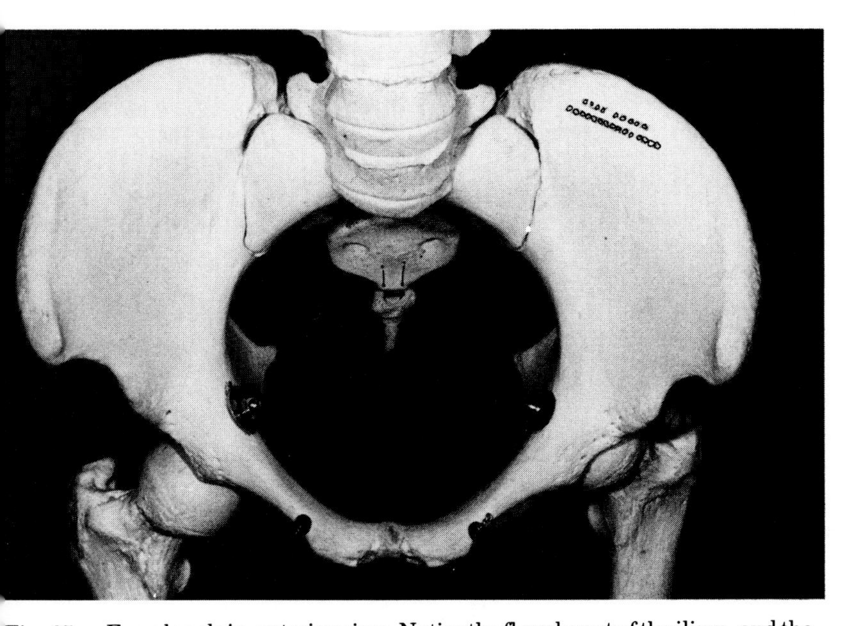

Fig. 95c Female pelvis, anterior view. Notice the flared crest of the ilium, and the larger opening in the center of the pelvis.

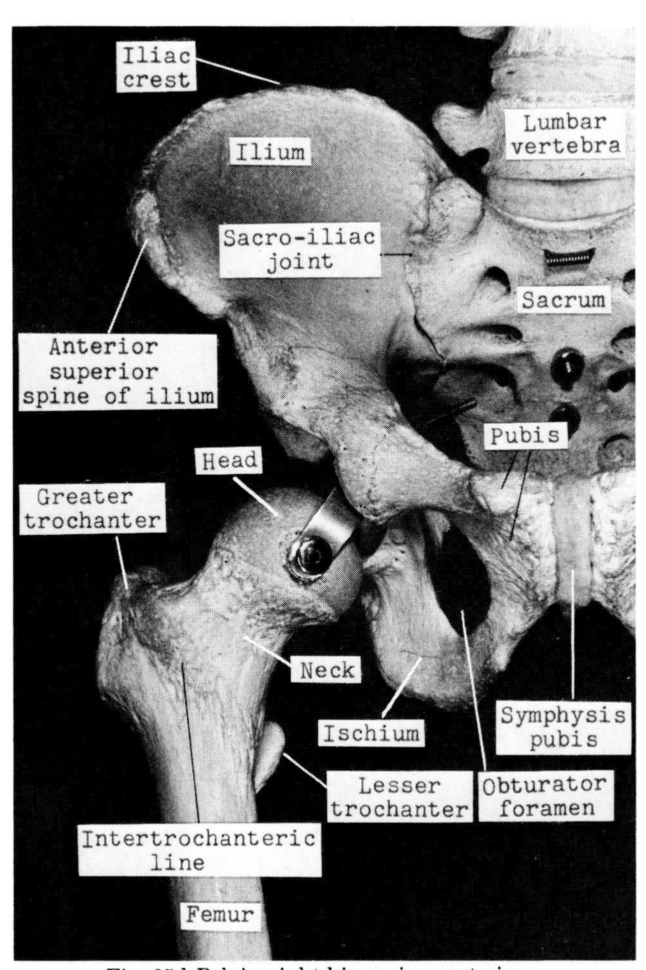

Fig. 95d Pelvis, right hip region, anterior.

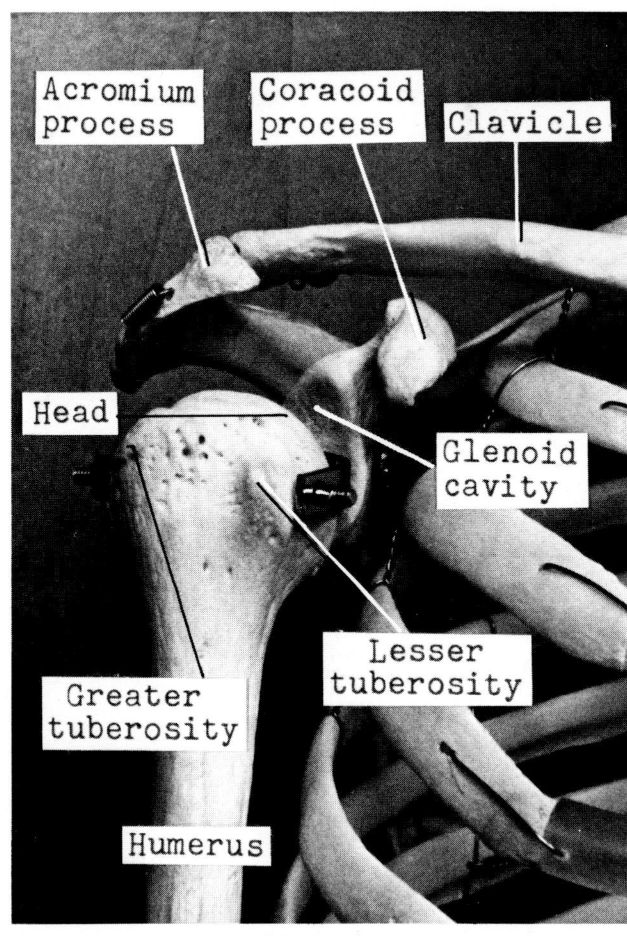

Fig. 95e Right shoulder region.

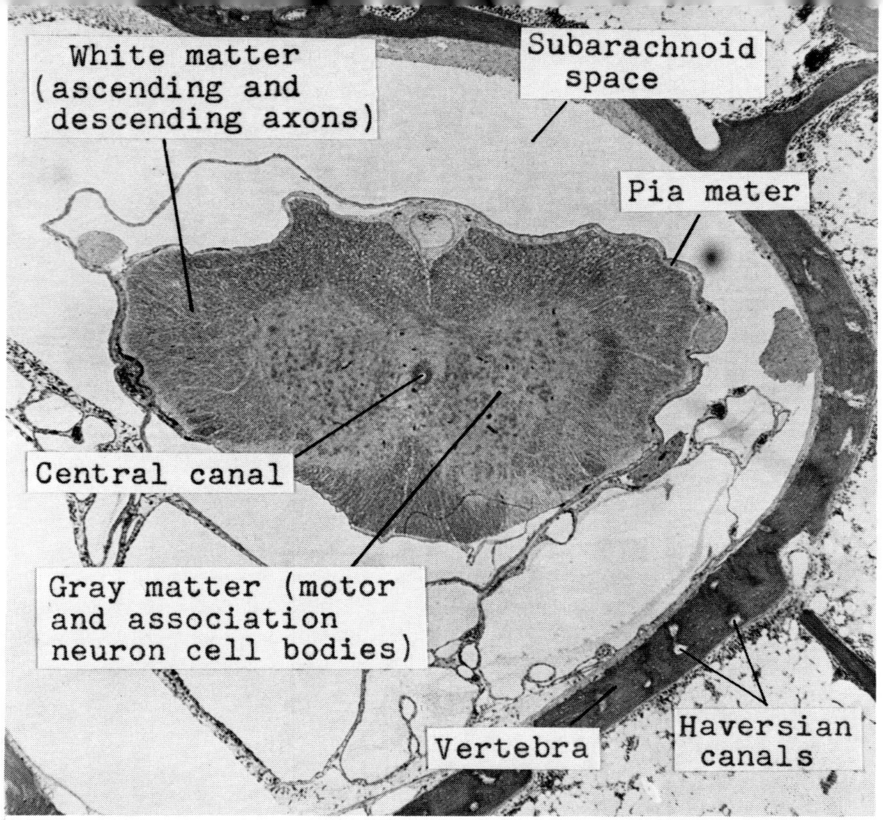

Fig. 96a Spinal cord (frog) in vertebra x.s. x40.

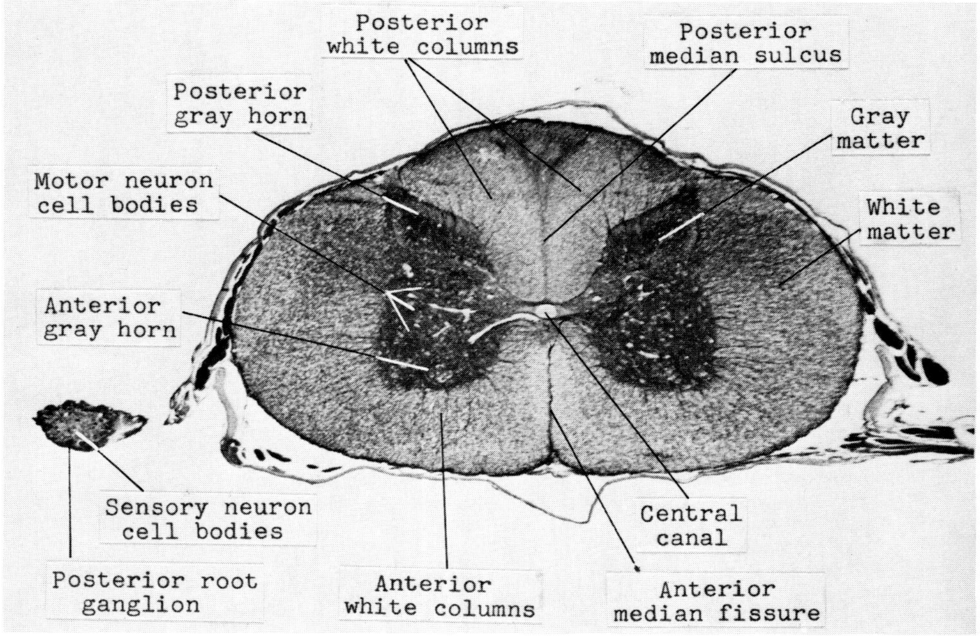

Fig. 96b Spinal cord x.s. x20.

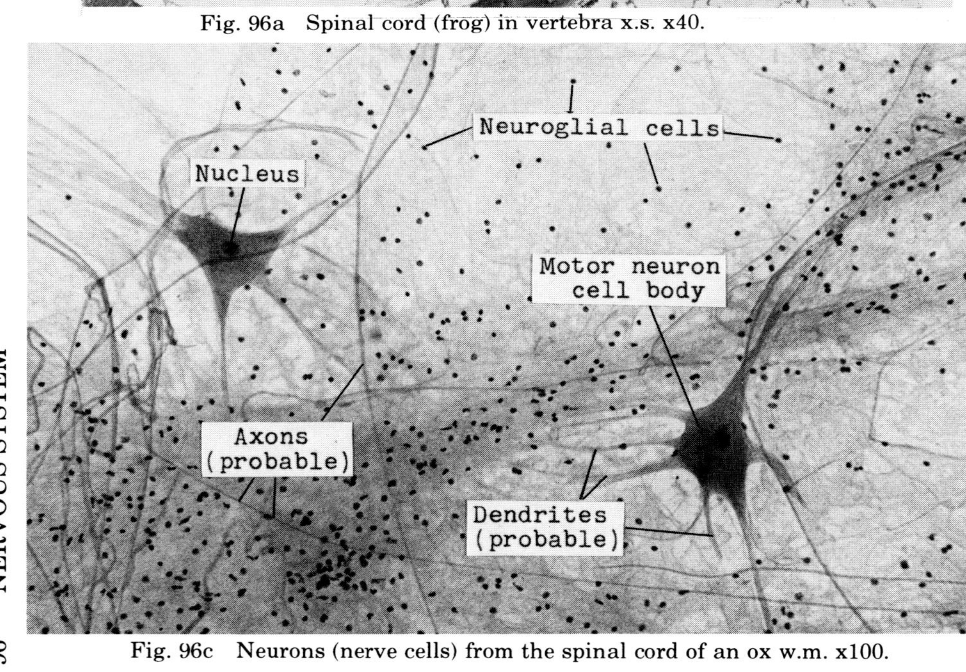

Fig. 96c Neurons (nerve cells) from the spinal cord of an ox w.m. x100.

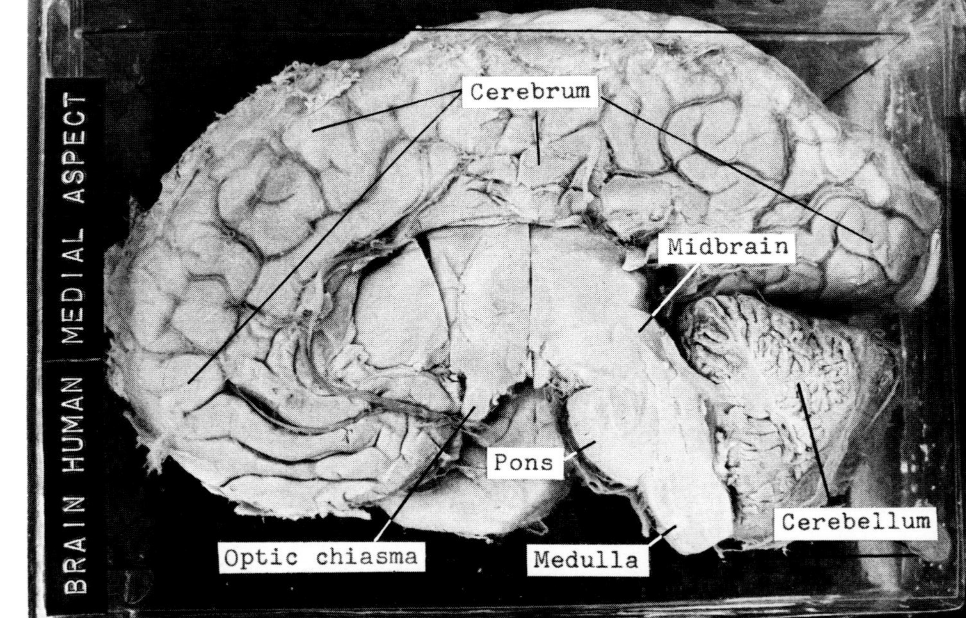

Fig. 96d Human brain l.s. x½.

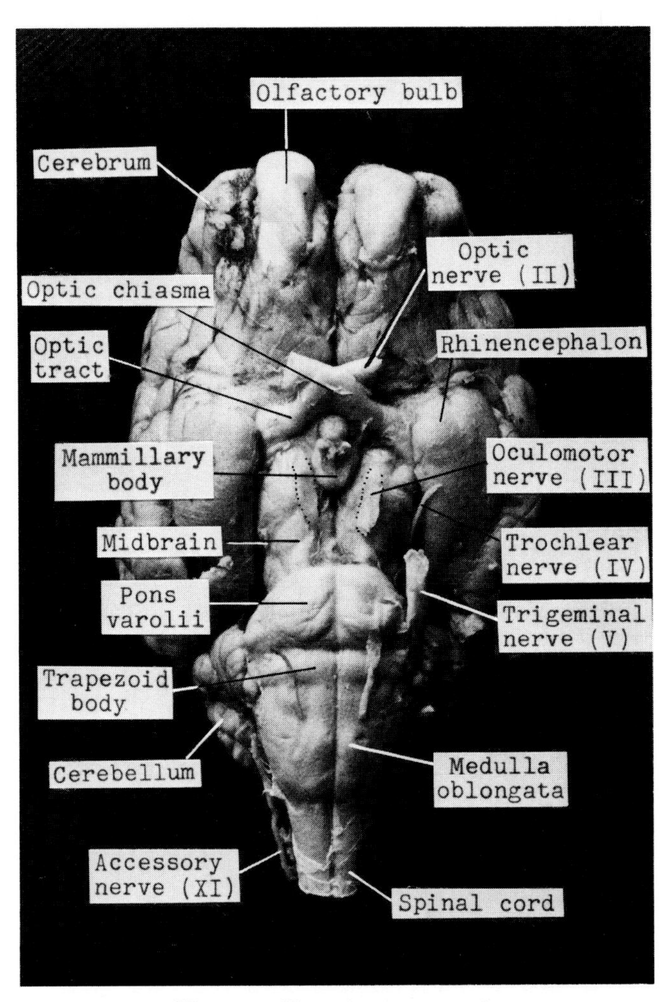

Fig. 97a Sheep brain, ventral view.

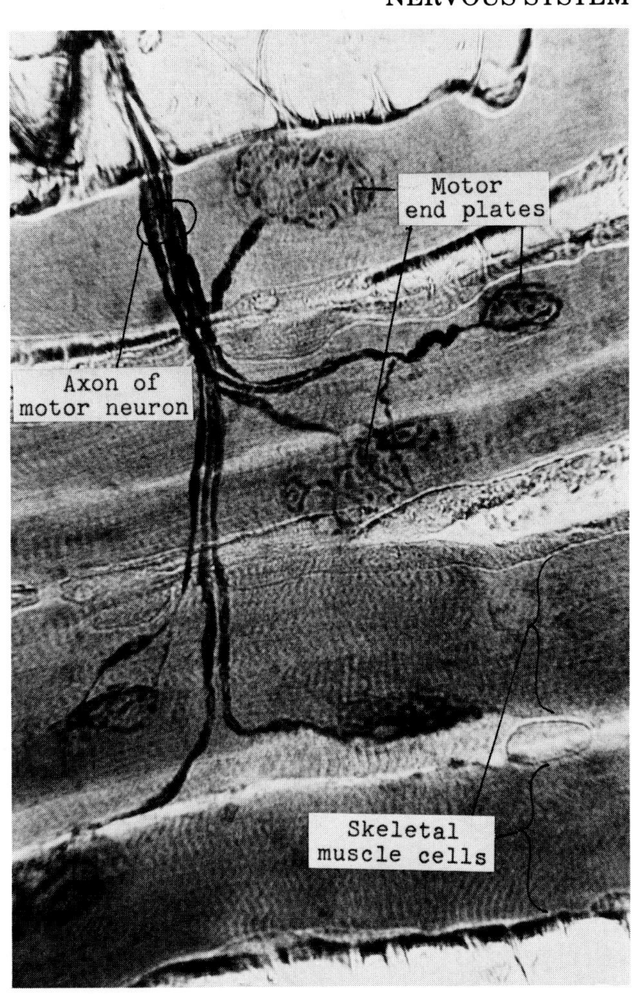

Fig. 97b Neuromuscular junction of motor end plates and striated muscle cells w.m. x430.

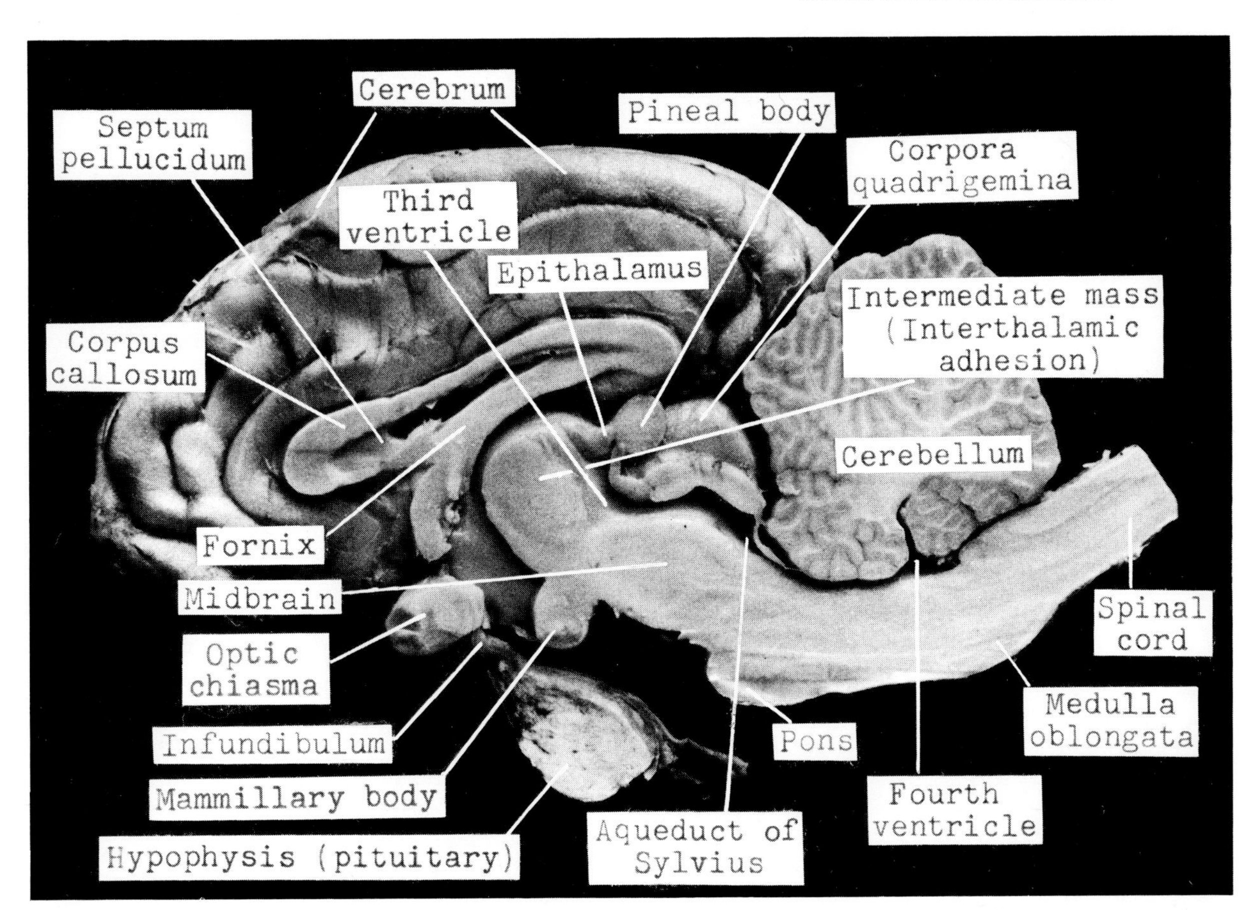

Fig. 97c Sheep brain dissection l.s.

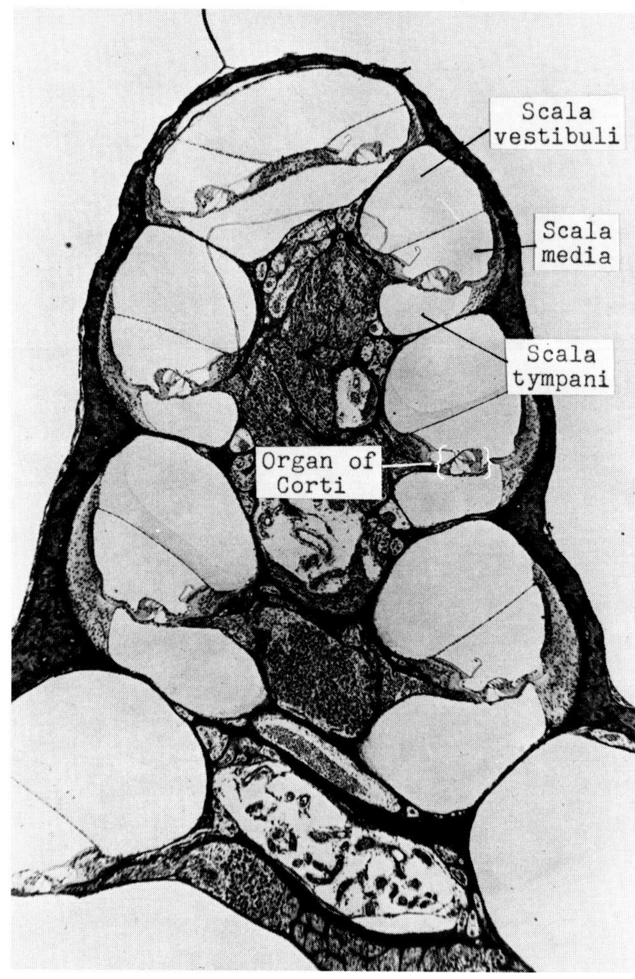

Fig. 98a Cochlea l.s. x40.

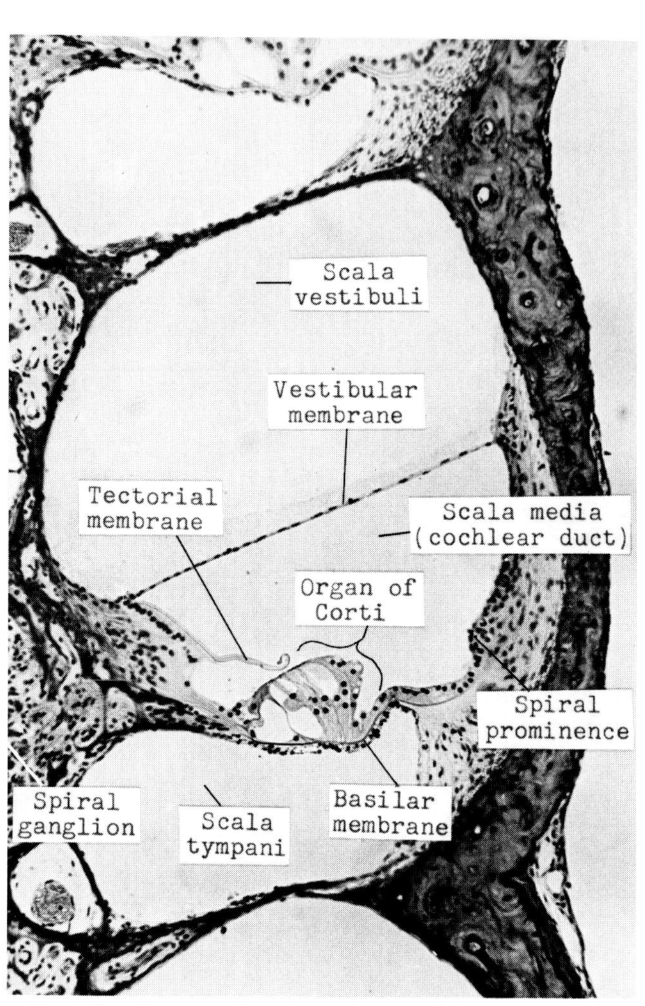

Fig. 98b Cross section through one of the turns of the cochlea l.s. x100.

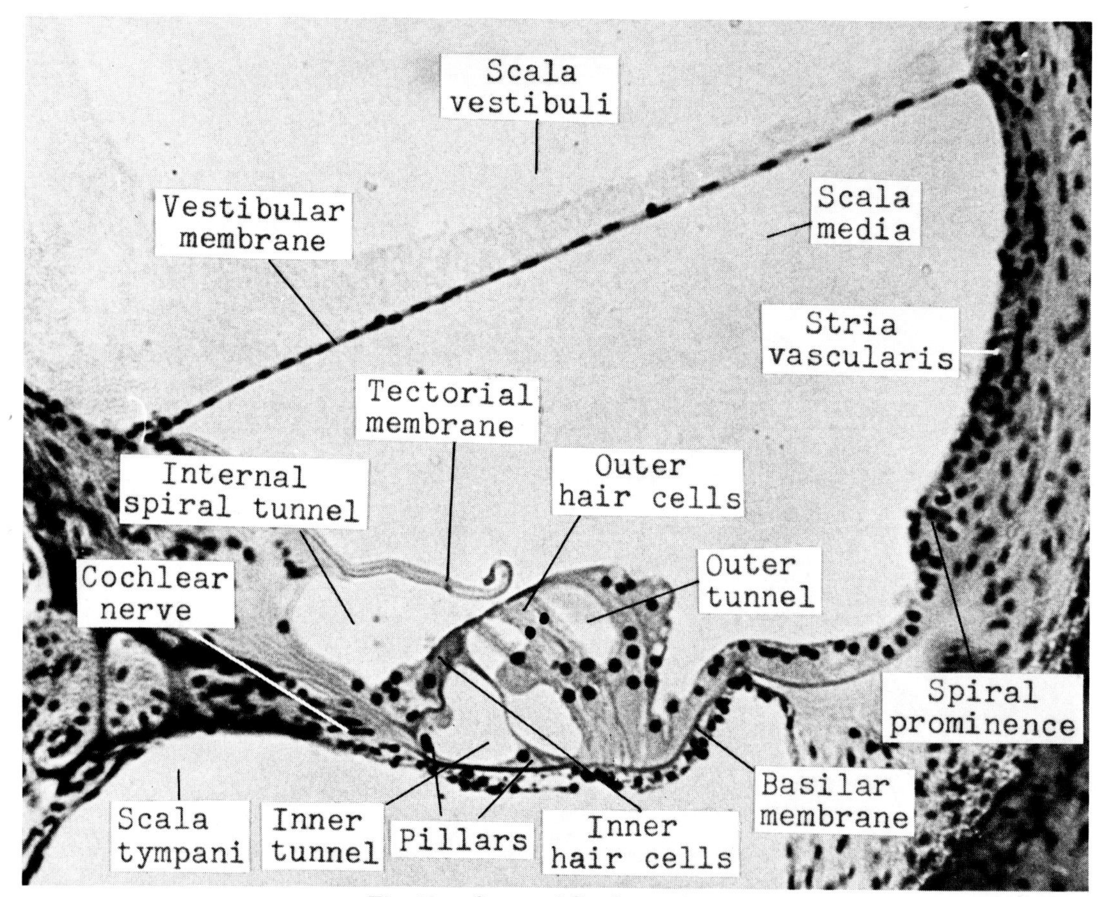

Fig. 98c Organ of Corti x.s. x200.

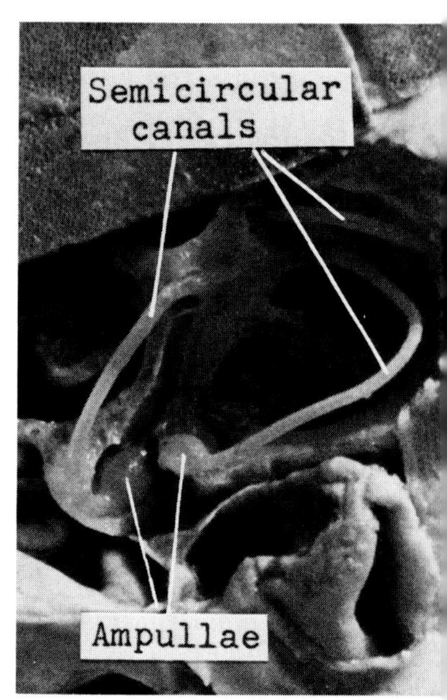

Fig. 98d Semicircular canals and ampullae of a shark x1.

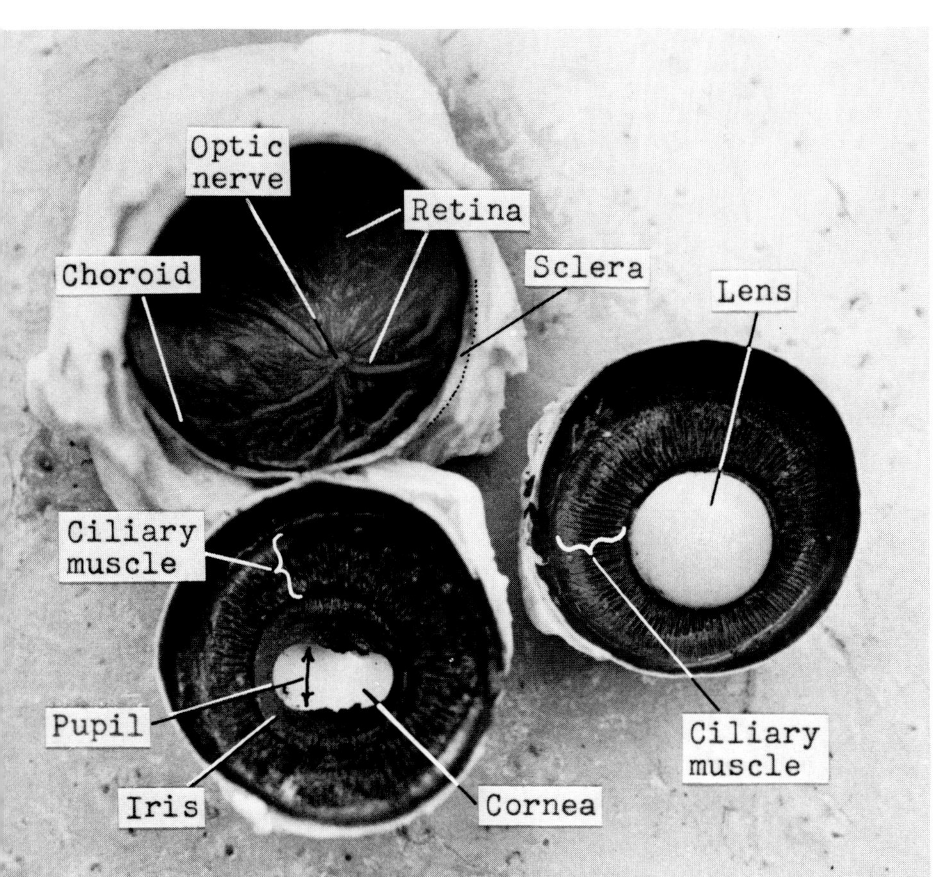

Fig. 99a Dissected sheep eyes x.s. The two left halves are of the same eye with the lens removed to show the pupil. The right eye shows the lens in place.

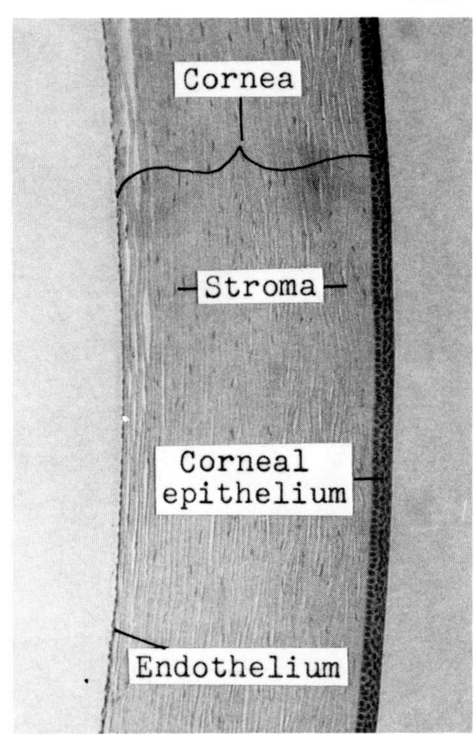

Fig. 99b Cornea l.s. x430.

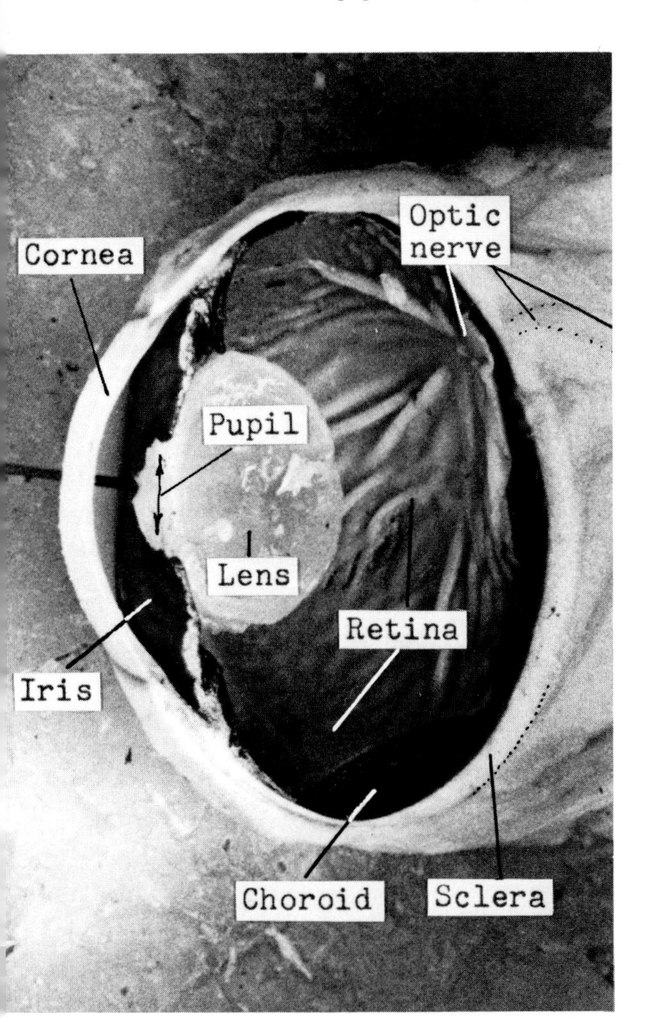

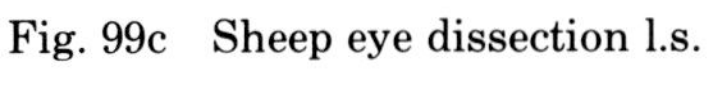

Fig. 99c Sheep eye dissection l.s.

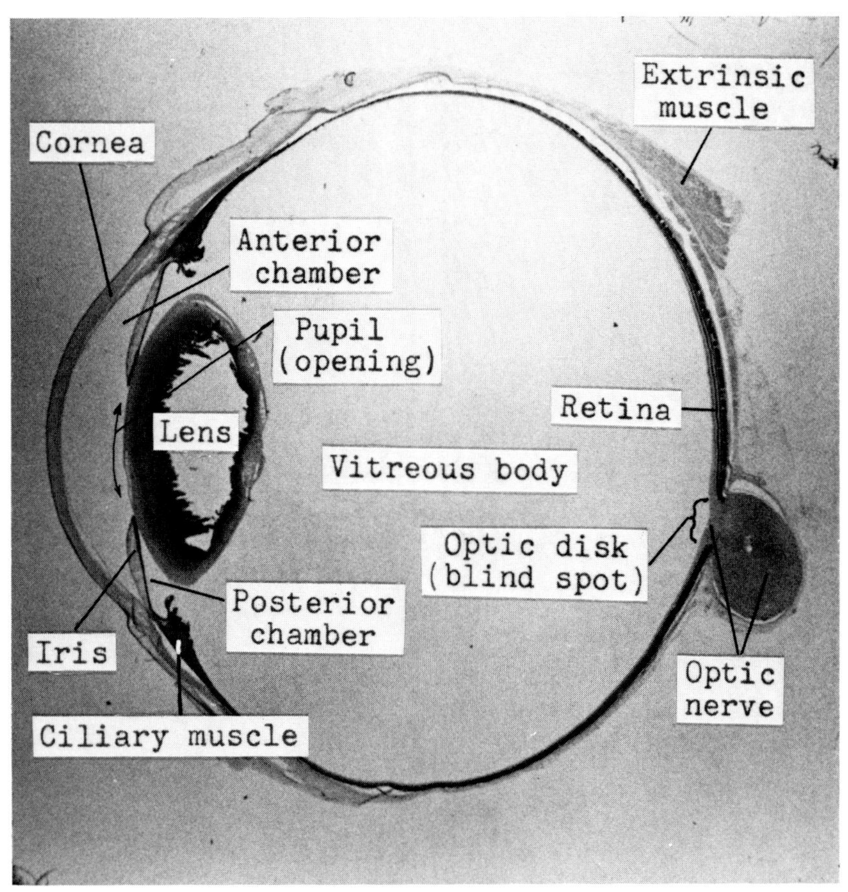

Fig. 99d Monkey eye l.s. x5.

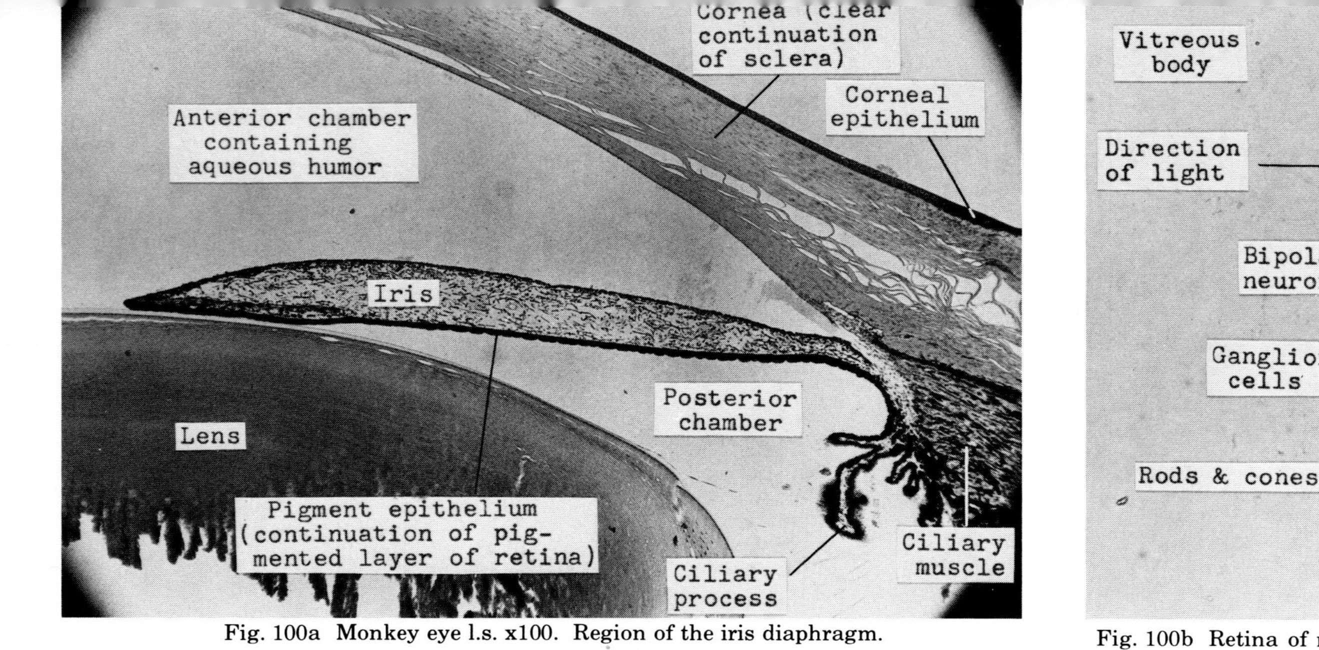

Fig. 100a Monkey eye l.s. x100. Region of the iris diaphragm.

Fig. 100b Retina of monkey eye l.s. x430. Notice that light must pass through several layers of nerve cells before it reaches the light-sensitive rods & cones. (Rods & cones are neurons also.)

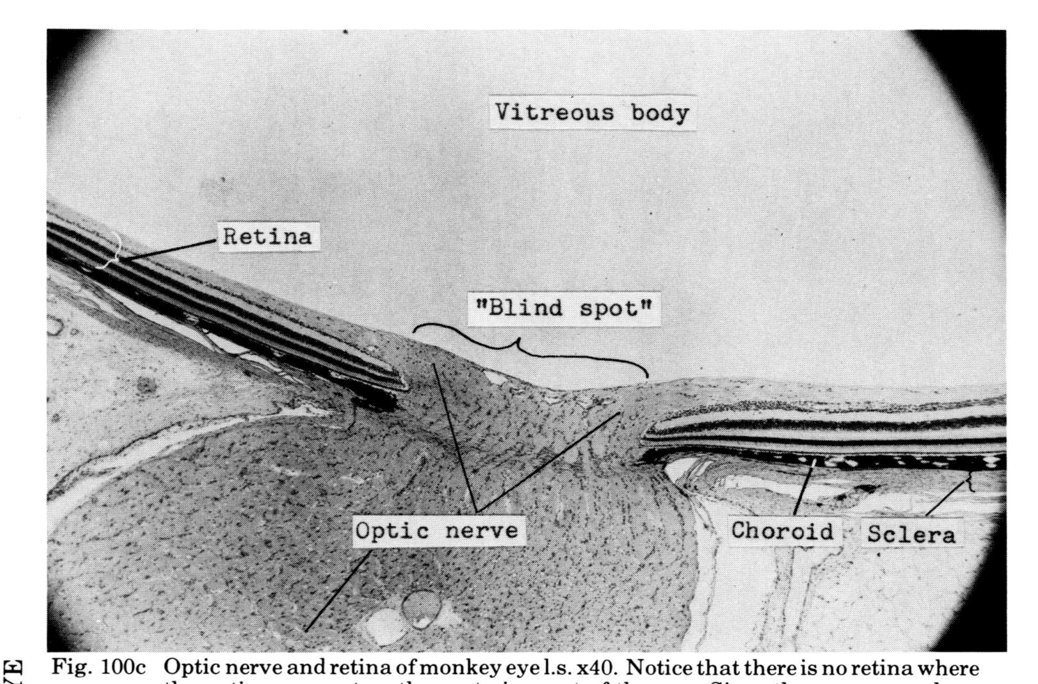

Fig. 100c Optic nerve and retina of monkey eye l.s. x40. Notice that there is no retina where the optic nerve enters the posterior part of the eye. Since there are no rods or cones in this area, no light is perceived and it is known as the blind spot.

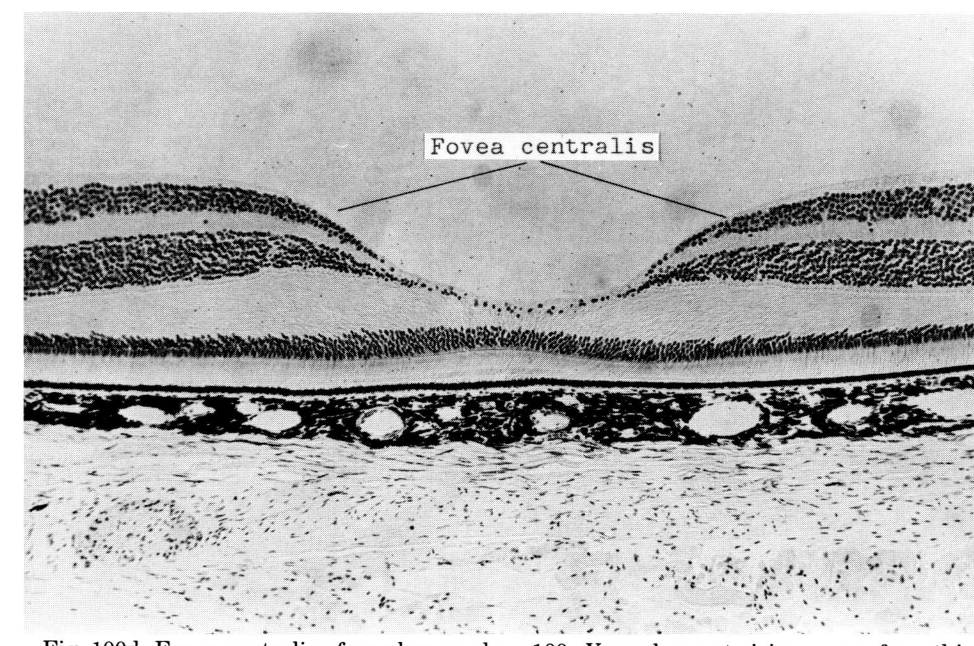

Fig. 100d Fovea centralis of monkey eye l.s. x100. Your sharpest vision comes from this area because: (a) the retina contains only cones in high concentration and (b) the reduction in overlying layers of other cells.

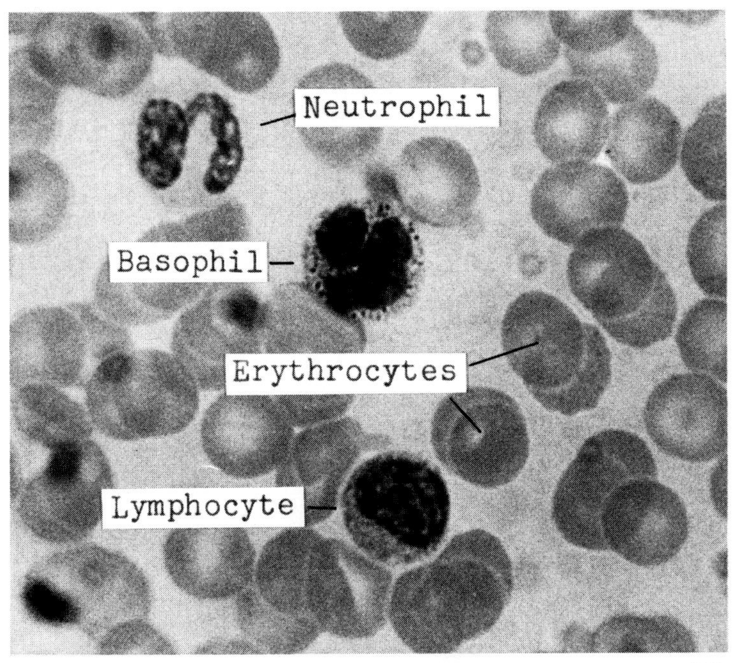

Fig. 101a Human blood w.m. x1000. The cytoplasm of the neutrophil stains only faintly and is difficult to see in this photo.

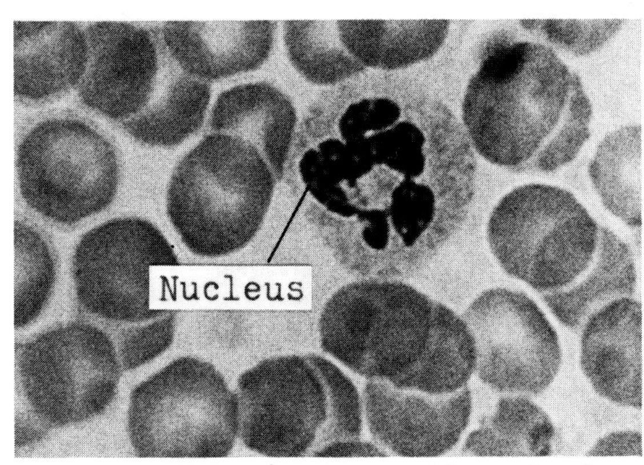

Fig. 101b Neutrophil w.m. x1000. The nucleus has many lobes and is polymorphic in these cells.

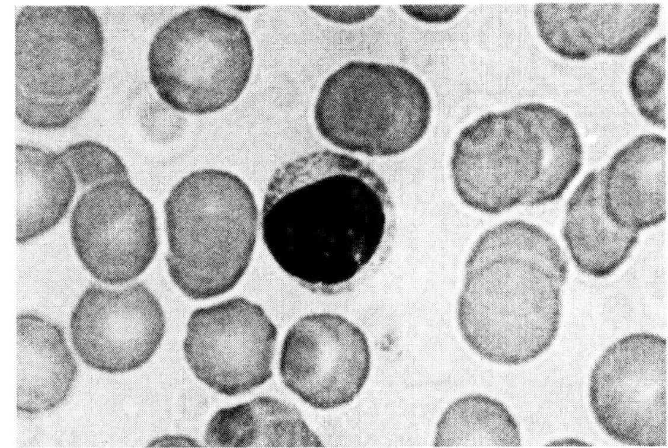

Fig. 101c Lymphocyte w.m. x1000. The nucleus usually takes up 80-90% of the cell.

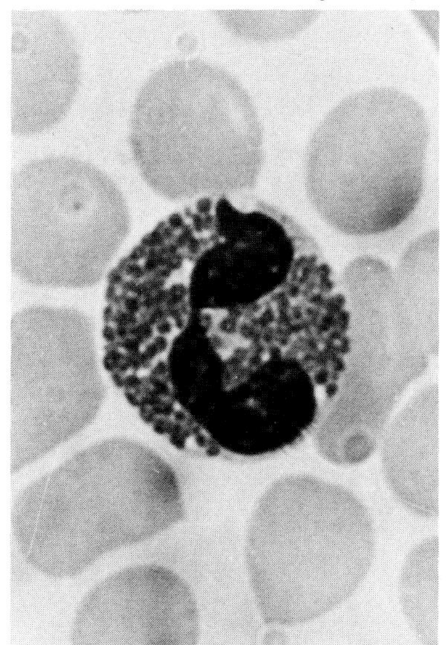

Fig. 101d Eosinophil w.m. x1000. The granules in the cytoplasm will stain red.

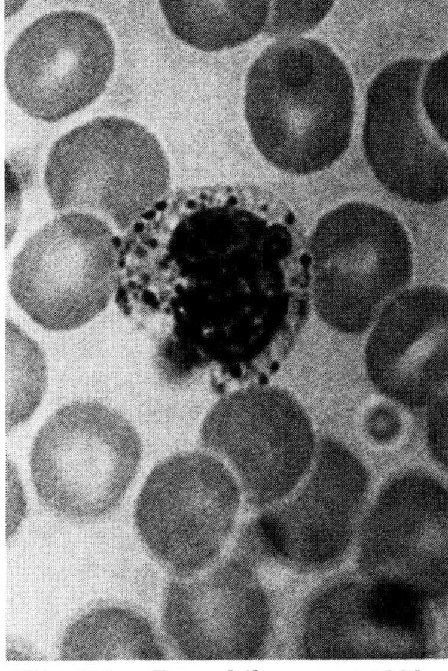

Fig. 101e Basophil w.m. x1000. The dark granules in the cytoplasm will stain a dark blue or purple.

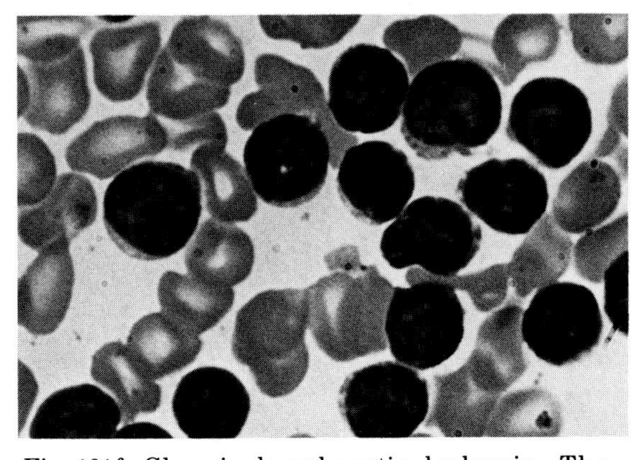

Fig. 101f Chronic lymphocytic leukemia. The patient died 4 days after this sample was taken. Notice the abnormally high number of lymphocytes.

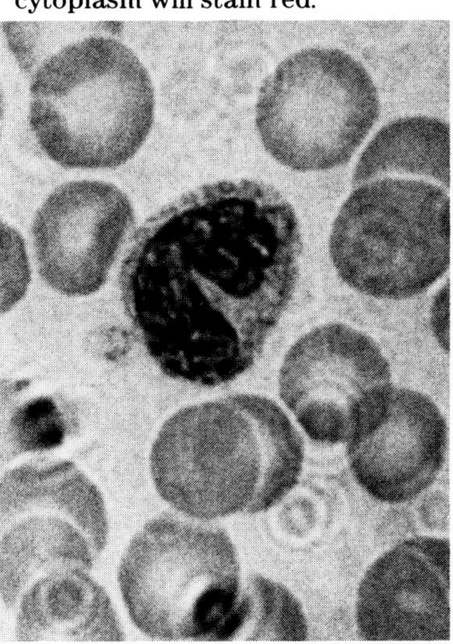

Fig. 101g Monocyte w.m. x1000. Monocytes are often up to 2x the size of other WBC's and commonly have a horse-shoe shaped nucleus.

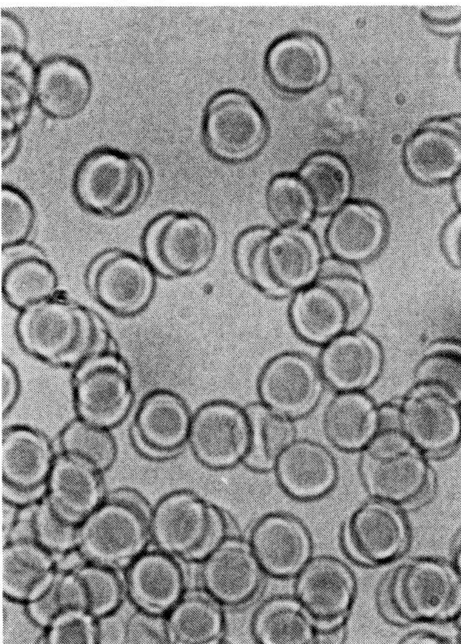

Fig. 101h Erythrocytes (live) w.m. x430. The biconcave shape is easily seen.

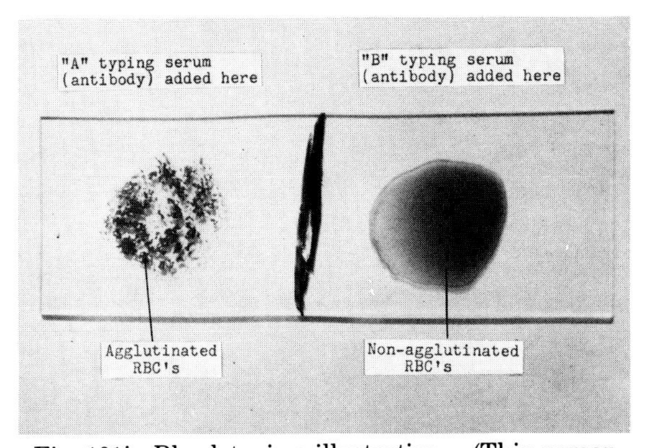

Fig. 101i Blood typing illustration. (This person had type "A" blood.) "A" antibody (typing serum) was added to the drop of blood on the left side and "B" antibody was added to the right. Agglutination occurred on the left side between the "A" antibodies and the "A" antigens on the person's RBC's.

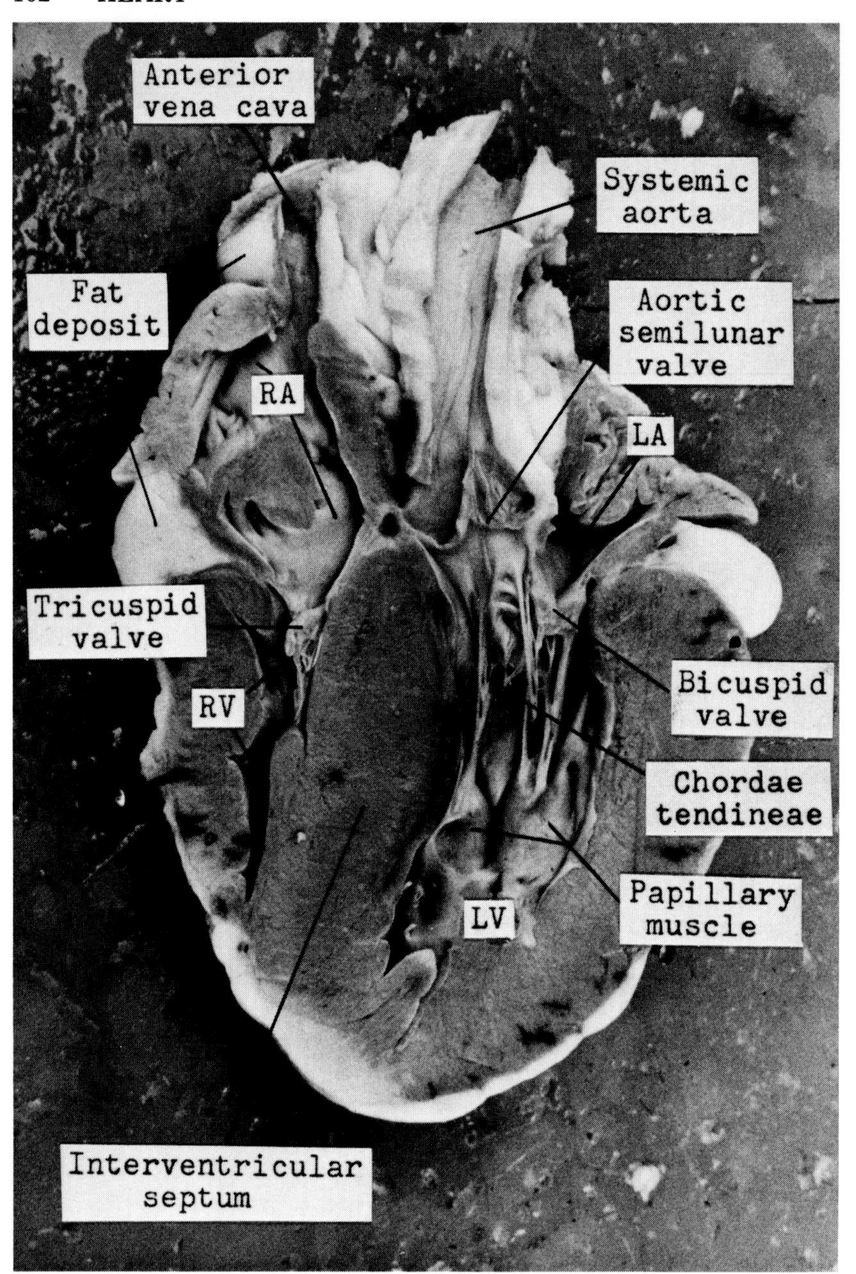

Fig. 102a Sheep heart dissection l.s. x1.

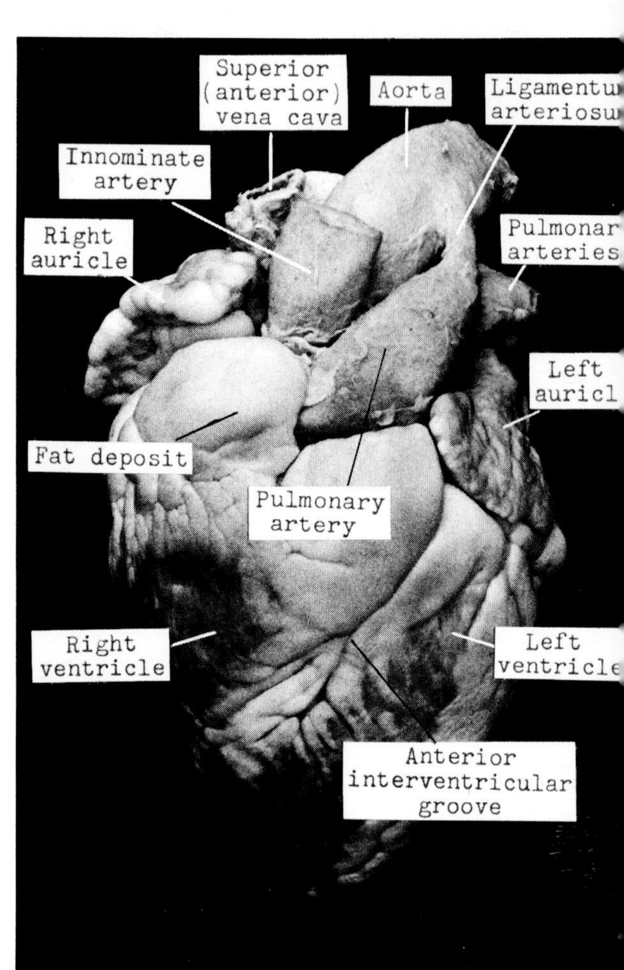

Fig. 102b Sheep heart, ventral view.

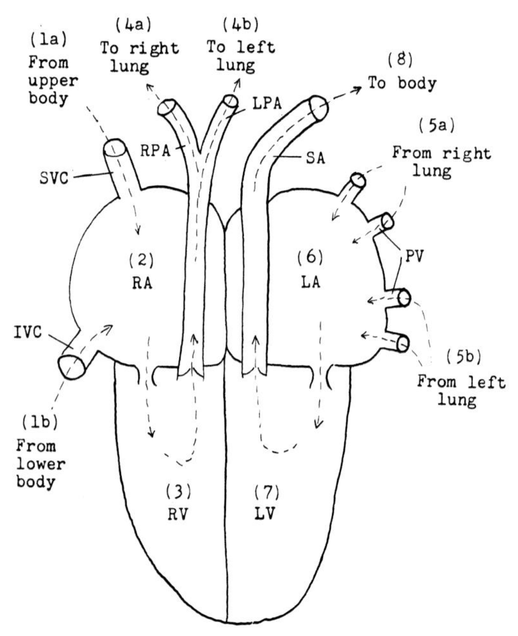

Fig. 102c Diagram of human heart circulation. (Use the numbers and follow the path of circulation.)

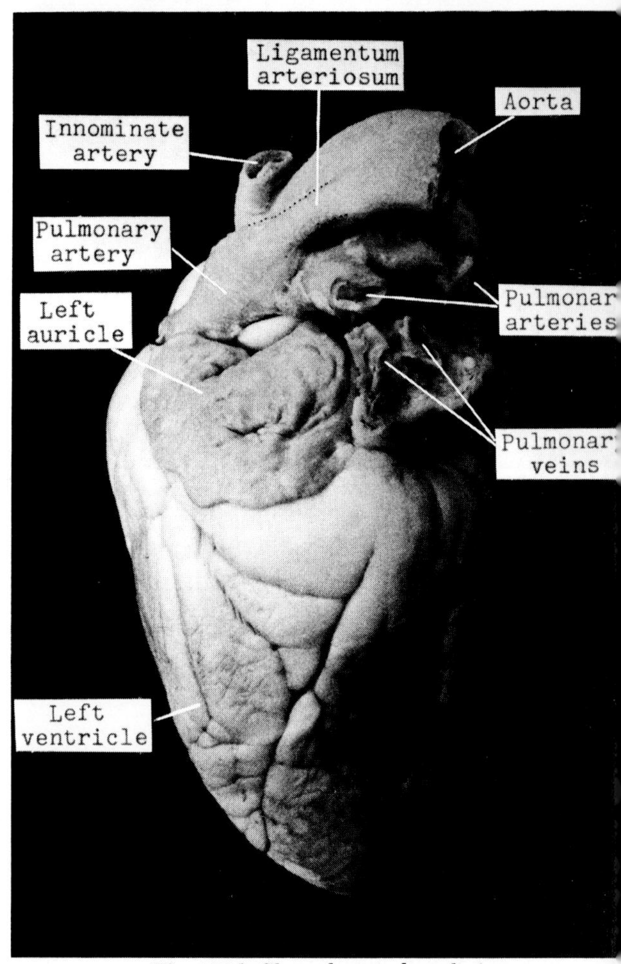

Fig. 102d Sheep heart, dorsal view.

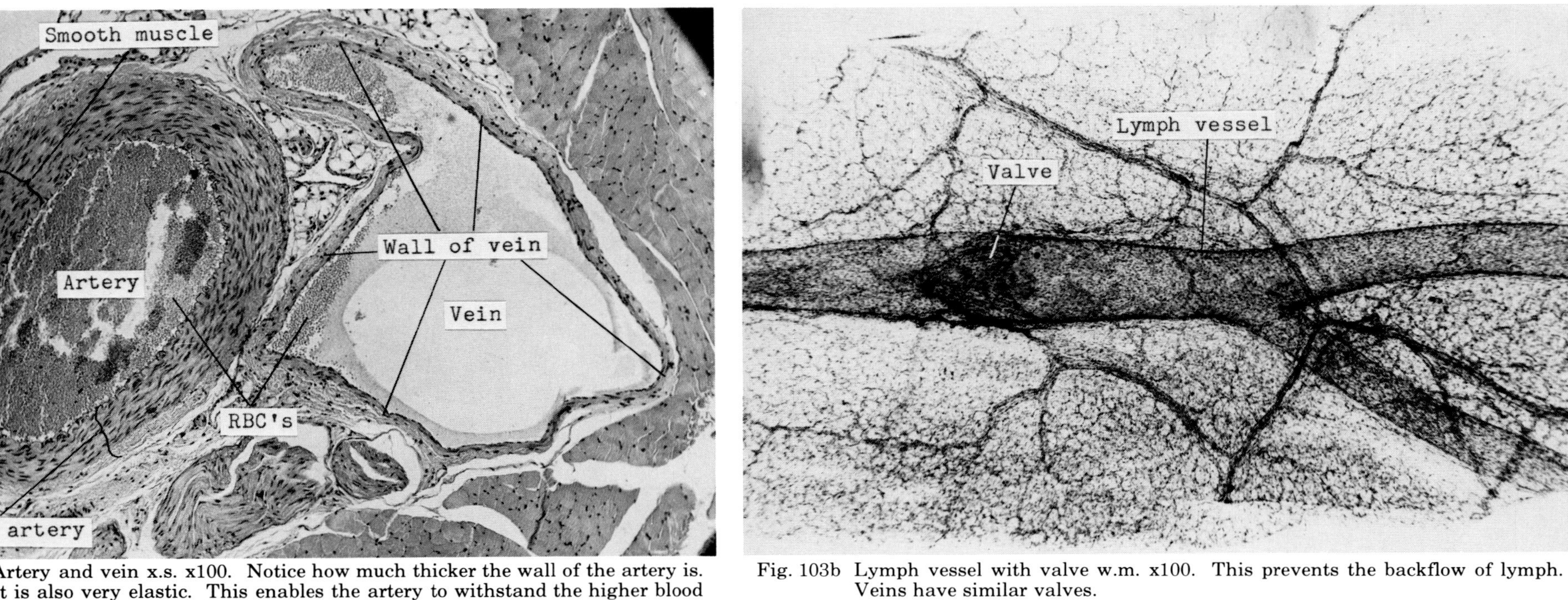

Fig. 103a Artery and vein x.s. x100. Notice how much thicker the wall of the artery is. It is also very elastic. This enables the artery to withstand the higher blood pressures coming directly from the ventricles.

Fig. 103b Lymph vessel with valve w.m. x100. This prevents the backflow of lymph. Veins have similar valves.

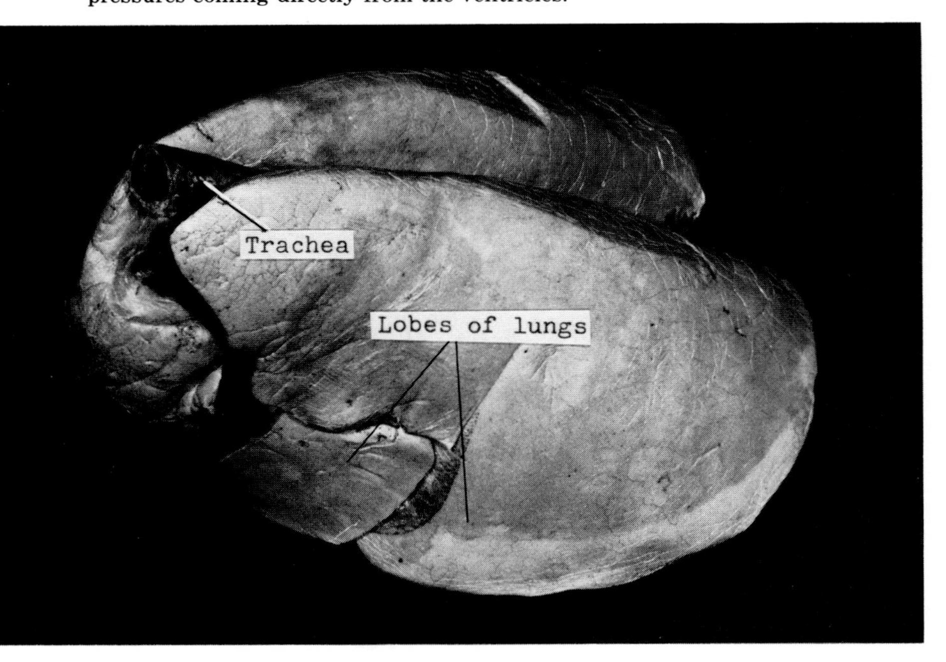

Fig. 103c Sheep lungs, inflated and dried.

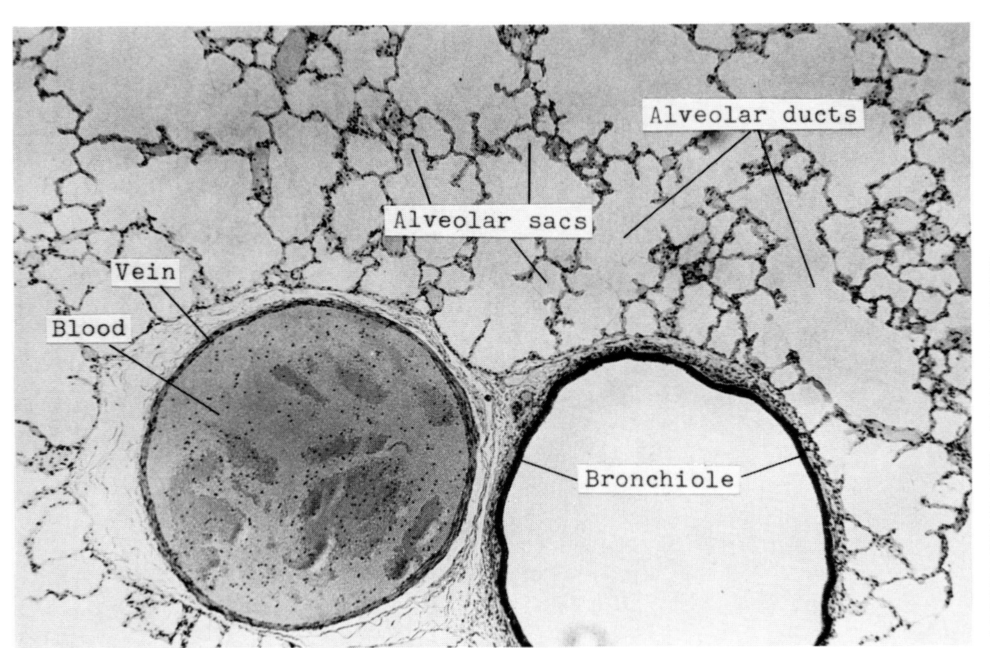

Fig. 103d Human lung x.s. x100.

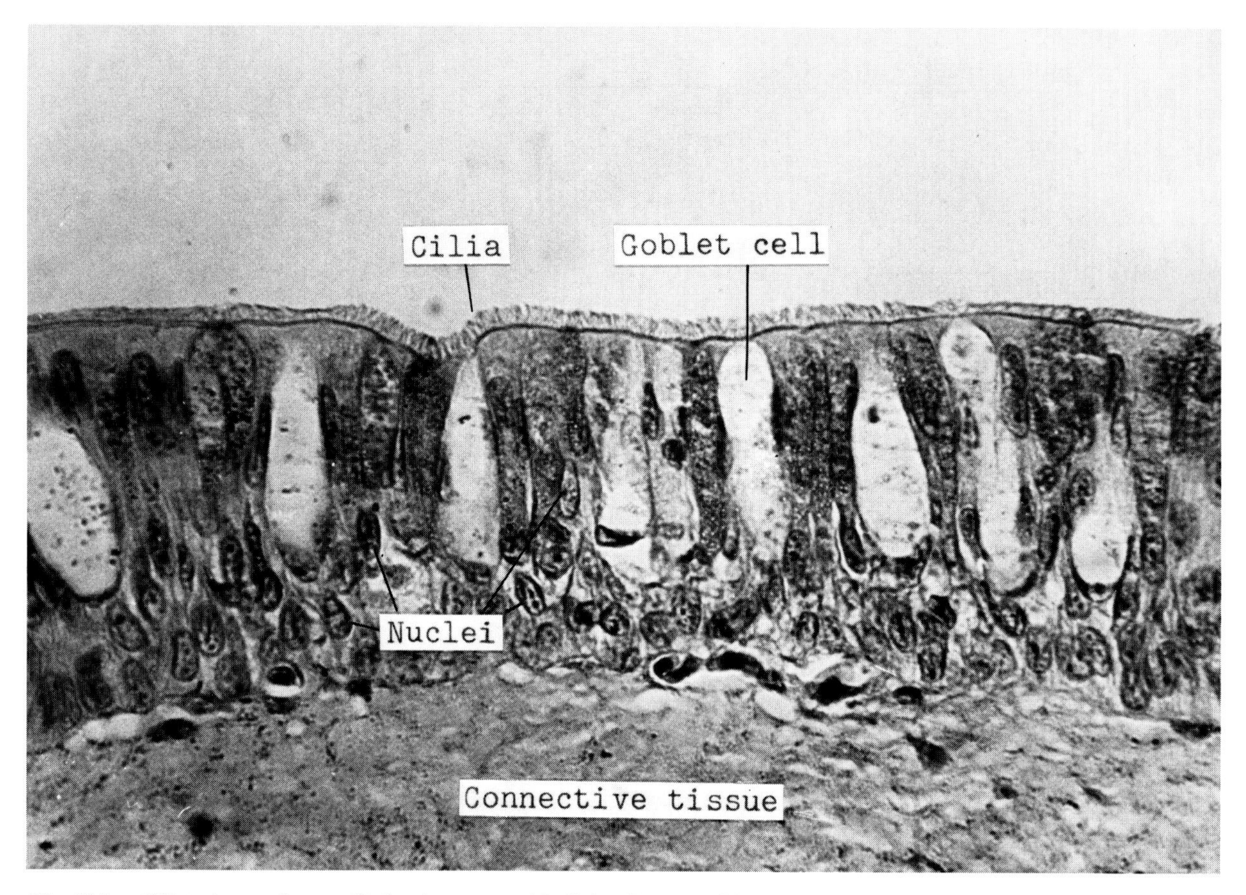

Fig. 104a Ciliated pseudostratified columnar epithelial cells l.s. x430. These cells are found lining the respiratory passages (mucous membranes of nose, trachea and bronchii). The goblet cells produce mucus which catches dirt particles and the cilia beat and drive the dirt-laden mucus to the back of the throat where it is swallowed. Smoking paralyzes these cilia. The smoker must then cough to remove the mucus in his lung passages.

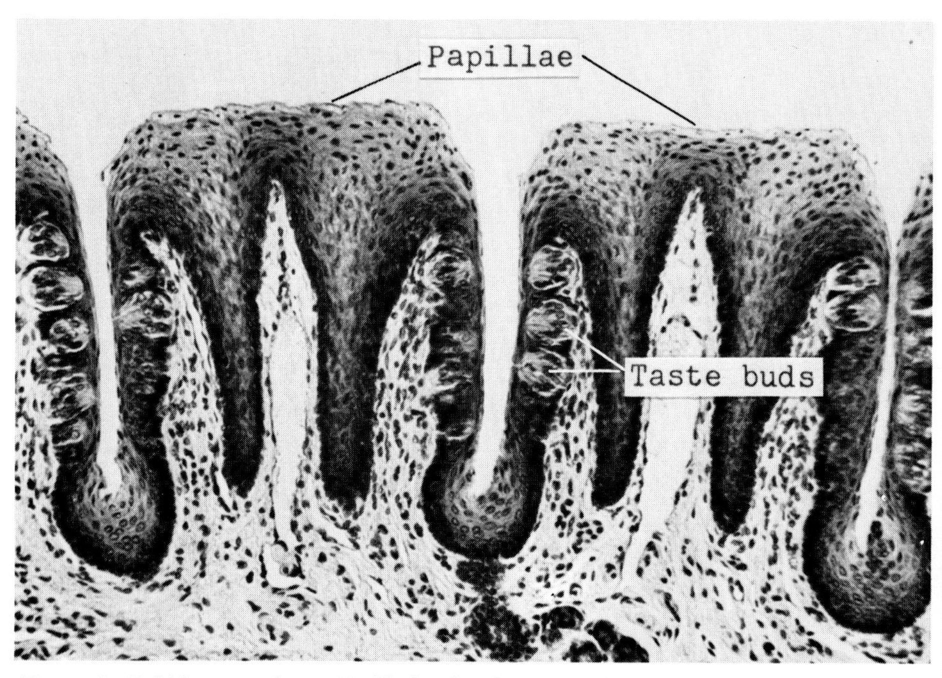

Fig. 104b Rabbit tongue l.s. x100. Notice the alternating deep clefts between the papillae containing the taste buds.

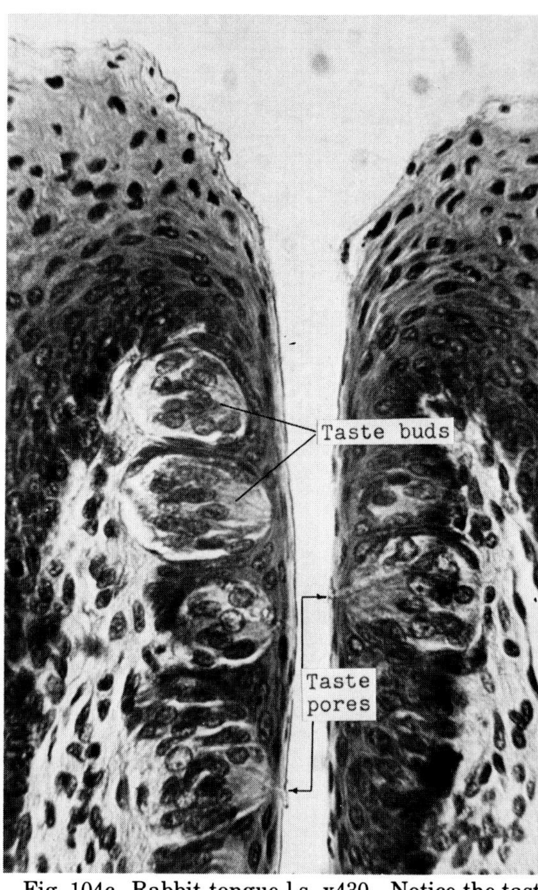

Fig. 104c Rabbit tongue l.s. x430. Notice the taste pores (entrances) into the taste buds.

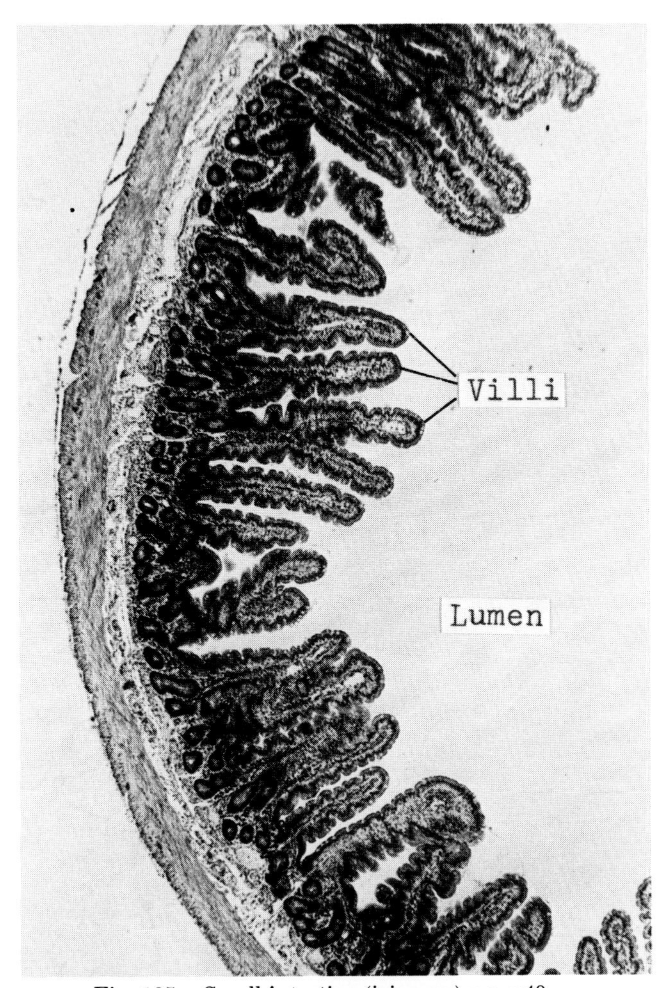

Fig. 105a Small intestine (jejunum) x.s. x40.

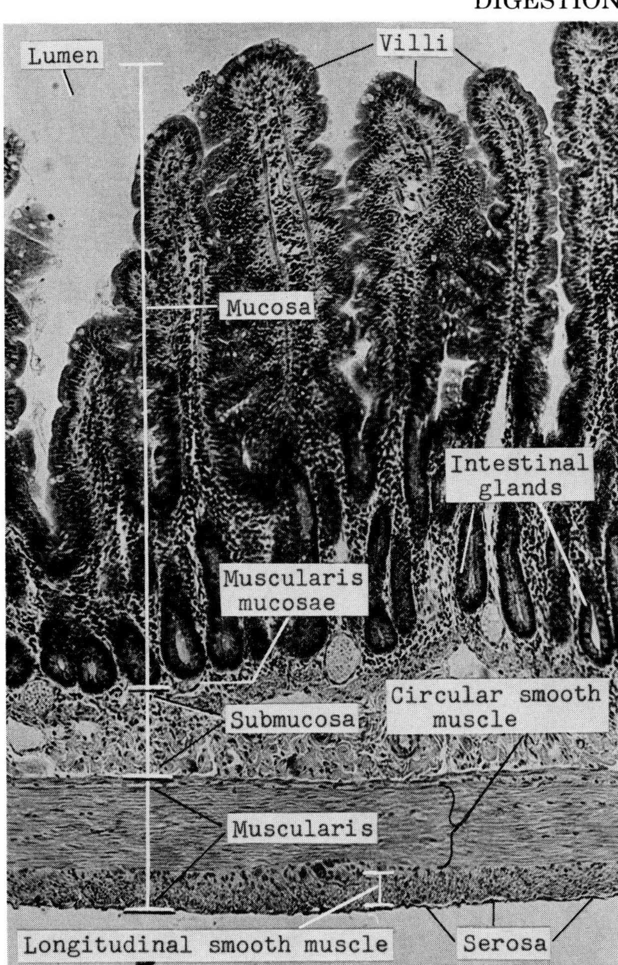

Fig. 105b Wall of the small intestine (jejunum) x.s. x100.

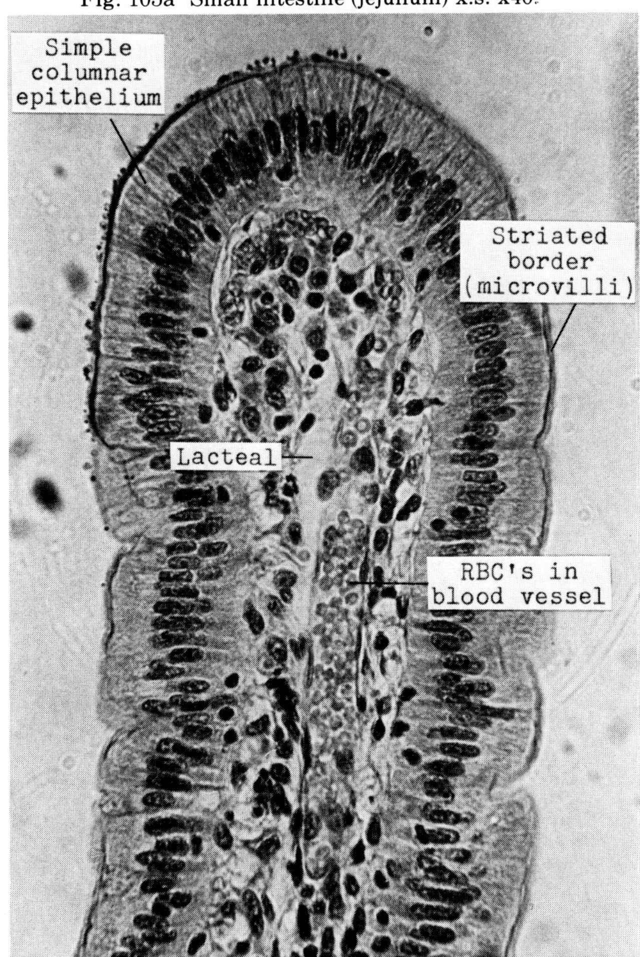

Fig. 105c Villus l.s. x430.

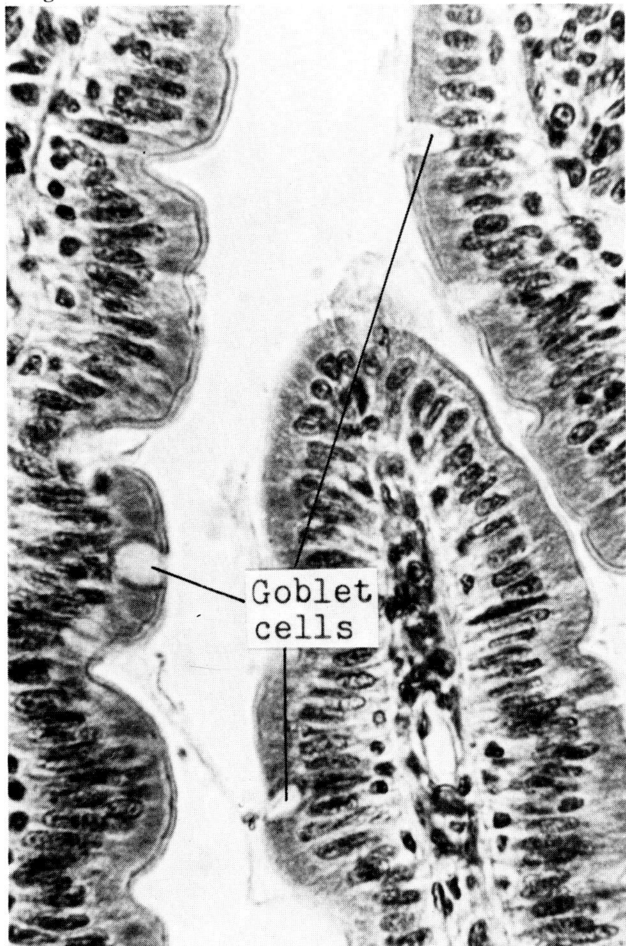

Fig. 105d Villi containing goblet cells l.s. x430. Goblet cells produce and secrete mucous for lubrication.

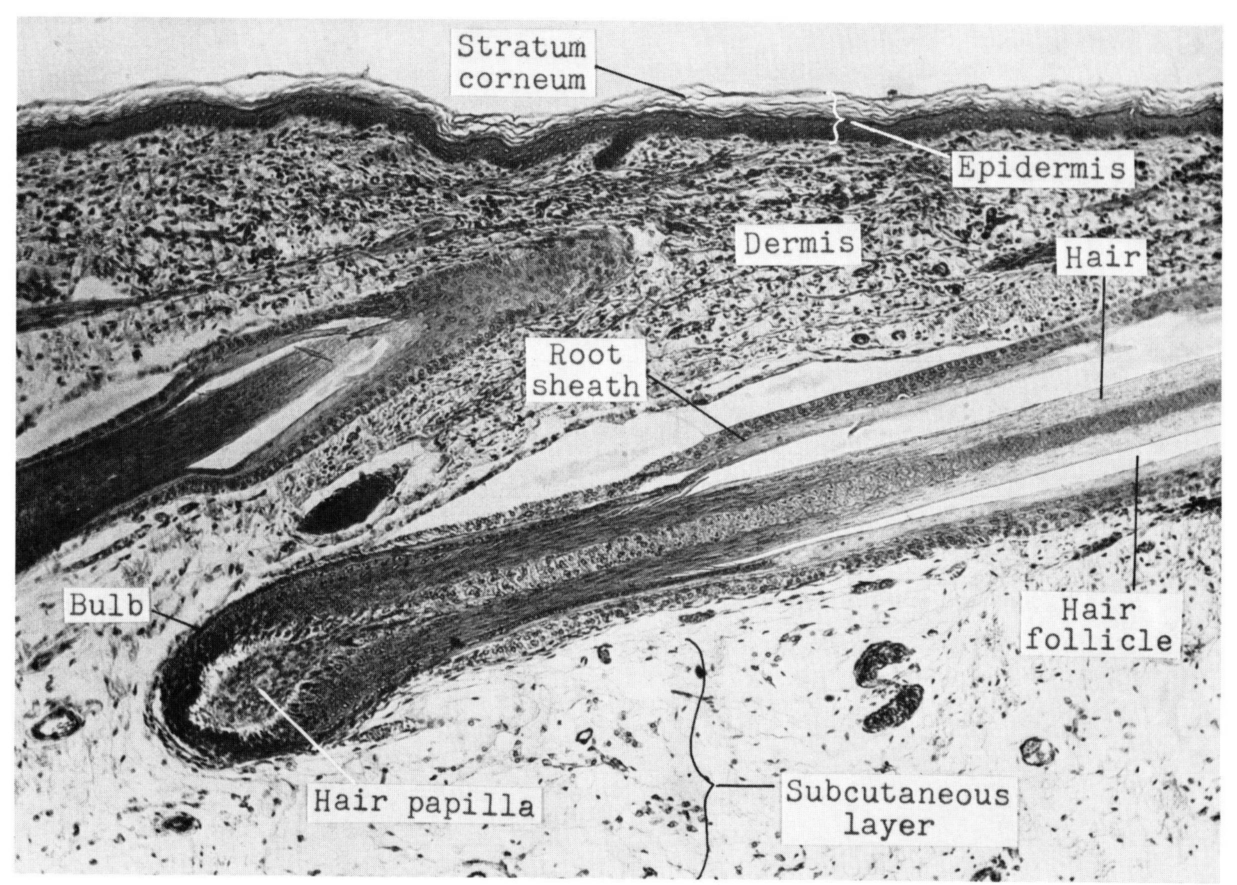

Fig. 106a Skin and hair l.s. x100.

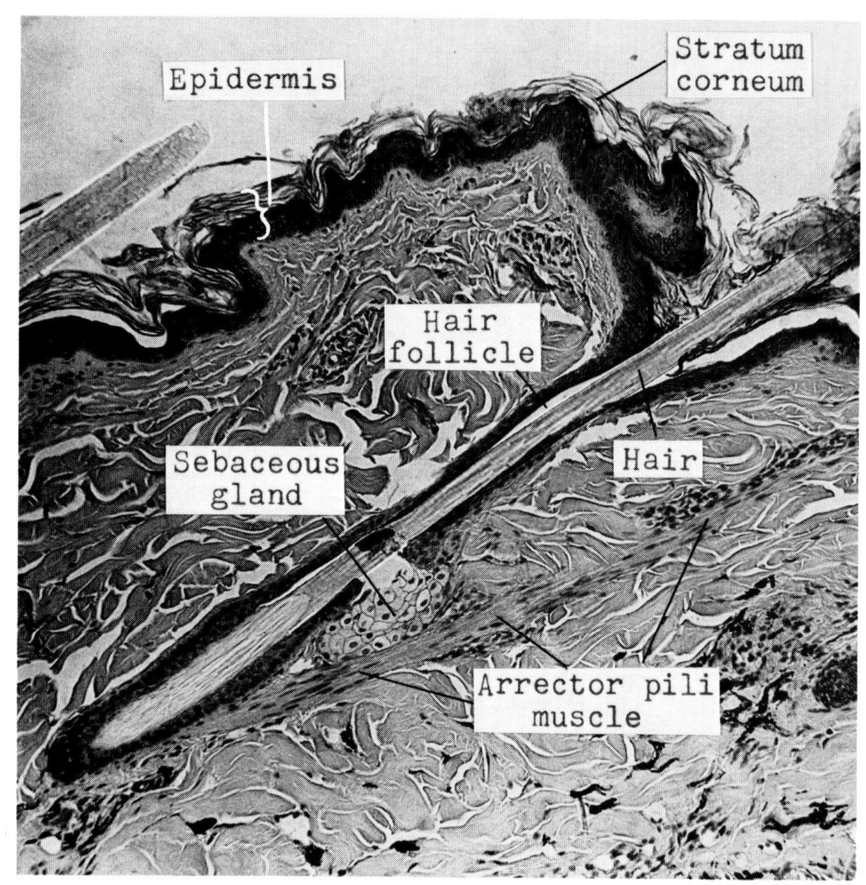

Fig. 106b Skin and hair l.s. x100. The arrector pili muscle "erects" the hair in animals for insulation against cold and/or to appear larger during fights. In humans, the erected hair pushes a fold of skin in front of it - the "goose bump."

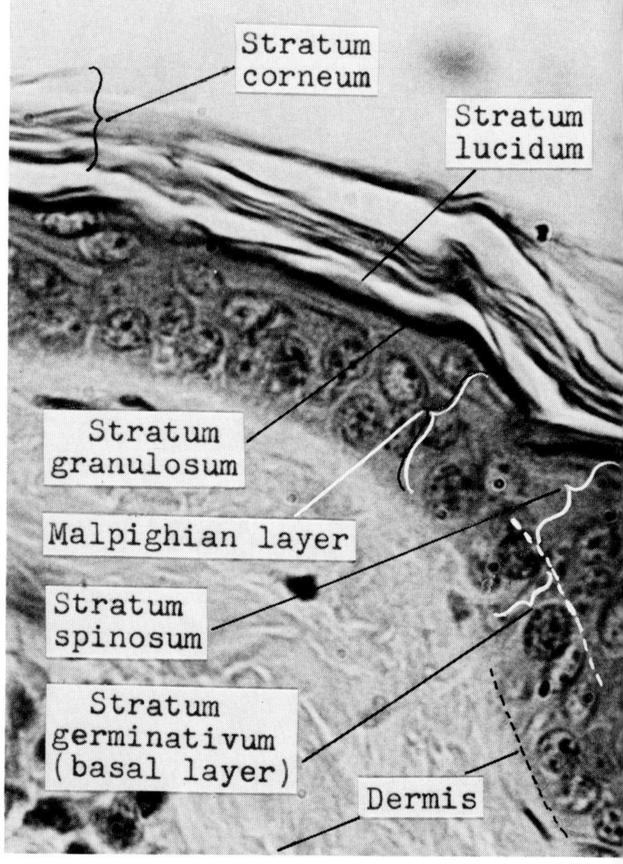

Fig. 106c Layers of the epidermis x.s. x430.

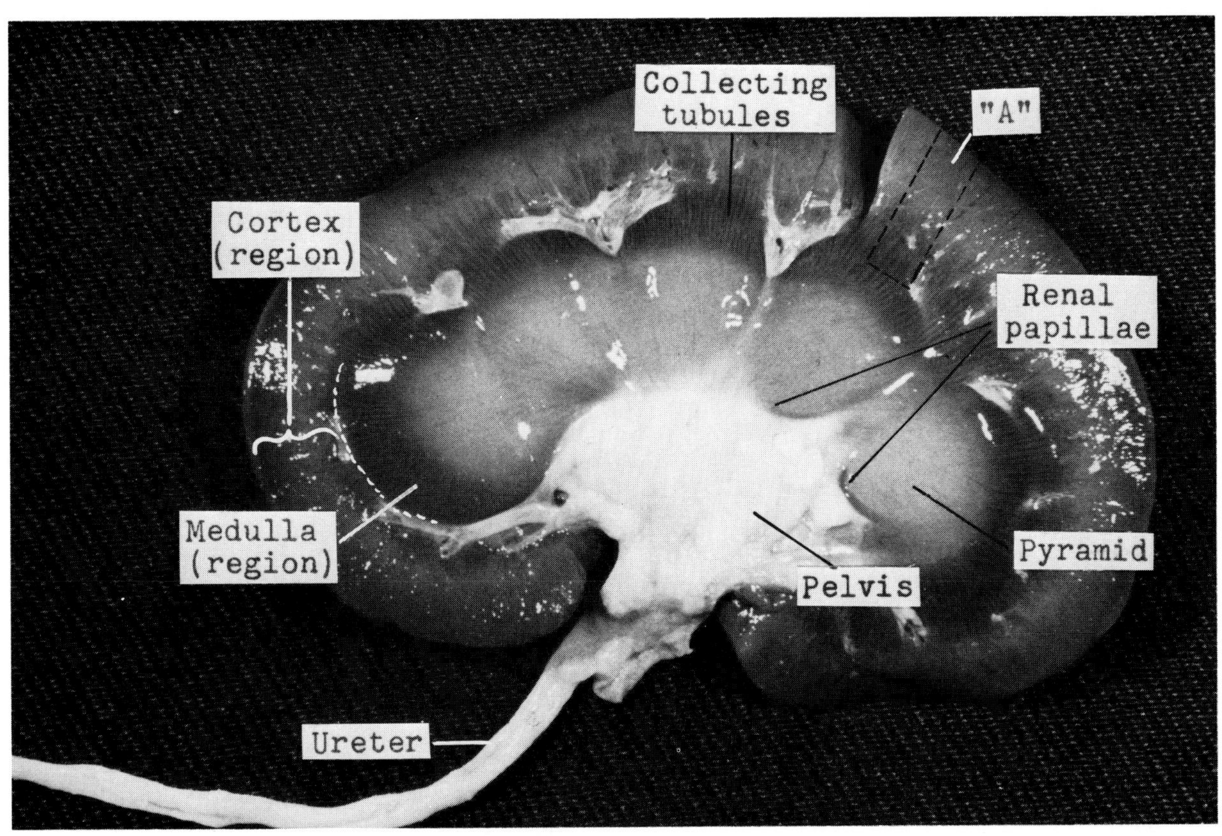

Fig. 107a Fresh lamb kidney dissection l.s. x1.

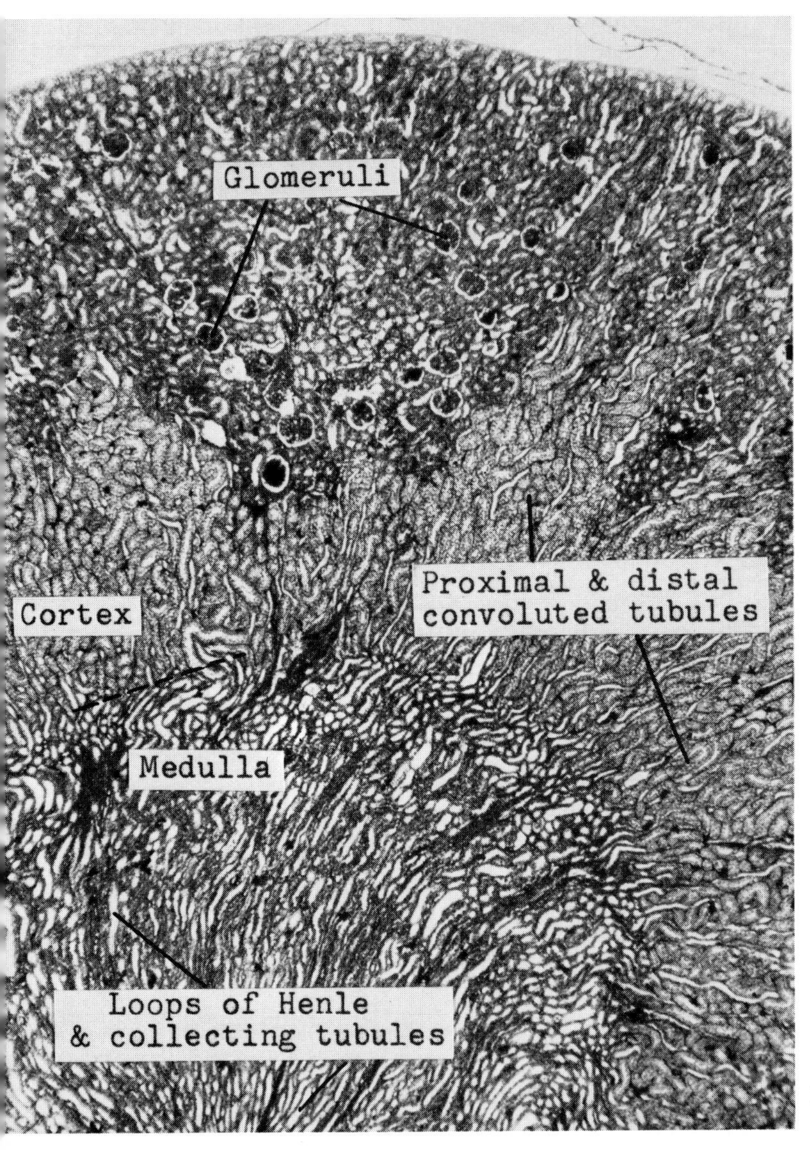

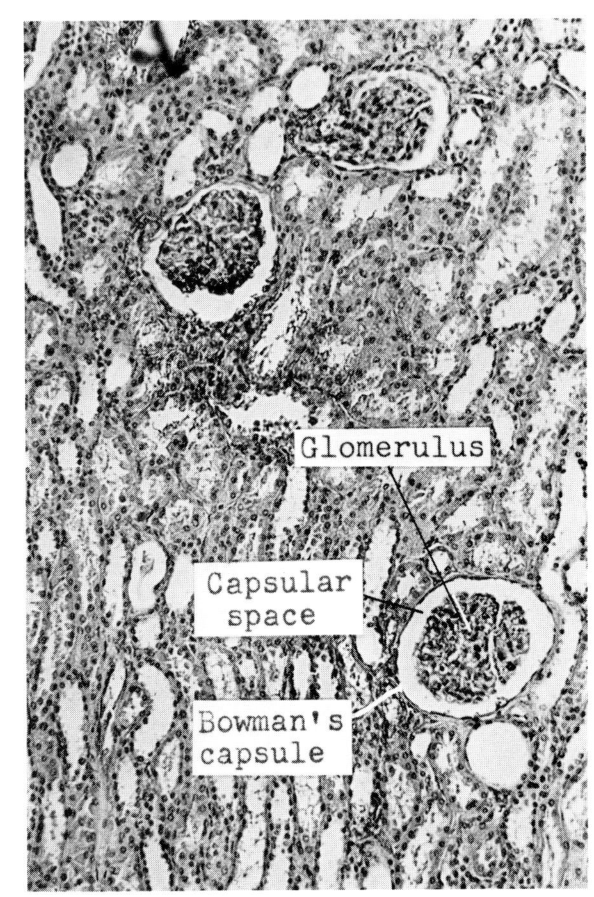

Fig. 107c Nephrons l.s. x430. (Cortex region.)

Fig. 107b Rat kidney l.s. x40. (From an area similar to region "A" in Fig. 107a.)

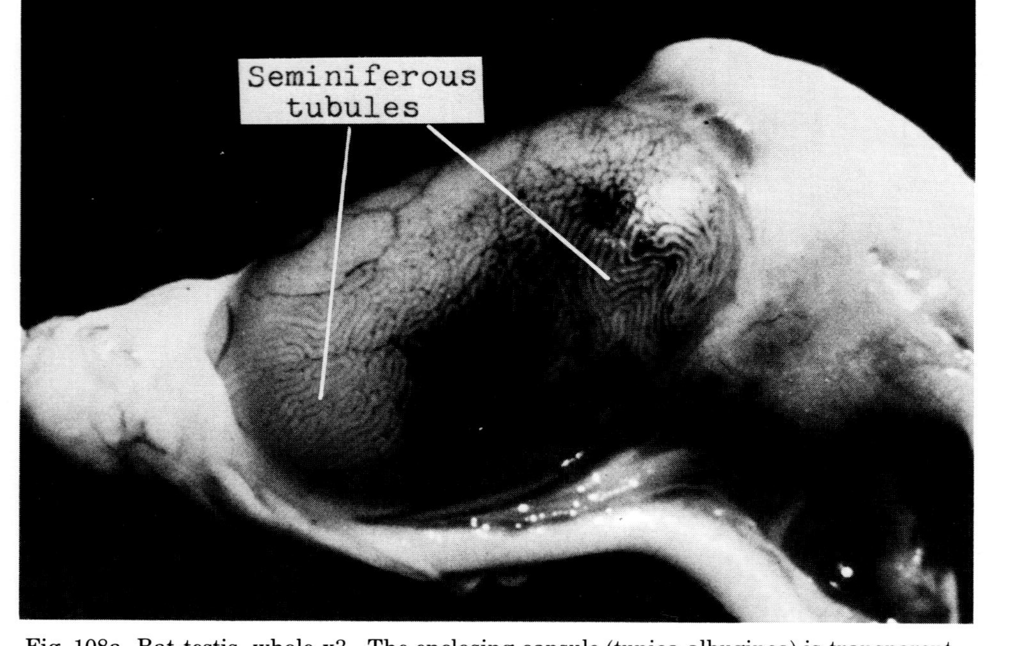

Fig. 108a Rat testis, whole x3. The enclosing capsule (tunica albuginea) is transparent enough to see the seminiferous tubules inside.

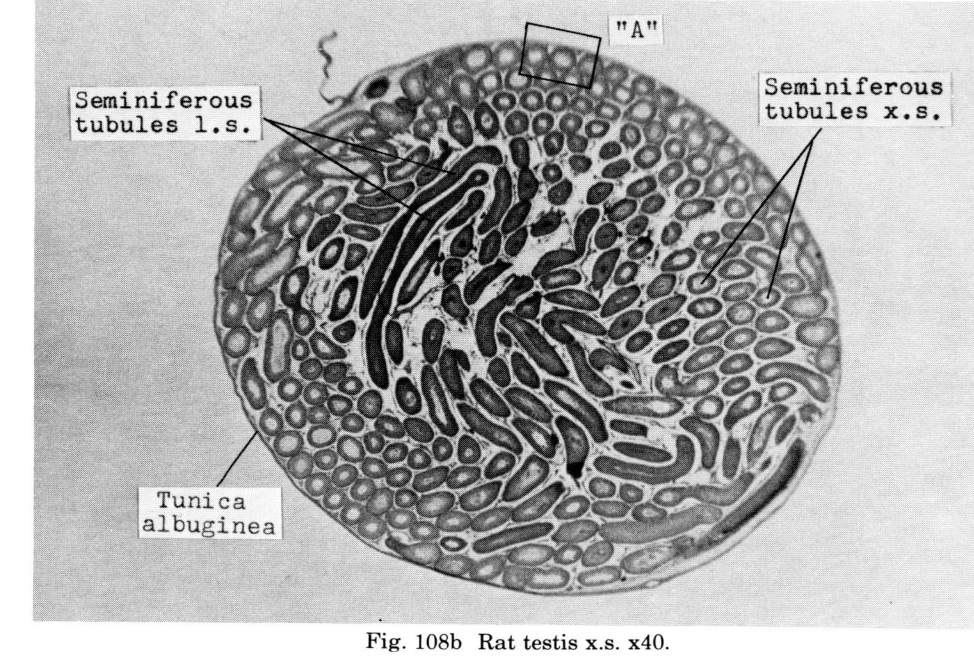

Fig. 108b Rat testis x.s. x40.

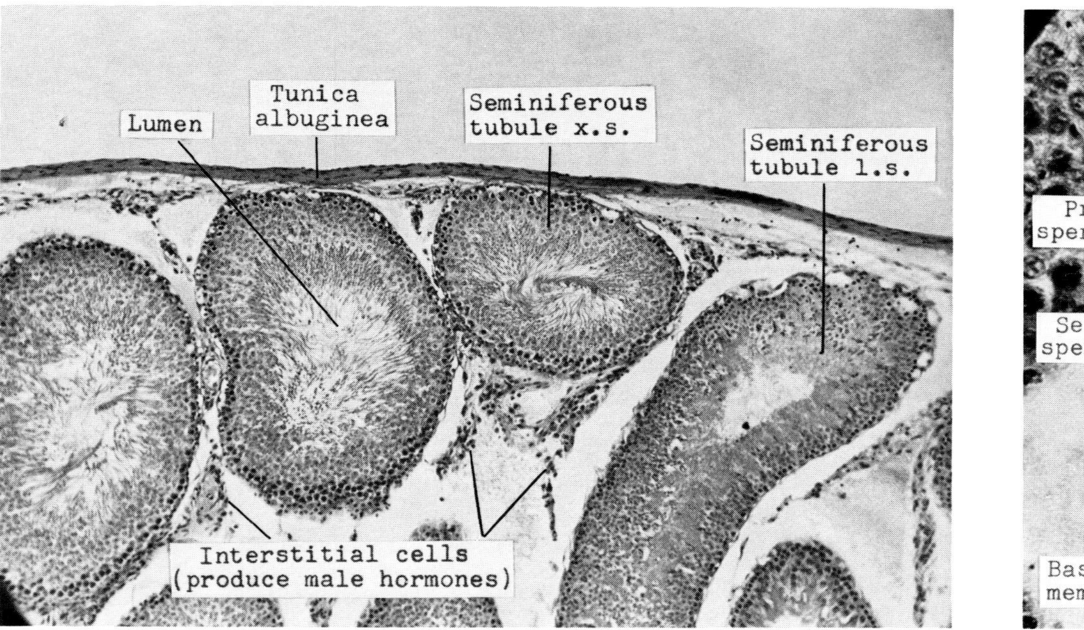

Fig. 108c Seminiferous tubules in a rat testis x.s. x100. (A higher magnification of a region like area "A" in Fig. 108b.)

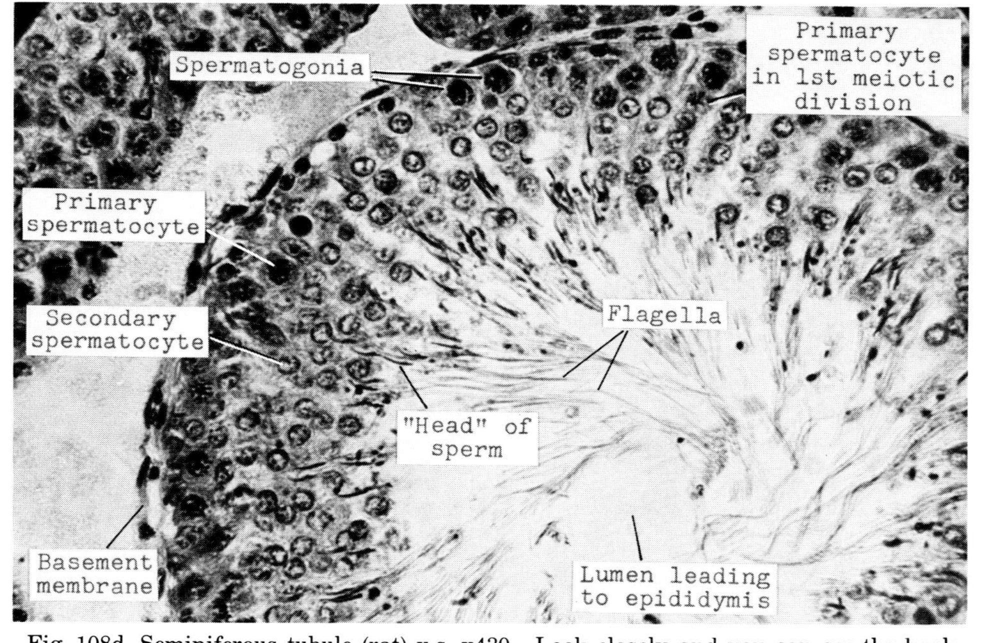

Fig. 108d Seminiferous tubule (rat) x.s. x430. Look closely and you can see the hook-shaped heads of the rat sperm. The cells around the edge of the tubule are dividing by meiosis and forming more sperm.

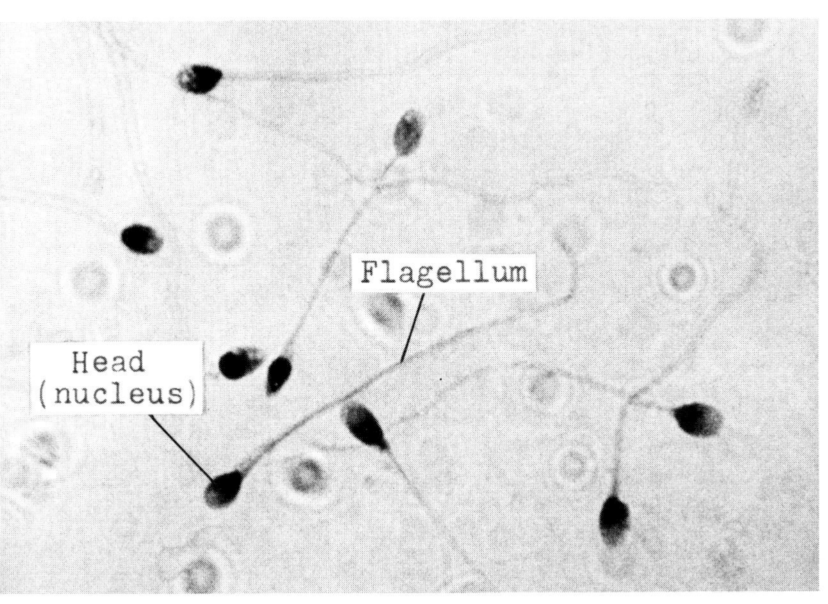

Fig. 109a Human sperm w.m. x1000.

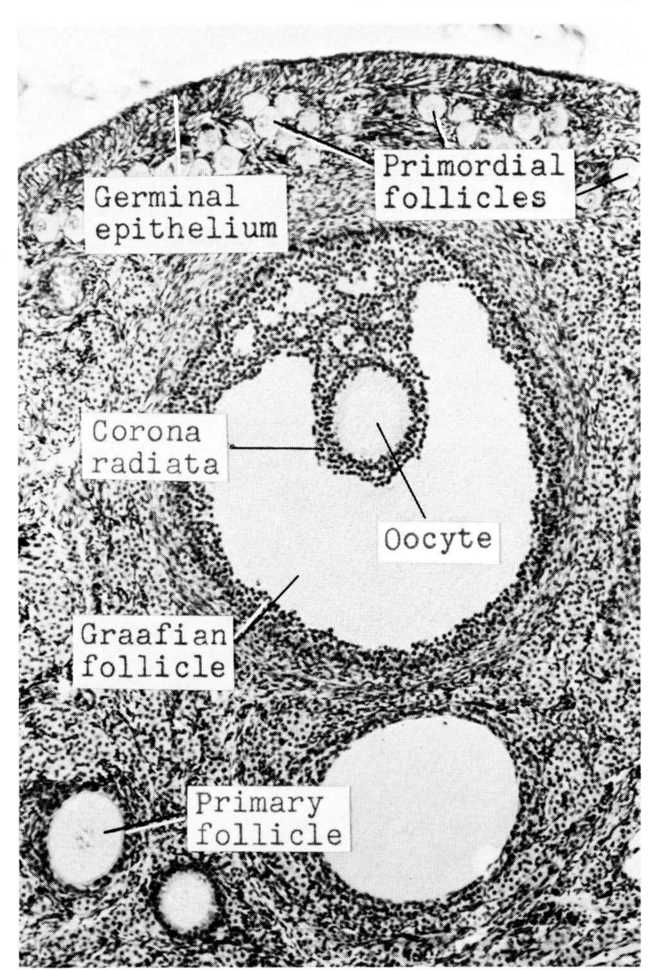

Fig. 109b Cat ovary with Graafian follicle x.s. x60.

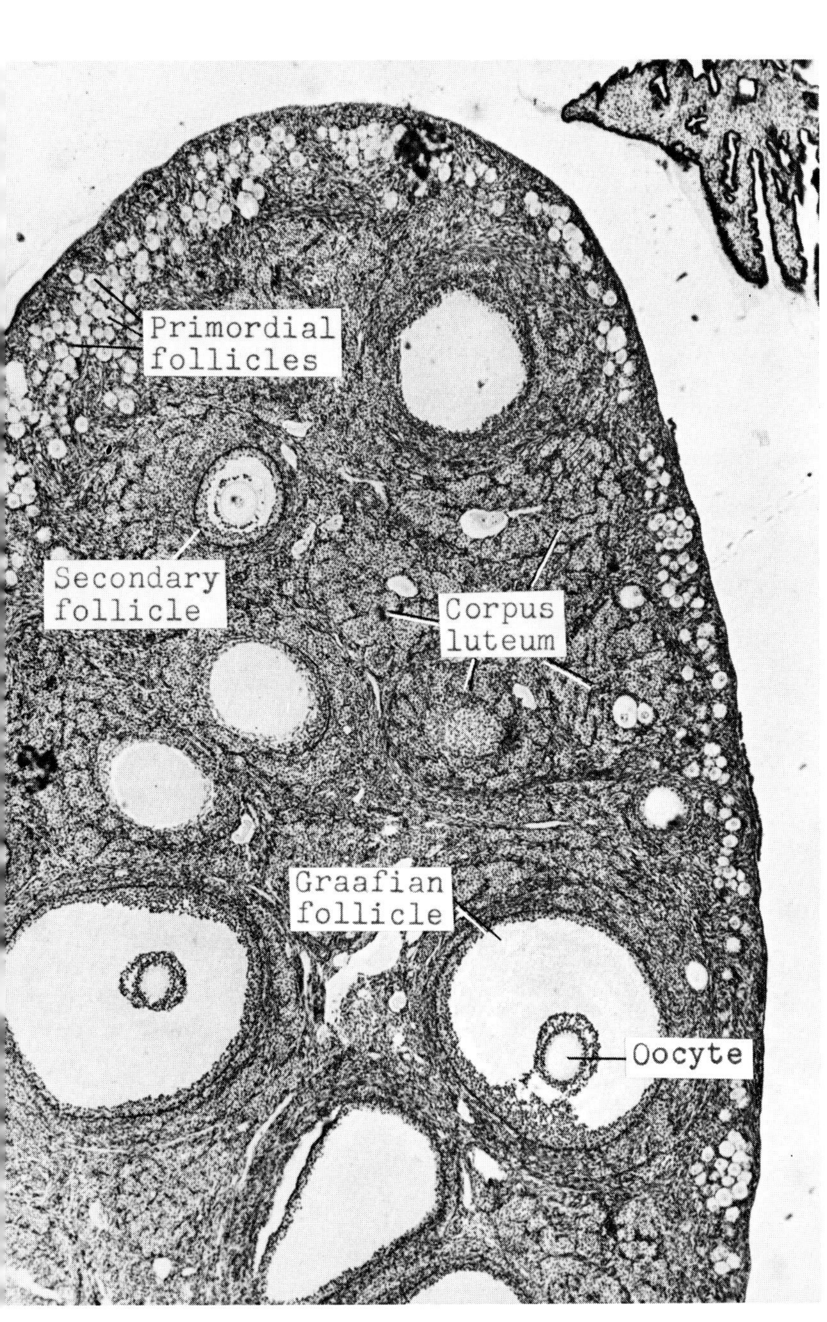

Fig. 109c Cat ovary l.s. x40.

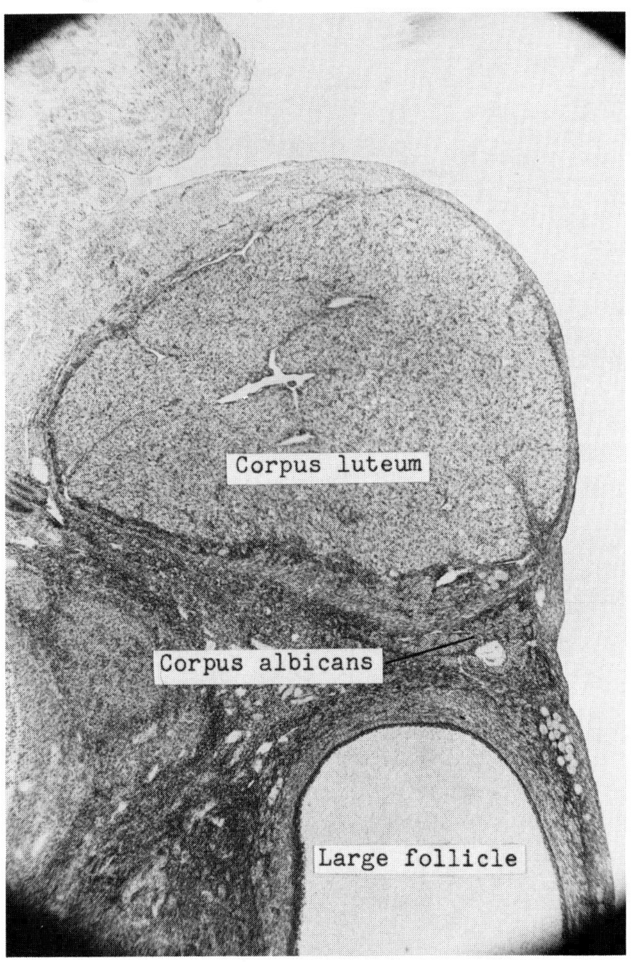

Fig 109d Corpus luteum in cat ovary x.s. x40.

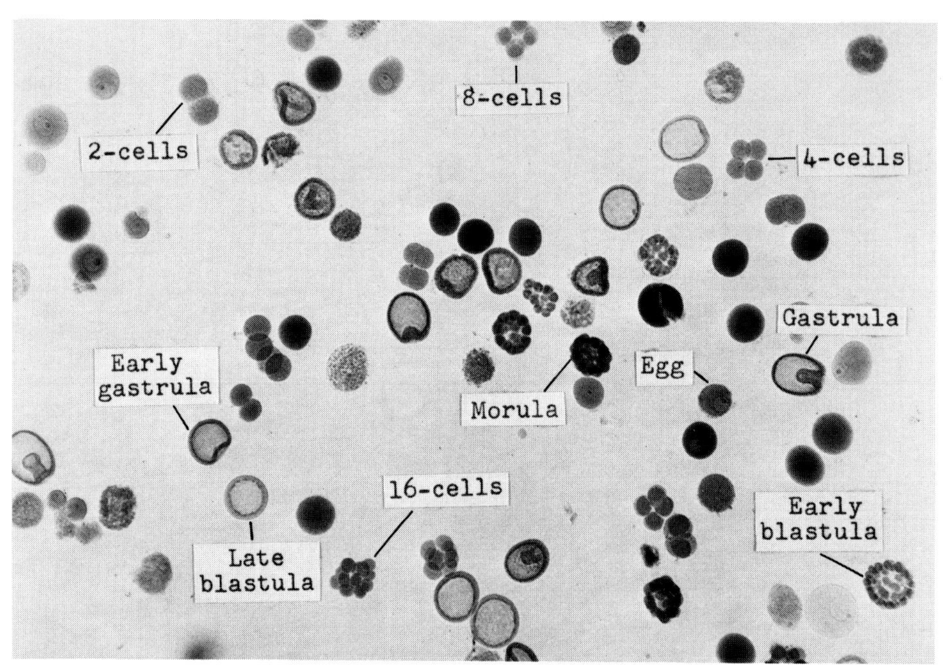

Fig. 110a Starfish embryos, various stages w.m. x40.

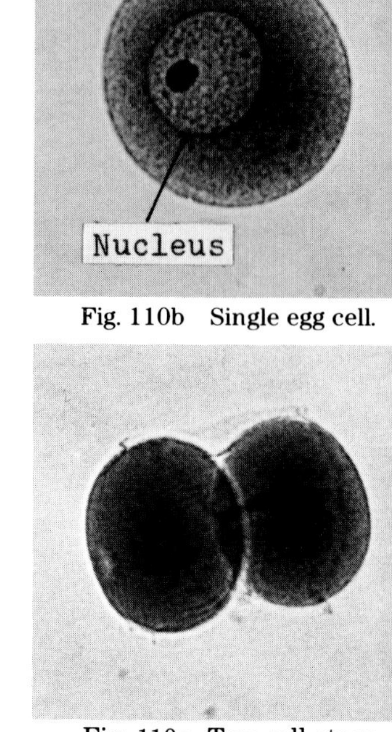

Fig. 110b Single egg cell.

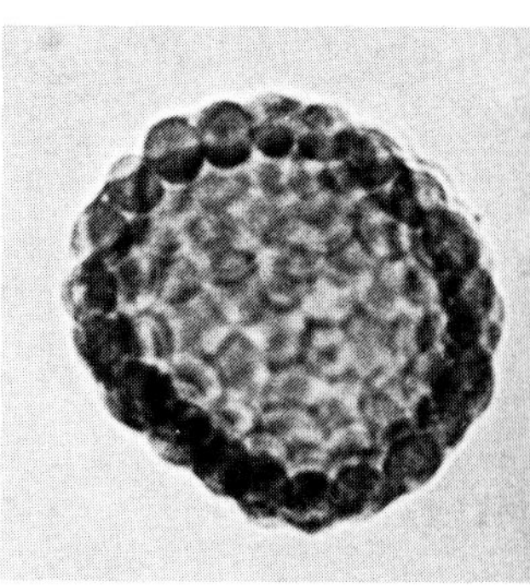

Fig. 110c Two cell stage

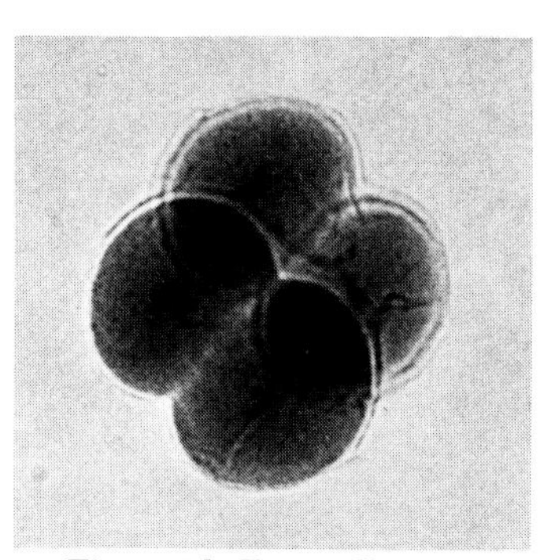

Fig. 110d Four cell stage

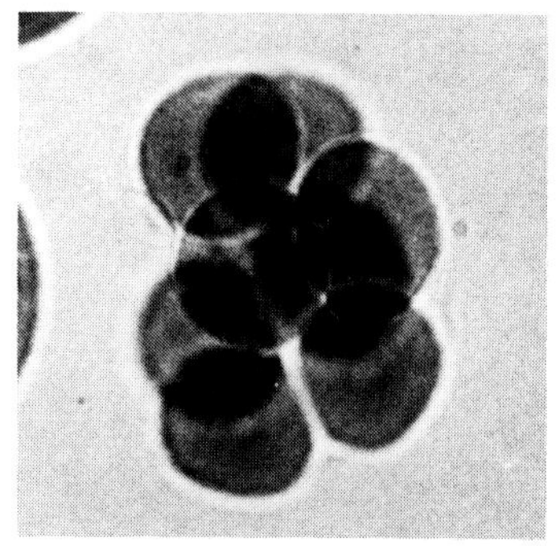

Fig. 110e Eight cell stage

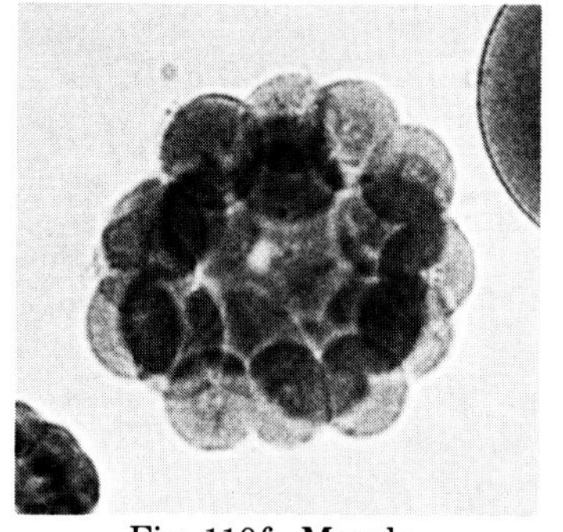

Fig. 110f Morula

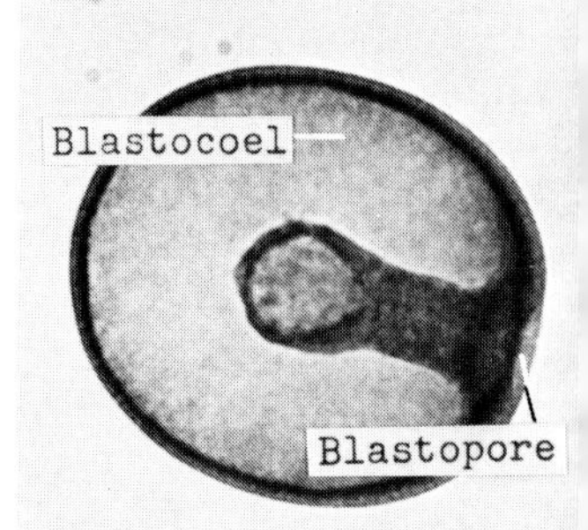

Fig. 110g Early blastula

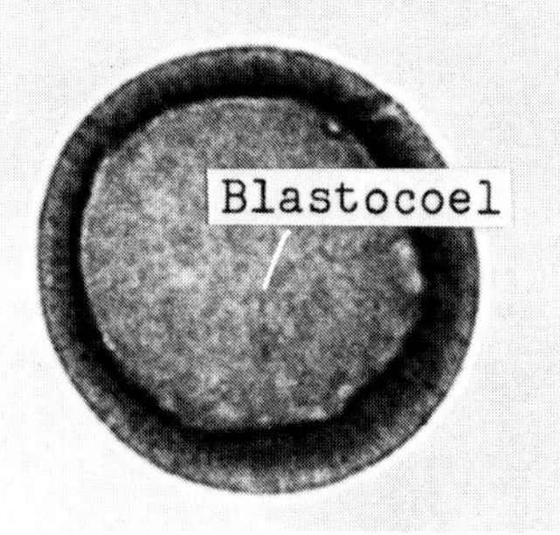

Fig. 110h Late blastula

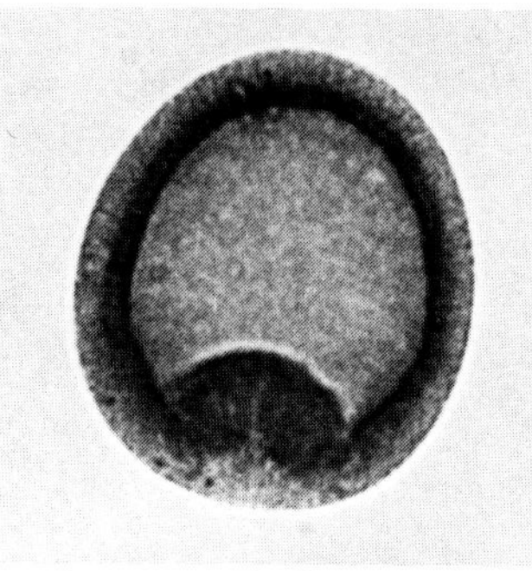

Fig. 110i Early gastrula

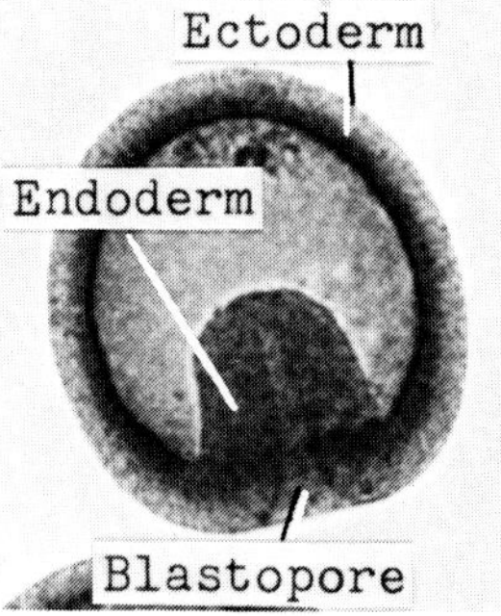

Fig. 110j Gastrula

Fig. 110k Gastrula

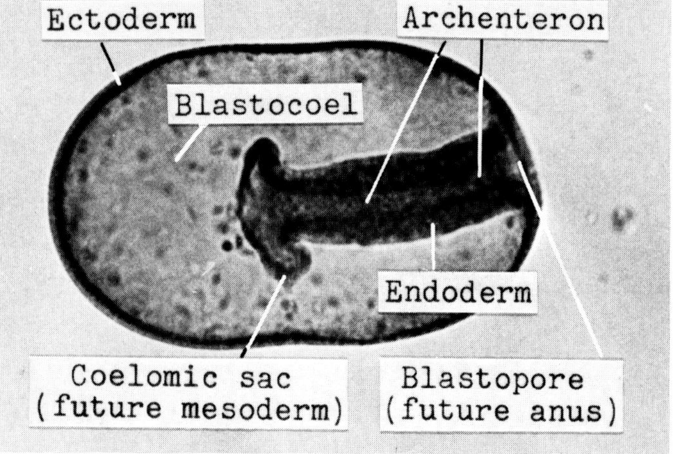

Fig. 110l Gastrula w.m. x100.

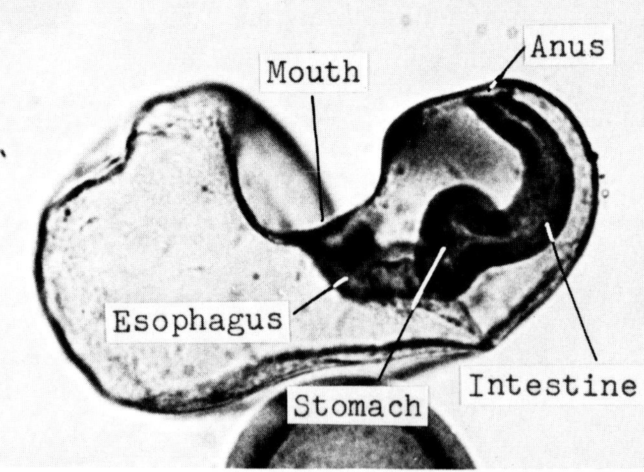

Fig. 110m Bipinnaria larva of starfish w.m. x100.

Marchantia 25-26

Marrow cavity 94
Marrow 89
Mastax 63
Mastigophora 13-14
Mastoid process 93
Matrix 89
Maturation region 38
Maxilla
crayfish 69
grasshopper 72
human 93
Maxillary palps 72
Maxillipeds 69
Medial malleolus 94
Medial 94
Medulla (kidney) 107
Medulla oblongota
frog 80
human 96
sheep 97
Medusa buds 55
Medusa 55
Megagametophyte 37
Megasporangium
pine 36-37
Selaginella 30
Megaspore mother cell
flower 47
pine 36-37
Megaspores 30
Megasporophyll 36-37
Megastrobilus 35-37
Mental foramen 93
Meristem (apical) 41
Merozoites 13
Mesenchyme
planaria 58
sponge 51
Mesenteric artery 78
Mesentery 77
Mesoderm 110
Mesoglea 53-54
Mesophyll 43
Mesothorax 72
Metacarpals 80, 92, 94
Metameres 66
Metaphase 3-4
Metapleural fold 76
Metatarsals 80, 94
Micrasterias 23
Micronucleus 9
Micropylar chamber 37
Micropyle
flower 47
pine 37
Microspores 30
Microsporophyll
flower 45, 48
pine 35
Selaginella 30
Microvilli 105
Midbrain
human 96
sheep 97
Midgut 76
Midvein 44
Miracidium 59
Mitosis
animal 3
plant 4
Mnium 28
Mollusca 64-65
Monera 5-6
Monkey face 40
Monocot
leaf 43
root 39
stem 40
Monocytes 13, 101
Morula 110
Mosses 27-29
Motor end plates 97
Motor neuron 91, 96-97
Mouth 80
Mucosa 105
Musci 27-29
Muscle
cardiac 91
circular 68, 105
frog 78
longitudinal 68, 105
mandibular 69-70
smooth 91
striated 90
Muscularis mucosae 105
Muscularis 105
Mushroom 18
Mycota 15-19
Myofibrils 90
Myotomes 76

N

Nares 80
Nasal 93
Navicula 7
Navicular 94
Neanthes 66
Necator americanus ... 63
Needle (pine) 35, 43
Nemathelminthes 61-63
Nematocyst 54
Nematodes 61-63
Nephridia
clam 64
earthworm 67-68
Nephridiopore 66
Nephrons 107
Nereis 66
Nereocystis 24
Neuroglial cells 91, 96
Neuromuscular
junction 97
Neurons 91, 96-97
Neutrophil 13, 101
Nostoc 6
Notochord 76
Nucellus 37
Nuclear membrane 1, 4
Nucleolus 2, 4, 21
Nucleus 1-4, 12, 14, 21, 22, 38, 91, 101
Nutrient foramina 94

O

Obelia 55
Obturator foramen 95
Occipital condyles 93
Occipital 93
Oculomotor nerve 97
Oedogonium 23
Oil droplets 2
Olfactory bulb 97
Olfactory lobes 80
Olfactory nerves 80
Oligochaeta 67
Onion
cells 2
mitosis 4
root 38
Oogonium
Fucus 24
Oedogonium 23
Vaucheria 8
Operculum 27-28
Opisthorchis 59
Optic chiasma
human 96
sheep 97
Optic disc 99
Optic lobe 80
Optic nerve II 97
Optic nerve 99-100
Optic tract 97
Oral groove 9-10
Oral hood 76
Oral spines 74
Oral sucker 59
Oral tentacles 76
Orbital surface of
sphenoid 93
Organ of Corti 98
Organelles 1
Os calcis 94
Oscillatoria 6
Osculum 50-52
Osteocyte 89
Ostia 70
Ostium 50-51
Ostracods 71
Outer hair cells 98
Ova *(Volvox)* 14
Ovary
Amphioxus 76
Ascaris 61-62
cat 109
crayfish 71
flower 45-47, 49
fluke 59
frog 77
hydra 54
pig 86
tapeworm 60
Oviducts
Ascaris 61-62
crayfish 69
earthworm 66
frog 77
grasshopper 72
Ovipositor 72
Ovulate cones 35-37
Ovules
flowers 45-47, 49
pine 36-37
Ovum (cat) 109

P

Palisade cells 43-44
Palps
grasshopper 72
Neanthes 66
Pancreas
frog 77
pig 81, 85
Papilla (hair) 106
Papillae 104
Pappas 46
Paramecium 9-11, 14
Paraphyses
Moss 29
Fucus 24
Peziza 17
Parapodia 66
Parenchyma
fern 32
root 39
stem 40-41
Parietal blood vessel 68
Parietal 93
Peanut 49
Pectoralis 79
Pedicel 94
Pedicellariae 73-75
Peduncle 49
Pelecypoda 64-65
Pellicle 9-10
Pelvis
hip 95
kidney 107
Penicillium 17
Penis
pig 86
squid 65
Peranema 14
Peridinium 7
Periosteum 89
Perisarc 55
Peristome 74
Perpendicular plate of
ethmoid 93
Petals 45-46
Peziza 17
Phaeophyta 24
Phalanges
frog 80
human 92, 94
Phalanx 94
Pharyngeal gill slits 76
Pharyngeal pouch 57-58
Pharynx
Amphioxus 76
earthworm 67
fluke 59
Neanthes 66
nematode 63
planaria 57-58
Phloem
fern 32
leaf 43-44
root 39
stem 40-42
Phloem ray 42
Photoreceptors 76
Physalia 56
Pia mater 96
Pig 81-86
Pileus 18
Pillars 98
Pine
reproduction 35-37
stem 42
needle x.s. 43
Pine needle x.s. 43
Pineal body 97
Pinnae 30-32
Pinnularia 7
Pistil 45-46, 49
Pistillate cone 35-37
Pith 41
Pituitary 97
Placenta 47
Planaria 57-58
Plants 21-49
Plasmodium vivax 13
Plasmolysis 1
Platyhelminthes 57-60
Pleopods 69
Pleural cavity 82
Plumule 48-49
Pneumatophore 56
Polar nuclei 47
Pollen
flower 45, 47
pine 35-37
Pollen chamber 37
Pollen sac
flower 45, 48
pine 36
Pollen tube 48
Polyps 55-56
Pons
human 96
sheep 97
Porifera 50-52
Portuguese man-o-war ... 56
Posterior adductor
muscle 64
Posterior chamber ... 99-100
Posterior gray horn 96
Posterior median
sulcus 96
Posterior mesenteric 84
Posterior retractor
muscle 64
Posterior root ganglion .. 96
Posterior vena cava
pig 85
squid 65
Posterior white
columns 96
Potato cells 2
Primary
spermatocyte 108
Primordial follicles 109
Proboscis 11
Progametangia 16
Proglottids 60
Pronotum 72
Prophase 3-4
Prostomial nerve 68
Prostomium 66
Prothallus 33-34
Prothorax 72
Protista 7-14
Protonema 28
Protozoa 9-14
Proximal convoluted
tubule 107
Pseudocoelom 62
Pseudopodia 12
Pubis
frog 80
human 92, 95
Puccinia graminis 19
Pulmocutaneous
artery 78
Pulmonary artery
frog 78
sheep 102
Pulmonary trunk 84
Pulmonary veins 102
Pupil 99
Pyloric caecum 73-75
Pyloric duct 73
Pyloric stomach 71
Pyramid 107
Pyrenoids 2, 21, 22

R

Radial canal
sponges 50-52
starfish 75
Radicle 48-49
Radio-ulna 80
Radiolaria 12
Radius 92, 94
Raphe 7
Ray flower 46
Rays 63
Receptacle
(flower) 45-46, 49
Receptacles
Fucus 24
Rectum
pig 81, 86
grasshopper 72
Rectus abdominis 79
Red blood
cells 101, 103, 105
Redia 59
Renal artery 81-84
Renal papillae 107
Renal vein 82-83, 85
Reproductive system
(pig) 86
Resin duct
pine needle 43
stem 42
Retina 99-100
Rhinencephalon 97
Rhizoids
fern 33-34
Rhizopus 15
Rhizome 32
Rhizopus 15-16
Ribs 92
Right atrium 102
Right subclavian
artery 84
Right ventricle 102
Rods 100
Romalea 72
Root cap 38
Root hairs 38
Root sheath (hair) 106
Roots 38-39
Rose 46
Rostellum 60
Rostrum 69
Rotifers 63
Roundworms 61-63
Rust (wheat) 19

S

Sac fungi 17
Sacro-iliac joint 95
Sacrum 92, 95
Sarcodina 12
Sarcolemma 90
Sartorius 79
Scala media 98
Scala tympani 98
Scala vestibuli 98
Scaphoid 94
Scapula 92
Schizophyta 5
Sclera 99-100
Sclerenchyma
fern 32
leaf 43
stem 40-41

Scolex 60
Scrotum 86
Scutellum 48
Scypha 50-52
Scyphistoma 55
Sea anemones 56
Sebaceous gland 106
Secondary follicle 109
Secondary
  spermatocyte 108
Seed coat 48-49
Seed leaf 48
Seeds
  angiosperm 45, 48, 49
  gymnosperm 35-37
Segments 66
Selaginella 30
Semicircular canals 98
Semimembranosus 79
Seminal receptacle
  crayfish 69
  earthworm 67
  fluke 59
Seminal vesicle
  Ascaris 62
  earthworm 67
Seminiferous tubules 108
Sensory neuron 96
Sepals 45-46, 49
Septa 67
Septum pellucidum 97
Serosa 105
Seta
  Marchantia 26
  moss 27-28
Setae 66, 68
Sheep liver fluke 59
Shell gland 59
Shell 64
Shoulder 95
Sieve tube 40
Siphon 65
Skeletal cup 56
Skeletal muscle 90
Skeletal plates 75
Skeleton
  frog 80
  human 92-95
Skin gills 75
Skin 106
Small intestine
  frog 77-78
  pig 81
  x.s. 91, 105
Smooth muscle
  artery 103
  tissue 91
Sori 31
Sorus 31-32
Sperm
  fern 33-34
  Fucus 24
  hydra 54
  human 109
  moss 29
  Oedogonium 23
  planaria 57-58
  rat 108
Sperm duct
  crayfish 69
  earthworm 66
  tapeworm 60
Sperm grooves 66
Spermatids 58
Spermatocytes 108
Spermatogonia 108
Sphenoid 93
Spicules 50-52
Spinal cord
  cat 96
  frog 80
  ox 91
  sheep 97
  x.s. 96
Spindle fibers 3
Spine 94
Spines 73-75
Spinous process 94
Spiracles 72
Spiral ganglion 98
Spiral tunnel 98
Spirilla 5
Spirochetes 14
Spirogyra 14, 21, 22
Spleen
  frog 77-78
  pig 81, 85
Sponges 50-52
Spongin 52
Spongocoel 50-52
Spongy bone 94
Spongy cells 43-44
Sporangia
  fern 31-32
  fungi 15-16
  Marchantia 26
Sporangiophore 15-16
Spores
  fern 30-32
  Marchantia 26
  moss 28
  Rhizopus 15-16
Sporophyte
  ferns 30-34
  Marchantia 26
  moss 27-28
Squamosal suture 93
Squid 65
Stamen 45-46, 49
Staminate cones 35-36
Staphlococci 5
Starch grains
  potato 2
  root 39
Starfish
  adult 73-75
  embryo 110
Stele 38-39
Stems 40-42
Stentor 11
Sterigma 18
Sternum
  grasshopper 72
  human 92
Sternum (frog) 80
Stigma 45-46, 49
Stipe
  fern 30
  mushroom 18
  seaweed 24
Stolon 15
Stoma 43-44
Stomach
  crayfish 70
  frog 77-78
  grasshopper 72
  pig 81, 85
  starfish 73
  starfish larva 110
Stone canal 74
Stratified squamous
  epithelium 87
Stratum corneum 106
Stratum
  germinativum 106
Stratum granulosum 106
Stratum lucidum 106
Stratum spinosum 106
Streptococci 5
Stria vascularis 98
Striated border 105
Striated muscle 90, 97
Striations 90
String bean 49
Strobila 55
Strobilus 30
Stroma (eye) 99
Style 45-46, 49
Styloid process 93, 94
Subarachnoid space 96
Subclavian arteries 84
Subcutaneous layer 106
Submucosa 105
Subneural blood
  vessel 68
Sucker
  fluke 59
  starfish 75
  tapeworm 60
Superior articular
  facet 94
Superior orbital
  fissure 93
Superior vena cava 102
Supraorbital foramen 93
Surirella 7
Suspensor cells 16
Sutures 93
Swimmeretts 69
Symphysis pubis 92, 95
Synergid cells 47
Systemic arch 78

T

Taenia 60
Talus 94
Tapeworm 60
Tarsals 80
Tarsus
  human 94
  grasshopper 72
Taste buds & pores 104
Tectorial membrane 98
Telia 42
Teliospores 19
Telophase 3-4
Telson 69
Temporal 93
Tendon of Achilles 79
Tentacles
  coelenterates 53-56
  Neanthes 66
  rotifer 63
Tergum
  crayfish 69
  grasshopper 72
Terminal vacuole 23
Termite protozoa 14
Test 12
Testes
  Ascaris 62
  crayfish 71
  fluke 59
  frog 78
  hydra 54
  pig 86
  planaria 57-58
  rat 108
  tapeworm 60
  x.s. 108
Testicular arteries 83, 86
Thallus
  Fucus 29
  lichen 20
  liverwort 25
Theca 56
Third ventricle 97
Thoracic vertebrae 92-93
Thorax
  grasshopper 72
  pig 82
Thymus 81-82
Thyrocervical trunk 84
Thyroid gland 81-82
Tibia
  human 94
  grasshopper 72
Tibio-fibula 80
Tissues 87-91
Tomato
  flower 45
  fruit 49
Tongue
  frog 80
  rabbit x.s. 104
Trachea
  grasshopper 72
  pig 83
  sheep 103
Tracheal rings 72
Tracheophyta 30-49
Tradescantia 44
Transverse process
  frog 80
  human 94
Trapezium 94
Trapezoid body 97
Trapezoid 94
Trematoda 59
Triceps brachii 79
Triceps femoris 79
Trichina 63
Trichinella spiralis 63
Trichocysts 10
Tricuspid valve 102
Trigeminal nerve 97
Triquetral 94
Trochanter 72
Trochlear nerve 97
Trophozoites 12
True ribs 92
Truncus arteriosus 78
Trypanosoma 13
Tube cell
  flower 49
  pine 36
Tube feet 74-75
Tube nucleus
  flower 49
  pine 36
Tunica albuginea 108
Turbatrix 63
Turbellaria 57-58
Typhlosole 68

U

Ulna 92-94
Umbilical
  arteries 81-84, 86
Uredospores 19
Ureter
  frog 77
  lamb 107
  pig 81-83, 85-86
Urethra 81, 86
Urinary bladder
  frog 77-78
  pig 81-82, 86
Urinary system (pig) 82
Uropod 69
Urostyle 80
Uterine horn 81
Uterus
  Ascaris 61-62
  fluke 59
  frog 77
  pig 81, 86
  tapeworm 60

V

Vacuole 1, 23
Vagina
  Ascaris 61-62
  pig 81, 86
  tapeworm 60
Vaginal vestibule 81, 86
Valve 7-8
Vas deferens
  Ascaris 62
  crayfish 71
Vascular bundle
  anther 48
  dicot stem 41
  fern 32
  leaf 43-44
  monocot stem 40
Vaucheria 8
Vegetative cells
  (volvox) 14
Vegetative filament 22
Veins
  leaf 43
  pig 85
  pulmonary 102
  x.s. 103
Velum 76
Venter 29
Ventral blood vessel 67
Ventral nerve cord
  Ascaris 62
  earthworm 67-68
  planaria 58
Ventral sucker 59
Ventricle
  clam 64
  human 102
  pig 81-83
  sheep 102
Vertebrae 92-95
Vertebral foramen 94
Vestibular membrane 95
Vestigal oviducts 78
Villi x.s. 105
Villus 87, 105
Vinegar eels 63
Visceral mass 65
Vitreous body 99-100
Vocal sac 80
Voluntary muscle 90
Volvox 14
Vomer 93
Vomerine teeth 80
Vorticella 11
Vulva 61

W

Walking legs 69
Water net 23
Wheat rust 19
Wheel organ 76
White blood cells 101
White matter 96
Wood ray 42
Wood 42
Woody dicot stem 42

X

Xanthophyceae 8
Xylem
  fern 32
  leaf 43-44
  root 39
  stem 40-42

Y

Yeast 10, 17
Yellow green algae 8
Yolk glands
  fluke 59
  tapeworm 60
Yucca flower 46

Z

Zoochlorellae 10
Zooids 14
Zygomatic process of
  temporal 93
Zygomatic 93
Zygomycetes 15-16
Zygospore
  Rhizopus 16
  Spirogyra 22
Zygote
  Rhizopus 16
  Spirogyra 22
  tapeworm 60
  Vaucheria 8